Smart & Sustainable Built Environments

Smart & Sustainable Built Environments

Edited by

J. Yang
P. S. Brandon
A. C. Sidwell

Blackwell
Publishing

© 2005 Blackwell Publishing Ltd

Editorial offices:
Blackwell Publishing Ltd, 9600 Garsington Road, Oxford OX4 2DQ, UK
   Tel: +44 (0) 1865 776868
Blackwell Publishing Inc., 350 Main Street, Malden, MA 02148-5020, USA
   Tel: +1 781 388 8250
Blackwell Publishing Asia Pty Ltd, 550 Swanston Street, Carlton, Victoria 3053, Australia
   Tel: +61 (0) 3 8359 1011

First published 2005 by Blackwell Publishing Ltd

Library of Congress Cataloging-in-Publication Data:

Smart & sustainable built environments / edited by J. Yang, P. S. Brandon, A. C. Sidwell.
     p.   cm.
   Includes bibliographical references and index.
   ISBN-10: 1-4051-2422-9 (hardback : alk. paper)
   ISBN-13: 978-1-4051-2422-5 (hardback : alk. paper)
   1. Sustainable buildings.   2. Sustainable architecture.   3. Ecological houses.   I. Title: Smart and sustainable built environments.   II. Yang, J. (Jay)   III. Brandon, P. S. (Peter S.)   IV. Sidwell, A. C.

TH880.S63   2005
720'.47–dc22
2004025945

ISBN-10: 1-4051-2422-9
ISBN-13: 978-14051-2422-5

A catalogue record for this title is available from the British Library

Set in 9.5/12 pt Palatino
by TechBooks
Printed and bound in India
by Replika Press Pvt. Ltd., Kundli

The publisher's policy is to use permanent paper from mills that operate a sustainable forestry policy, and which has been manufactured from pulp processed using acid-free and elementary chlorine-free practices. Furthermore, the publisher ensures that the text paper and cover board used have met acceptable environmental accreditation standards.

For further information on Blackwell Publishing, visit our website:
www.blackwellpublishing.com

Contents

v

Preface

This book is a fresh attempt at demonstrating our current and potential capabilities of R&D in sustainable development in the built environments. More importantly, it hopes to bring about a more philosophical consideration of our approaches to sustainability amidst the present 'action-packed', technology-driven solutions to the issues and concerns.

The book provides a snapshot of existing methods and technologies as well as potential opportunities and future strategies to develop smart and sustainable built environments, presented through the eyes of international experts who have contributed each of the chapters in the book. The selection of topics and contributions was based on past achievements in the field and was a result of the networking, debates, information exchange and convergence following the First CIB international Conference on Smart and Sustainable Built Environments, which was held in Australia in November 2003. The book presents a further opportunity to disseminate information and enrich the literature of relevant sustainable initiatives in a global perspective, through the specific expertise, performances, case studies, commentaries and future overviews covered in the chapters.

The book has six parts, each addressing a specific area of concern for the development of smart and sustainable built environments. Individual chapters cover a broad spectrum of issues ranging from technological advancement, through the assessment and evaluation of past experiences, to communication, education requirements and future strategies. With the book's strong focus on integration among the various sustainability initiatives, and on disciplinary priorities, innovative products and technological advancement, we hope to put human values and aspirations in the right perspective by promoting a better identification of needs among society and industries, more effective communication between stakeholders, and increased education and feedback to the general public. We believe that innovations and solutions, as demonstrated through examples in this book, will come forward when our clients – that is, our industries, businesses and the public alike – become more accustomed to the sustainability phenomenon and can relate it to their way of life.

We acknowledge all contributors in this book, not only for their expert views and contributions here, but also for their pioneering spirit, dedication and persistence and for their visions of the future of the complex issues of sustainable development.

Jay Yang
Peter S. Brandon
Anthony C. Sidwell

Introduction – bridging the gaps in smart and sustainable built environments

J. Yang, P. S. Brandon and A. C. Sidwell

The evolving nature and constraints of sustainable development

Sustainable development is quickly becoming a global phenomenon, focused by many people who represent a wide range of professions and interests. In the field of the built environment, sustainability issues are being highlighted, debated and explored in many parts of the world as they seem to bring the promise of solving many problems. Despite its current hype, sustainable development has had a changing nature during its relatively short period of development. It has identified many constraints and gaps that need to be properly dealt with.

The term 'sustainability' was first used in the United States in its present context in the National Energy Policy Act of 1969. According to Kibert (1997), the idea of *intertemporal choice*, a precursor to the notion of sustainability, was strictly based on economic argument in the 1970s. The real watershed of global interest in sustainable development emerged from the World Commission on Environment and Development (WCED, 1987) in its Brudtland's report, which became frequently quoted for its definition of sustainable development as:

> 'the development that meets the needs of the present without compromising the ability of future generations to meet their own needs'.

The above concept was further expanded at the Earth Summit held in Rio de Janeiro in 1992 (UNCED, 1992) in Agenda 21 (Policy plan for environment and sustainable development in the 21st century). In the final declaration, 27 principles were agreed upon; these interweave political, economic, legal, social and environmental dimensions (Bentivegna *et al.*, 2002). This has evolved significantly, from the conventional view of being 'green' and reducing the use of natural resources, to the linking of social and economic dimensions with ecological protection goals (Kohler, 2002).

Sustainability in now about balancing the Earth's physical resources with the social, economic and environmental needs of our societies (Marsh, 2000). In the construction field, sustainable development has also been referred to as 'the creation and responsible maintenance of a healthy built environment based on resource efficient and ecological principles' (Grey and Halliday, 1997). In addition, it can be seen as providing a contribution to alleviating poverty, creating a healthy and safe working environment, distributing social costs and benefits of construction equitably, facilitating employment creation, developing human resources, and acquiring financial benefits and uplift for the community (CIB, 1999).

Within a relatively short time span of 35 years, the concept and practice of sustainable development has grown quickly from an inexplicit reference in a national energy policy matter to a truly global phenomenon. Its evolution has coincided with the rapid expansion in information, knowledge, technology, products and practices, covering a broad spectrum of the architectural, engineering and construction (AEC) industries. As the definitions and approaches evolve, sustainable development also faces an increasing number of constraints on many fronts. An example of some of the constraints, priorities and complexities faced in solving the problems is given in Table I.1.

More recently, literature began to cover specific arguments about the varying starting points, processes and end results of sustainability for people, depending on the many different priorities and backgrounds such as social, environmental and economic perspectives from which they come (Mawhinney, 2002). The views of the economists differ from those of environmentalists in that the former place reliance on the status quo of the economy and the central measure of GDP as the proxy for development, while the latter consider the implications for environmental damage, especially irreversible damage, as the basis of all measures. Researchers also warned that the multifacetted nature of sustainable development involves more than energy consumption and water conservation, considered in isolation, as some earlier developments had tended to focus on. There is an increasing recognition of the 'softer' targets of social, institutional, psychological and other agendas, particular within an integrated approach. Adding to the complexity, innovative and smart technological developments for built facilities tend to promise many environmentally friendly outcomes but these often come at the cost of huge financial burdens up front. Affordability is no longer a fashionable topic for research but it is the cold reality for stagnation in some aspects of sustainable development. The issue of 'time' of development is poorly served despite recognition that it will impact on the evaluation of sustainability. Moreover, the processes of change, adaptation and growth, and the management of these processes are also defining elements in the evolution and development of smart and sustainable built environments (Yang *et al.*, 2003). Along the way, sustainable development will also need to:

- understand natural systems and their ways of interaction with the human environment
- identify the priority needs of human development and the aspirations of the general public, which are not adequately identified at present
- deal with additional risks as well as the potential gains of adopting new materials, technologies and practices
- provide the institutional flexibility and mechanism required for changing legislation, standards, codes and guidelines that are already taking shape at regional and federal levels in more and more countries
- improve the technical know-how of professionals

Table I.1 Examples of constraints, priorities and complexities in Sustainable Construction.

Constraints	Potential priorities	Typical complexities
Resource depletion	Reduction of energy consumption during construction and use Conservation of water resources Development of alternative materials	• Lack of awareness and sharing of knowledge and experiences among professionals and trades/people • Incompatible methods of procurement and construction • Inefficiency in process modelling
Financial target	Lean construction Target setting, information sharing and benchmarking Technology innovation	• Dependency on multi-level coordination, and government incentive • The conservative nature of the construction business • Inability to assess and handle risks • Input/Benefit analysis
Environmental damage	Design for minimum waste Reduction in construction waste and usage waste Minimization of pollution through efficient operation Maintenance and improvement of biodiversity	• Deficiency in comprehending natural systems and phenomena • Inability of design tools • Consumer habits • Legislation and governance • General public awareness
Social context and political stance	Respect for people and nature Health and safety principles Legislation and codes Implementation and incentives Education of professionals and the community	• Lack of competence in managing the processes of changing attitude of people and institutions • Lack of appropriate education channels • Inability to establish 'best practice'

• develop more united and consolidated approaches amid the differences caused by varying interpretations, preferences, priorities and prejudice towards sustainability principles
• respond to the difficulties in defining universally acceptable role models, caused by the differences cited above and changing public expectations.

As the global community increases the common awareness of sustainability issues, it is necessary for all involved to respond quickly to these changes and constraints.

Current responses from the industry

Over the last few years, specific references to sustainable development by the stake-holders of the construction and property industries have become increasingly common. According to the Green Building Council, Australia, while there are pioneers and 'activists' who whole-heartedly embrace the principles in their practice or services, the majority of practitioners are just becoming accustomed to relating to the 'Triple Bottom Line' (TBL) theory, or social, economic and environmental targets, in developing business strategies and formulating long term goals. However as the community, or the 'clients', develop more awareness and interest in the relatively 'novice' issue of sustainability, our industries have responded in a more 'practical' fashion and are keen to 'move on' in search of the seemingly right mechanisms for delivering a final outcome that is supposedly environmentally sustainable. In contrast to the wide array of technical 'innovations', there have been few coordinated, industry level and strategic efforts to consider the priorities, drivers, stimuli, obstacles and constraints described in the last section.

There are varying levels of attitudes towards the adoption of sustainability principles in practice. From simple and individual measures such as sediment control on a housing construction block to grander and multinational research schemes such as adaptive buildings and intelligent cities, the AEC industry has made significant progress in technological advancement in this field. A substantial amount of information and knowledge has been gained from practical experiences over the last few decades. Some of these advancements, information and knowledge have indeed made a big difference. Perhaps because of this, the final measurable outcome had often been a prime concern for architects, engineers and builders alike.

Developers often claim the importance of providing 'what the customers want' without acknowledging their own responsibilities to educate the clients and change the way that industries operate. Some are more interested in promotional and marketing gains with unsophisticated and isolated products and services. In keeping the balance between issues of business survival, local industry growth, public acceptability and political motivations, governmental interventions have also been mixed. In Australia, Brazil and China, for example, while there have been initiatives to take on the sustainability challenge at federal, state and local government levels, the upgrading and expansion of relevant building codes, standards and regulations have been slow. Many changes to date tend to include only the minimum requirements, rather than more comprehensive measures or targets related to sustainability. Without adequate upgrading of skills, a considerable number of practitioners are struggling with the new concepts and will be reluctant to change the way they conduct their professional routines. At the bottom end of the scale, some operators are treading a fine line between meeting the minimum requirements and illegal practice.

The overall result of such responses and attitudes presents difficulties that inhibit more effective implementation of sustainable strategies in our whole built environments. Many of the efforts of past research and development, often seen as vision provider for the industry, contemplated technical innovation on isolated issues that were applicable only within certain localities concerned with immediate outcomes. This has resulted in gaps in sustainable development such as the difficulties in identifying

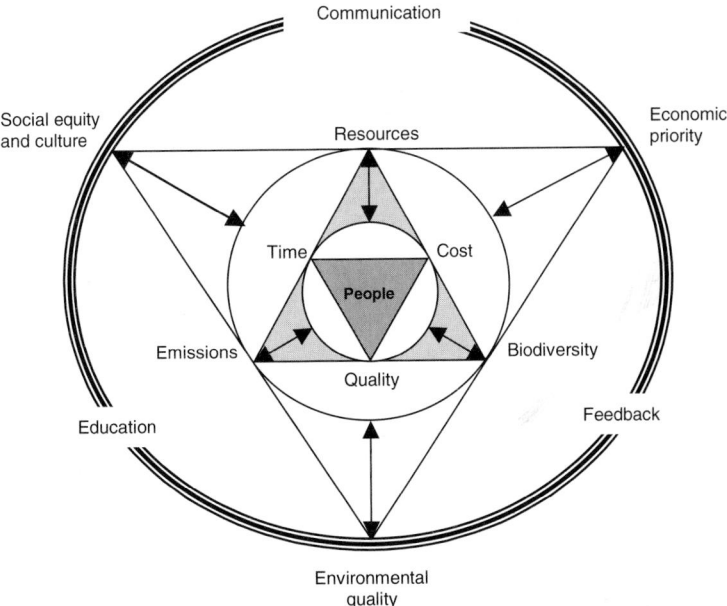

Figure I.1   An integrated approach to key issues of smart and sustainable built environments.

common priorities, defining appropriate action plans for industry sectors or levels of governments, managing information as well as knowledge, and coordinating global, regional and disciplinary activities of professional and general public interest. A mechanism for integration facilitated by an over-arching approach is needed.

The 'bigger' picture

The evolving definition and constraints of sustainable development concepts exhibit a close correlation with key management principles. Sustainable development centres on people and will have the same underpinning of time, quality and cost (i.e traditional management parameters). As a new and evolving paradigm, it will add new dimensions such as resources, emissions and biodiversity, as justified in the CIB Agenda 21 report (CIB, 1999). Extending such a philosophy to a global context will take in further aspects of social equity, cultural issues, governance, environmental quality and economic constraints. Another area of future focus is the education, communication and feedback between stakeholders. These dimensions and aspects must be treated on an integral basis, sharing information with each other, and should be jointly explored as foundational support to sustainable built environments in a 'big picture', as illustrated in Figure I.1.

For many years, the TBL approach has been the guiding principle of sustainable development in many countries and sectors of the AEC industry. To encourage 'buy-ins' of all facets of life and wider ranged participation in pursuing the sustainability

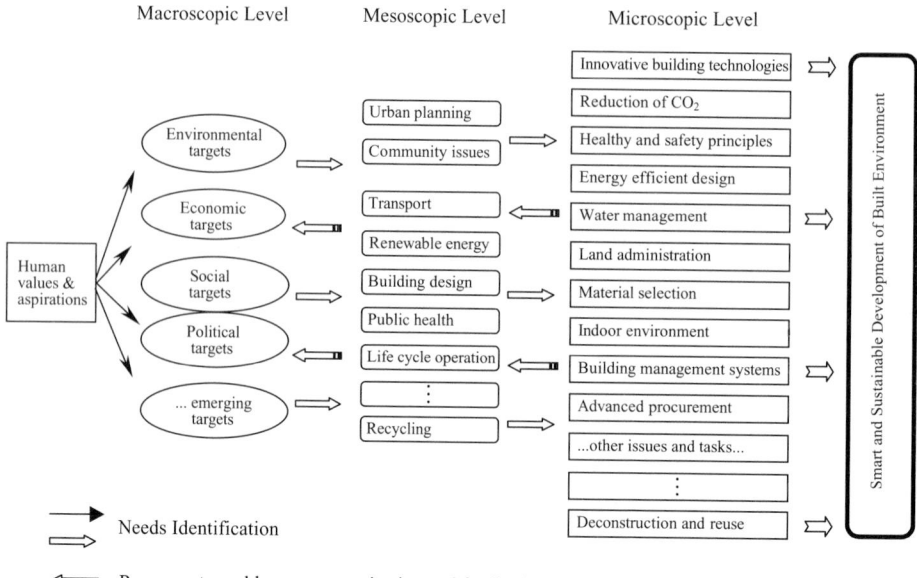

Macroscopic Level   Mesoscopic Level   Microscopic Level

Figure I.2 The 'bigger' picture of smart and sustainable development of the built environment.

goals, there need to be expanding dimensions. A more complete picture capable of representing the overall philosophy of sustainability and facilitating integrated solutions to the issues and problems needs to be drawn. Figure I.2 is an attempt at describing such an expanded approach, as the 'bigger picture'.

In contrast to many current technically focused solutions that adopt the top-down approach, the 'bigger picture' promotes a bottom-up approach starting with human aspirations and essential values. These values will be clarified through the identification of need and become expanded TBL targets at the macroscopic level. These targets will then link with the current physical and social infrastructures that are operated at the mesoscopic level, and expand to cover interrelated domains such as transport, resources use, and public health, which are all integral parts of our built environments. Detailed and specific responses to technical issues and tasks can then be managed and acted upon at the microscopic level. Going up levels (from left to right in Figure I.2) identifies essential needs. The opposite action will provide responses to problems and means of communication and feedback that are currently missing in most sustainable development themes. The internal linking of different levels and between elements of the levels is also of great importance.

Integrated development

To bridge the current gaps, the next stages of development must incorporate a more integrated approach. It also requires major effort on behalf of industry and public education and the identification and promotion of exemplar projects.

Smart features and sustainable outcomes

The two extremes of promoting 'green' labels only or chasing the ultimate 'technology wonders' at all cost are things of the past. Before R&D work begins, the economic justification, the environmental consequences, and the readiness of the general public for acceptance must always be borne in mind. It is almost a common expectation for intelligent technologies to respond to sustainability issues and address environmental concerns. Smarter ways of working, in addition to computer facilitated automation or intelligence, should also be identified and encouraged. Successful examples of incorporating smart technologies in pursuing sustainable goals will become new breeds of exemplars worldwide.

Integration between project stakeholders

Professional institutions have an important role to play in stocktaking technology inventories and equipping practitioners with the latest information on the sustainability aspects required for their jobs. All sectors of the industry, not just architects, environmentalists and consulting engineers, need to assume the responsibility of educating the whole society, as well as individual practitioners, towards sustainability.

At the industry level, communication between professional groups, especially those who can exercise substantial influence over clients, must be improved. In an office building refit project consulted by the Center for Building Performance and Diagnostics of the Carnegie Mellon University, USA, it was found that the mechanical engineers could not comprehend the philosophy of a hybrid HVAC system incorporating the use of desiccant air handlers, as required by the design consultants. A recent Australian study on the adaptation of raised floor systems (RFS) and technologies in office buildings identified that the lack of understanding of commercial estate agents had a major impact on the slow take-up of this new technology. Up to now, estate agents have never been considered, as all the previous effort by design consultants and building system suppliers to promote RFS adoption had been directed towards corporate clients and facility managers.

An ever increasing volume of varying and competing building codes, energy standards, development guidelines, and environmental rating tools have been put out by individual trade bodies and different levels of governments. Without proper coordination and control, this will lead to a large degree of duplication, chaos and a waste of resources.

Professional institutions and governance

Previously, one of the obvious missing links in the whole picture of developing sustainable built environments had been the inability to integrate processes, starting points, bottom-lines, targets, variations and optimization, and to provide a hierarchical coordination between professional institutions, the industries, government agencies and society as a whole. This problem is now being addressed, with a range of professional institutions taking on the initiatives to lead the way. For instance, the Green

Building Challenge, coordinated from Canada, incorporates the national Green Building Councils of more than fifteen countries in developing and testing commonly acceptable environmental performance assessment systems that are capable of reflecting varying priorities, technologies, building norms and culture related issues exhibited in these countries (Larsson, 2004). Many of these Green Building councils are quasi-governmental bodies that have facilitated the necessary coordination between public and private organizations. In the Netherlands, the various sectors of the construction industry joined forces recently to assess and specify the environmental impact of their products and services, with the aid of the Dutch Federation of Suppliers to the Construction Industry and the establishment of MRPI, an Environmental Product Declaration Scheme.

As the world's premium professional association providing a global network for international building research and construction innovation, the CIB has made significant efforts over the last decade to pursue sustainability research and promote best practices. Through the work of its many working commissions and member organizations, the CIB now covers a wide range of interests from urban sustainability to future studies in construction, and has produced a wealth of literature such as the *Agenda 21 for Sustainable Construction in Developing countries*. In response to the integration challenge, the CIB has facilitated joint projects and case studies among some of its working commissions and has established new task groups such as the CIB TG55, specializing on integration issues between smart features and sustainable outcomes in developing the built environment (CIB, 2004).

Future development will see international bodies taking the lead in working with national professional institutions and government agencies to encourage, direct, facilitate and manage the changes in industry at regional or even global levels, rather than at a local scale.

The role of information technology

Given these arguments, it can be envisaged that the definition of 'sustainable development' will continue to evolve. Central to sustainability problems is the prediction, direction and control of current, evolving and future technology and knowledge development. As clients, developers, contractors, consultants and the public begin to define guidelines and regulations in response, the industry will have the major task of dealing with the new information and knowledge that has accumulated through practice and research and development. This will quickly become mandatory, given the rate of information growth. The handling of these complex issues and activities can be aided by information technology, a proven tool for managing projects and developing the construction industry.

To date, the use of IT was limited to fields such as the control of intelligent building systems, the computation of embodied energy of building materials and environmental rating tools (ITCBP, 2004). With recognized significance to the construction industry over the last three decades, IT possesses key characteristics that are capable of scoping, benchmarking, modelling, simulating, communicating and finally integrating processes and routines for sustainable development. There are areas that can be dealt with directly and immediately by the deployment of IT tools. Dedicated research will identify future roles and priorities for IT in sustainability equations.

Exemplar projects

Successful sustainable development must respond to the changing nature and constraints in the scope of work, increasing public awareness and improving professional recognition. As the construction industry is known for its reluctance to change, any genuine transformation in our way of working will need to focus on the adoption of sustainability principles by the building sector and public as a whole. To educate the vast majority, it will be vitally important to identify and promote exemplar projects worldwide.

In the global context of natural resource and environment, European countries that have observed tangible improvement through earlier work on sustainable development should lead the way for less privileged countries. Recently the Sustainable Building Symposium has shown a very positive move in this regard by holding regional conferences (SB04) in Brazil, South Africa, China and Malaysia, as a prelude to the main global event, Sustainable Building 2005, in Japan. The success stories of developing countries, such as the sustainable urban settlement schemes developed by the CSIR of South Africa and the Ecological Building and Eco-homes by SRIBS of China, may provide good examples for countries that are still developing economies as the first priority. The dissemination of information on the exemplar projects through news coverage, print media and the Internet will be essential.

The way forward

The way forward will require a new philosophical and comprehensive approach, as opposed to existing technical approach to the phenomena of sustainability. A rationalized, structured, and integrated approach should replace the ad hoc applications and impatient deployment of technologies aimed at immediate benefits and returns. Along the way, practitioners require a knowledge upgrade. Stakeholders require communication. The public requires education.

This book provides an example of such an integrated approach by demonstrating the possibilities in six key development areas of smart and sustainable built environments, from continuing studies on our sustainable living environments (Part 1), emerging tools of the trade (Part 2), and design, construction and operational issues of built assets (Part 3), to the much needed new focus on the stocktake and evaluation of our past performances (Part 4), the management and transfer of sustainability knowledge (Part 5), and the education and process control of sustainable development (Part 6). The chapters include discussions on legal systems and implications of sustainable development, the roles of local councils' actions towards sustainability initiatives, and the issue of time in the lifespan consideration of building performance. There are presentations on international exemplar projects in different geographical regions and countries reflecting varying levels of economic development status and maturity of sustainable initiatives. Much effort has also been directed towards the integration issues between 'smart' and 'green' and the roles and significance of various technologies, including IT, towards such integration. Development processes and needs for effective communication, training and evaluation of experiences and strategies are also amplified through dedicated discussions.

The chapters in the book paint only a small picture of the wonderful work that is taking place around the world. They are mere examples of what international researchers and leading practitioners are capable of in terms of theories, techniques, products, services and information development. Whenever there is sufficient need, knowledge will be made available and technologies will be developed. The crux of the issue for future phases of sustainable development, however, will be moving from an arbitrary, hasty and top-down approach in our thinking to the adoption of a bottom-up need-identification that originates from essential human aspirations.

References

Bentivegna, V., Curwell, S., Deakin, M., Lombardi, P., Mitchell, G. and Nijkamp, P. (2002) A vision and methodology for integrated sustainable urban development: BEQUEST, *Building Research & Information*, 30(2), 83–94.

CIB (1999) *Agenda 21 on Sustainable Construction*, International Council for Research and Innovation in Building and Construction Report Publication 237, CIB, Rotterdam.

CIB (2004) International Council for Research and Innovation in Building and Construction website, http://www.cibworld.nl/pages/ib/0402/TG55Report.html

Grey, R. W. and Halliday, S. P. (1997) Designing and revitalizing communities, in Brandon. P., Lombardi. P. L. and Bentivegna, V. (Eds) *Evaluation of the Built Environment For Sustainability*, E & FN Spon, London, 168–179.

ITCBP (2004) IT & Sustainability, The IT Construction Best Practice Programme website, http://www.itcbp.org.uk/itsustainability

Kibert, C. J. (1997) Environmental cost internalization for sustainable construction, in Brandon. P., Lombardi. P. L. and Bentivegna, V. (Eds) *Evaluation of the Built Environment for Sustainability*, E & FN Spon, London.

Kohler, N. (2002) The relevance of BEQUEST: an observer's perspective, *Building Research & Information*, 30(2), 130–138.

Larsson, N. (2004) Government and private sector actions in improving the environmental performance of buildings, *Proceedings of the 2004 Sustainable Building Conference*, Shanghai (SB04 China), Volume 1, 18–32.

Marsh, C. (2000) Sustainable construction, *Building Services*, September.

Mawhinney, M. (2002) *Sustainable Development – Understanding the Green Debate*, Blackwell Science.

UNCED (United Nations Conference on Environment and Development) (1992) *Earth Summit 92*, Regency Press, London.

WCED (World Commission on Environment and Development) (1987) *Our Common Future*, United Nations, New York.

Yang, J., Guan, L. and Brandon, P. (2003) Decision support to knowledge management of sustainable construction development, *Proceedings of the Joint International Symposium of CIB Working Commissions W55, W65 and W107*, Singapore, Volume 2, 249–57.

Part 1
Creating smart
and sustainable cities

1 Smart sustainable office design – effective technological solutions, based on typology and case studies

A. A. J. F. van den Dobbelsteen[1], *M. J. P. Arets*[1]
and A. C. van der Linden[1,2]

Summary

In order to achieve global sustainability, an environmental improvement factor of 20 is necessary. Assessments indicate that there has been only a marginal improvement in office buildings. A more effective approach is therefore required.

In 2002, we assessed 12 Dutch 'normal' government office buildings on environmental criteria, e.g. the use of building materials, energy and water consumption. This case study identified the most important parameters and building elements that contribute to sustainability.

In order to compare the environmental impact of different design alternatives, to determine the improvement potential, and to find effective solutions, over the last three years, typology studies have been conducted, e.g. of the building geometry, height, bearing structure and the facade. These studies provide directions for smart sustainable technological design solutions, effectively leading to an improvement factor of 20. In this chapter, we also discuss the limitations of technological solutions for sustainability.

Introduction: sustainability in numbers

In order to enable *sustainable development*, by the year 2040, an environmental improvement factor of 20 needs to be achieved. The progress of sustainable building towards this goal can be determined through LCA (life cycle approach)-based tools and by comparison with a general reference for the year 1990, and can be expressed in *improvement factors* or *environmental indices* [1].

Environmental assessment demonstrated that government office buildings, with no particularly sustainable design, on average, achieved an improvement factor of

---

[1] Delft University of Technology, Faculty of Architecture, PO Box 5044, 2600 GA Delft, The Netherlands.
[2] Aacee Bouwen en Milieu, Jan Ligthartplein 39, 3706 VE Zeist.

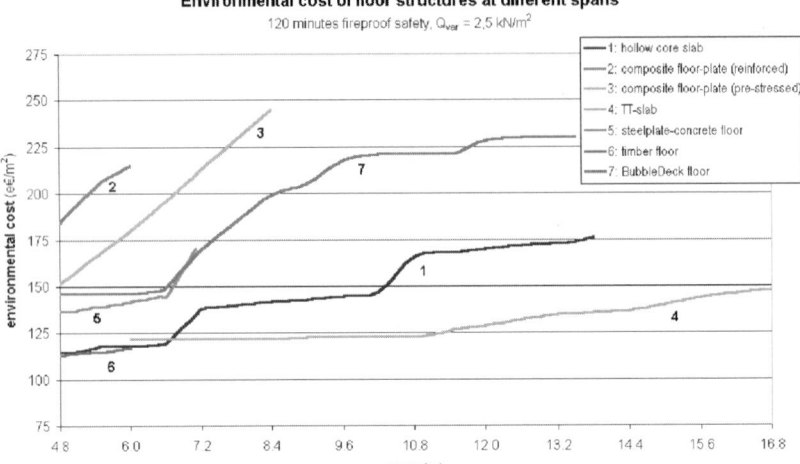

Figure 1.1   Environmental load of different floor structure alternatives at different spans [5].

1.1 to 1.2 [2]. As Figure 1.1 demonstrates, relating the performance achieved to the date of construction shows no clear correlation, indicating marginal environmental improvement over the years.

Buildings that are widely considered sustainable achieve factor 1.4 to 2.5 [3], but, by the year 2000, a factor of 5 would have been necessary to keep a linear target line towards 2040. Therefore, we can conclude that sustainable building has not been as effective as it should have been. *Effectiveness equals importance times improvement potential.* The proportional improvement of the measures applied has therefore been insufficient, and/or building design has been concentrated on relatively unimportant aspects or building elements. For effective sustainable building, solutions need to be found that are related to important elements and imply a large improvement.

Important aspects and elements in buildings

A first step in determining effective improvements is finding the main sources of the environmental load on a building. Taking into account a reference lifespan of 75 years for building materials, 77.5% of the annual environmental load of a building is due to energy consumption during occupation: 19.5% to the building materials, and 3.1% to water consumption.

The predominance of the load due to energy consumption emphasises the importance of energy-efficient technologies and, in particular, sustainable energy sources. Further division of energy consumption reveals three relatively important sources, each accounting for approximately 30% of the environmental load: lighting, heating (and cooling), and equipment.

As a result of these findings, a study was conducted of the building facade, aimed at a simultaneous environmental improvement of energy consumption and the use of building materials [4]. Both existing and new solutions for a facade that is adaptable to changing weather conditions and time of day were analysed. The facades with separate windows for daylight access and outside viewing involved elements for dynamic control of sunlight and daylight and integrated heat and cold storage systems.

Most of the adaptable building facades achieved environmental improvement factors of between 1.2 and 1.4. A combination of roller blinds, high-performance glass panels, and a water-filled heat and cold storage system achieves a factor of 1.8. Note that this relates to approximately 70% of the environmental energy load. The greater part of the environmental load of the building materials (almost 60%) is caused by the bearing structure [2]. Following these findings, a further study was conducted on the bearing structures, focusing on the most important element: floor structure [5]. Figure 1.1 shows the results of the assessment of seven different floor types with different structural spans, all meeting the same functional and technical requirements. The graph shows a maximum improvement factor of 2.0 between the sustainably best and worst solutions at a certain span. The timber floor, TT-slabs and hollow core slabs prove most sustainable.

If the required load-bearing capacity were to be increased, for instance in order to provide a more flexibly usable office floor, the timber floor would no longer be an option. In this case, the relationships between the other alternatives would be similar to Figure 1.1.

Efficient building shape: a theoretical comparison

A typology study of basic building shapes with the same floor space required (see Figure 1.2) was conducted to determine the environmental impact of different

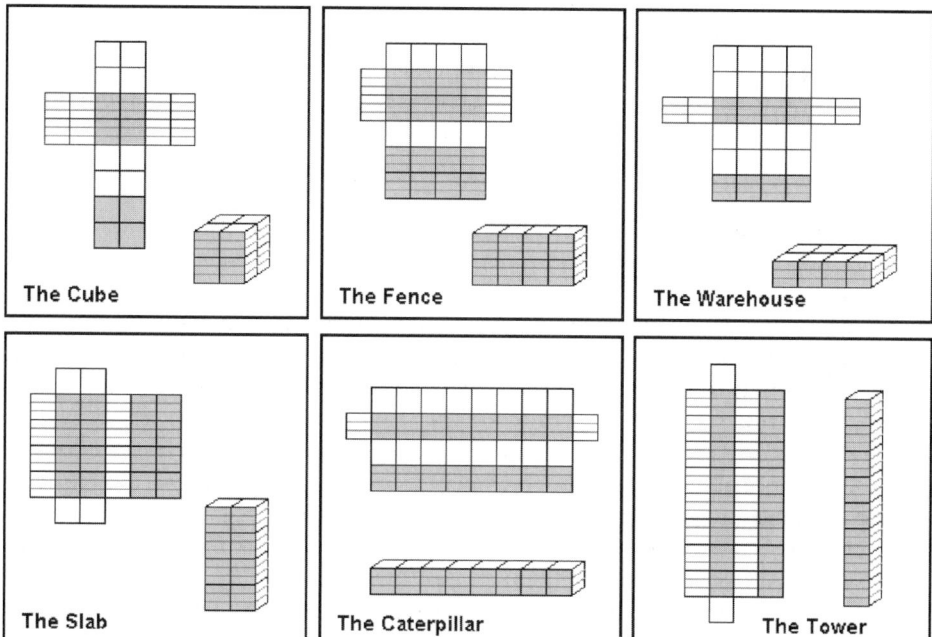

Figure 1.2   Different building geometry typologies with the same floor surface.

Table 1.1 Annual environmental load per m² (in environmental cost) of the main building elements and their importance, based on a lifespan of 75 years.

Element	Environmental cost	Importance	Including:
Ground floor	€1,472.03/m²	2.674	foundation, ground floor structure, floor decks, and floor and ceiling finishing
Roof	€465.29/m²	0.845	roof structure, beams and finishing
Storey floors	€606.54/m²	1.102	storey floor structures, beams, floor decks, stairs and balustrades, and floor and ceiling finishing
Facades	€365.57/m²	0.664	facade structures and cladding
Vertical elements	€373.75/m² GFA	0.679	columns, bearing and non-bearing walls, inner doors and windows, and wall finishing
Terrain paving	€20.27/m² GFA	0.037	terrain paving
Average	**€550.57/m²**	**1.000**	

geometric designs. For each of these shapes, specific geometric characteristics were determined.

In order to determine the importance of different building elements, Table 1.1 shows the environmental load (in environmental cost) per m² *gross floor area (GFA)* of building elements, including finishings, deduced from the study previously presented [2]. *Environmental costs* are societal costs, relating the prevention and defrayal of environmental damage, expressing the extent of environmental damage.

When the factors of Table 1.1 are related to the basic shapes of Figure 1.2, an indication is obtained for the environmental material performance of different building shapes (see Table 1.2). The improvement factors of Table 1.2 show that, theoretically, long (either vertically or horizontally) buildings and those with a large claim on ground area ('The Tower' and 'The Caterpillar'), have a shape that is unfavourable for the environmental impact of building materials. Even taking into account the relative environmental importance from Table 1.1, the most favourable building shape is 'The Cube'.

The impact of the building geometry on energy consumption can be determined accordingly, relating energy for heating and cooling completely, and lighting partly to the building envelope. Lighting was also taken as being partly dependent on the floor depth. All other parameters influencing energy consumption were assumed in relation to the floor area and were therefore equal in all cases. Table 1.3 gives the results.

Table 1.3 expectably demonstrates that the less compact the building is, the greater is its energy consumption. The cubic shape however does not lead to the lowest energy consumption as it has a greater floor depth than 'The Slab' and 'The Fence'. Again, a long shape (lying down or standing up) is least favourable.

In a more accurate comparison model, the alternatives had a cellular office layout with a basic *net floor area (NFA)*, but a GFA defined by the *gross/net ratio*, which is greater

Table 1.2 Importance-corrected impact of different building shapes on the use of building materials.

Shape type Importance >	Ground floor 2.674		Roof 0.845		Storey floors 1.102		Facades 0.664		Others 0.716		Total 1.000		
	Area	Rel.	Area	Rel.	Area	Rel.	Area	Rel.	Area	Rel.	Area	Rel.	Factor
Cube	4	10.7	4	3.4	20	22.0	16	10.6	24	17.2	68	57.3	1.00
Warehouse	8	21.4	8	6.8	16	17.6	12	8.0	24	17.2	68	61.8	0.93
Fence	4	10.7	4	3.4	20	22.0	20	13.3	24	17.2	72	62.7	0.91
Slab	2	5.3	2	1.7	22	24.2	24	15.9	24	17.2	74	63.0	0.91
Caterpillar	8	21.4	8	6.8	16	17.6	18	12.0	24	17.2	74	69.8	0.82
Tower	1	2.7	1	0.8	23	25.3	32	21.2	24	17.2	81	71.2	0.80

Table 1.3 Importance-corrected impact of different building shapes on energy consumption.

Shape type Importance >	Heating & cooling 44.3%			Depth	Lighting 37.6%			Others 18.1%		Total 100%	
	Envel.	Resp.	Relative		Resp.	Relative		Relative		Relative	Factor
Slab	28	100%	0.52	1	50%	0.28		0.18		0.98	1.02
Fence	28	100%	0.52	1	50%	0.28		0.18		0.98	1.02
Cube	24	100%	0.44	2	50%	0.38		0.18		1.00	1.00
Warehouse	28	100%	0.52	2	50%	0.38		0.18		1.07	0.93
Tower	34	100%	0.63	1	50%	0.28		0.18		1.09	0.92
Caterpillar	34	100%	0.63	1	50%	0.28		0.18		1.09	0.92

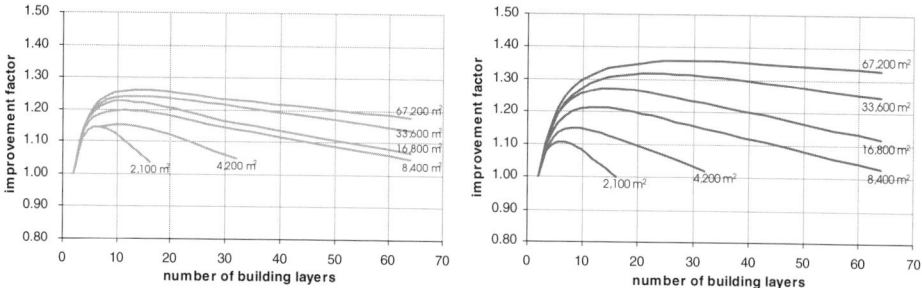

Figure 1.3   Relationship of environmental improvement factors for the use of (a) building materials and (b) energy consumption with the building height, for six net floor areas and a floor depth of 12.6 m.

in the case of high buildings [6]. Applying this model, an optimal number of building layers could be determined with regard to the use of building materials (see Figure 1.3a) or energy consumption (see Figure 1.3b), depending on the NFA required and based on a floor depth of 12.6 m. For the improvement factors, all alternatives were related to the reference of two storeys. When the results for materials and energy were integrated with the non-altering consumption of water, the optimum lies between 6 and 26 storeys, depending on which NFA is applied.

Geometry impact of real buildings

Assessment of real office buildings of moderate height [2] suggests a limited increase of the environmental load of building materials and decrease of energy consumption for higher buildings. However, as this case study was limited to twelve office blocks, the majority of which are less than 10 storeys, the validity of these results is limited.

In order to determine the impact of high-rises, Colaleo [7] compared Europe's tallest building, the Commerzbank Headquarters in Frankfurt, Germany (structural height: 259 m; see Figure 1.4), with a fictitious reference based on an equal number of users and

Figure 1.4   The Central Business District of Frankfurt, Germany, with the Commerzbank Headquarters building at right. Photograph by Andy van den Dobbelsteen.

Table 1.4 Environmental comparison of the Commerzbank Headquarters in Frankfurt, Germany, with two fictitious references [7].

Criteria	Annual environmental cost						Improvement factor	
	Commerzbank	Unit	GFA-reference	Unit	FTE-reference	Unit	GFA-ref	FTE-ref
Building materials	1 902 380	e€	694 690	e€	516 688	e€	0.37	0.27
Energy consumption	3 687 102	e€	2 931 998	e€	2 305 623	e€	0.80	0.63
Water consumption	78 292	e€	100 978	e€	100 978	e€	1.29	1.29
Total	**5 667 774**	e€	**3 947 867**	e€	**3 143 490**	e€	**0.70**	**0.55**
Per employee	2362	e€	1645	e€	1310	e€		
Per m² GFA	58	e€	41	e€	47	e€		

GFA (the *GFA-reference*) and a reference with an equal number of users, however with a referential personal use of space (the *FTE-reference*). The references had a maximum of 24 storeys.

Colaleo drew the conclusion that, in spite of the image of being ecologically sound, the environmental load of the Commerzbank is greater than lower buildings for the same number of people and/or with an equal floor space (see Table 1.4).

Also compared with two lower-rise Dutch ministries that have a comparable number of occupants and GFA, the Commerzbank has a relatively great environmental load. This is mainly due the inefficient use of floor space in the Commerzbank: it has a very high gross/net ratio because of heavy bearing structures and hanging gardens that are part of the climate system. Furthermore, the bearing structure of high-rise buildings imposes a greater environmental load than for average office blocks. In the case of the Commerzbank this is magnified by the choice of maintaining the dimensions of the three main bearing poles of concrete from bottom to top. Only the proportion of re-enforcement steel was reduced at higher levels.

Conclusions and discussion

A selection of the conclusions from the office building assessments and cases studies are:

- Use of sustainable, renewable energy sources and energy-efficient design are the most effective measures for sustainability.
- Facades that adapt to the time of day and changing weather conditions can lead to a factor 1.8 improvement in the use of energy and building materials.
- Bearing structures are important; efficient, relatively light structures are therefore effective. The preservation of the bearing structure is also an important measure to avoid great environmental damage. In order to achieve this, over-sizing of bearing structures seems allowable.

The geometry typology studies led to the following conclusions:

- A cubic shape is both material and energy efficient. For energy efficient consumption, shallow, compact structures give the best results.
- In terms of the environmental load, an optimum number of storeys can be determined, based on the chosen floor area and floor depth.

The Commerzbank study demonstrates the relatively unfavourable environmental impact of high-rise blocks and the importance of an efficient use of internal space (a low gross/net ratio).

The geometry typology studies were based on fictitious, simple offices with cellular floor plans. In the case of more tower-like designs, an open layout with a floor plan deeper than 12.6 m may be expected. For the gross/net ratio of the 32- and 64-storey alternatives, extrapolation of the values by Gerritse [6] was applied, yet a more dramatic increase of the indicator would be realistic. Furthermore, the same material use was assumed for the bearing structure, whereas the variants of 32 and 64 layers would need a heavier bearing structure. Energy consumption will also be greater: for instance, the increased airflow at high altitudes plays a significant role in the heat transmission of high-rise buildings. For high-rise alternatives, the results will therefore be less favourable than presented, as the Commerzbank assessment demonstrated. This outcome, however, does not imply that the Commerzbank is an environmentally unfavourable high-rise design, because a comparison with other high-rise buildings was not conducted. For a more accurate picture, more case studies are required.

More importantly, some advantages of high-rise buildings and densely built urban areas were not included in the environmental performance. In the case of high-rise buildings, there is a great reduction of land use, so building in ecologically more important rural areas is avoided. Furthermore, dense urban areas with high-rise buildings stimulate the use of public transport and thereby improve the transport system, as demonstrated by Newman and Kenworthy [8, 9]. These aspects were not taken into account in the building assessments presented in this section. They were, however, taken into account in the case study of European urban redevelopment by Dobbelsteen and Wilde [10] and demonstrated to be of great importance.

The time factor was also ignored. As with the use of space, a building's lifespan is an important aspect, because the eventual environmental load will depend on it. This subject is discussed in Chapter 6 of the book by Dobbelsteen and Linden. The functional quality, aesthetical value, and cultural meaning of a building, more than its technical quality, are decisive for its eventual lifespan and thereby for the environmental performance.

The studies presented mainly focused on the technological aspects of buildings. Based on findings so far, we think that the greatest effect on environmental sustainability will be achieved through an integrated approach to solutions for the factors of technology, space and time, as well as the initial concept of office work organization that precedes the design and use of office accommodation.

References

[1] Dobbelsteen, A. A. J. F. van den, Linden, A. C. van der and Ravesloot, C. M. (2002) Defining the reference for environmental performance. Anson M., Ko J. M. and Lam E. S. S., *Advances in Building Technology*, 1509–1516. Oxford, UK: Elsevier Science.

[2] Dobbelsteen, A. A. J. F. van den, Linden, A. C. van der and Klaase, D. (2002) Sustainability needs more than just smart technology. Anson M., Ko J. M. and Lam E. S. S., *Advances in Building Technology*, 1501–1508. Oxford, UK: Elsevier Science.

[3] Croes, H., Dewever, M. and Haas, M. (2001) *One number says it all* (poster). The Hague, Netherlands: Dutch Government Buildings Agency.

[4] Bokel, R., Jansen, S., Dobbelsteen, A. van den and Voorden, M. van der (2003) The development of an energy-saving office facade based on adaptation to outside weather conditions in the Netherlands. *Proc. Int. Conf. on Building Simulation*. Eindhoven, Netherlands: IBPSA.

[5] Arets, M. J. P. and Dobbelsteen, A. A. J. F. van den (2002) Sustainable bearing structures. Anson M., Ko J. M. and Lam E. S. S., *Advances in Building Technology*, 1449–1456. Oxford, UK: Elsevier Science.

[6] Gerritse, C. (1995) *Stapeling*. Delft, Netherlands: Publikatieburo Bouwkunde.

[7] Colaleo V. (2003) *Sustainability in Numbers for Sustainable Building*. Turin, Italy: Politecnico di Torino, Facoltà di Ingegneria.

[8] Newman, P. W. G. and Kenworthy, J. R. (1987) Gasoline consumption and cities – A comparison of U.S. cities with a global survey and some implications. Murdoch, W.A., USA: Murdoch University.

[9] Newman, P. and Kenworthy, J. (2001) Sustainable urban form: The big picture. Williams K., Burton E. and Jenks M., *Achieving Sustainable Urban Form*, 109–120. London, UK/New York, USA: Spon Press.

[10] Dobbelsteen, A. A. J. F. van den and Wilde, Th. S. de (2003) Space use and sustainability – environmental assessment and comparison of urban cases of optimised use of space, *Proc. Int. Conf. on Smart and Sustainable Built Environment*. Brisbane, Australia: Queensland University of Technology.

2 Sustainable building: perspectives for implementation in Latin America

V. G. Silva[1] and M. G. Silva[2]

Summary

Sustainable building in Latin America extends beyond the use of renewable materials, energy efficiency and low impact construction. Additional components are fitness for use, durability and adaptability over time, quality of indoor and outdoor areas, use of local materials, and social and economic development including employment, poverty eradication, improvement of income distribution and promotion of regional production.

This text discusses strategies and barriers for the implementation of sustainable building in the region based on four focal points:

(1) long- and short-term balance between building quality expectations at low environmental impact and the need to satisfy basic needs for large proportions of the population
(2) development of sustainable building regional parameters, which can be significantly different from those found in developed countries
(3) difficulties posed by formal and informal construction to sustainable building implementation
(4) introduction of sustainable building in professional education.

The industrialized countries taking part in the first rounds of the Green Building Challenge process (GBC) have developed environmental policies and finalized construction-oriented research investment. This solid foundation facilitated and allowed for immediate work on environmental assessment of buildings. Argentina, Brazil, Chile and Mexico are now part of the GBC. Although it is clear that they

---

[1] Brazilian Green Building Challenge (GBC) Team Leader and Member of International Initiative for a Sustainable Building Environment (iiSBE), Board of Directors. Faculty of Civil Engineering, State University of Campinas, Brazil.
[2] Brazilian GBC Team Member. Technology Center, Federal University of Espirito Santo, Brazil.

cannot replicate methods based on the success they had in other regions, GBC can be a valuable means of introducing concepts and raising awareness.

Introduction

In the southern Latin American region, sustainable building extends beyond the environmental concerns of use of renewable materials, energy efficiency, and low impact construction. Among other essential components are factors such as fitness for use, durability and adaptability over time, quality of indoor and outdoor living conditions, use of local building materials and craftsmanship, enhancement of employment and income distribution, promotion of regional production, poverty eradication and social and economic development. The new political situation in Brazil, the painful evolution in Argentina and the promotion of the Mercosur as a focus for regional development, provide a framework for sustainable building criteria and implementation.

This chapter presents four focal points:

(1) the long- and short-term balance between the expectations of building quality with low environmental impact and the need to confront poverty and satisfy the basic needs of a large proportions of the population
(2) the development of sustainable building regional parameters, related to local materials, climate, cultural factors and living conditions, which are significantly different from those found in developed countries
(3) stages of sustainable building development face difficulties posed by a wide range of variables of formal and informal construction, in order to reduce contrasting standards and improving quality, through both gradual and parallel implementation
(4) the introduction of sustainable building in professional education, including the practice of environmental evaluation of buildings as an integral part of the design process.

Contextualization of sustainable building in Latin America

Latin American countries face massive problems in keeping up the economy, enhancing income distribution, coping with unplanned growth of cities, and satisfying the basic needs of large proportions of their populations. The economic crisis in Argentina revealed the fragility in supplying imported building materials and devastated local materials industries and related employment [1]. Brazil struggles to overcome the reflections of global economic recession and stabilize its currency. Chile has historically concentrated efforts on providing the basic needs of the population, with little room left for environmental considerations [2].

The above-mentioned examples demonstrate how difficult it is to put the environment as a national priority in a region marked by poverty and economic instability. Consequently, the sustainability actions are focused on eradicating poverty, ensuring democracy, improving human settlements and preserving natural resources. Chile is probably the only country in Latin America and the Caribbean that is currently developing sustainability policies for the different industrial sectors [3].

A continuous and active government involvement is essential for the establishment of a sustainable built environment. Public procurement is a basic and important instrument for sustainable building implementation in any economy. In many Latin American countries, however, sustainable building is not a priority. The lack of effective power of the official environmental institutions, and the low degree of environmental concern among citizens were cited as the two major barriers for sustainability in the UN LAC diagnostic on policies and governmental institutions [4]. Without government leadership, lack of funding and of building regulations are major difficulties. On the client side, there is no demand for a sustainable built environment. People are unaware of the impact of buildings, of the meaning of sustainable building, or of the sustainable practices that are already part of the building culture and traditions.

The pressures on energy vary from one country to the other, as do the energy matrices, but in general terms, energy efficiency is not subject to legislation in Latin America, with little awareness of the economic and environmental benefits of building with better thermal quality. In Brazil, a national bioclimatic standard applicable to housing units is about to be launched. A similar instrument is found in Argentina, referring to thermal insulation, though it is applicable only to social housing financed by the government. Conversely, non-residential constructions do not incorporate mandatory standards or even the economic optimum of thermal insulation [1].

In terms of R&D investment, the research community in Latin America is relatively weak: national investment is less than 0.8% of the GNP [5]. Nevertheless, there are active research groups in the region covering most topics of a possible sustainable building Agenda. Reinforcement of this research network and the establishment of partnerships between countries can reduce both the cost and time required to solve the main technical problems, and provide a more appropriate experience exchange.

In the following sections, the discussion is mainly based on the Brazilian context, while establishing parallels to similar situations in Argentina, Chile, Mexico and other Latin American countries.

The Brazilian perspective

The Brazilian Agenda 21 is clearly concerned with urban problems such as transportation, the need for increasing sanitation services and low-cost houses, and with poverty alleviation. As a general document, the agenda does not address tasks for the construction sector and, in general terms, neither Brazilian society, NGOs nor the different governmental levels are really aware of the impact that the construction industry poses on the environment [3].

Sustainability is not currently used as a key criterion in decision making. The competitive factors in the construction sector are centred on the traditional view of cost, time and quality. *Quality*, particularly, is a theme that has gained importance in the last decade, and in some cases it includes *environmental quality* aspects, often understood in the form of energy consumption and reduction of materials wastage. There are several products on the Brazilian market that help to protect the environment, but little offer of global solutions. Also, the lack of basic data, such as climatic information to support building simulation, environmental loads of materials production and use, durability of materials and components in different applications and exposition conditions, is a major barrier to sustainable construction research and policy implementation.

Independent activities on recycling, improvement of energy and water efficiency, and urbanization of shanty towns date back to the early 1980s, but coordinated discussions on sustainability of the construction sector and the built environment are quite recent introductions in Brazil. Unlike in most developed countries, the leadership in sustainable building promotion has not been taken by governmental agencies. In its turn, the private sector is concerned about the cost that is supposedly embodied in 'green' developments that could erode the already-low profit margin for investors and contractors. The universities have been left with the role of demonstrating that, even though the benefits of sustainable construction may result in higher initial costs, they certainly lead to a high quality built environment that provides a better quality of life and a larger financial return in the long run.

Despite several good examples of bioclimatic architecture, and the general use of environment-friendly Portland cement and of efficient devices to reduce water and electricity consumption, so far it is not a tradition in Brazil to design buildings that aim from the outset to reduce environmental loads.

Nevertheless, in the last two years sustainable construction has attracted increasing attention from leading construction industry stakeholders, who were previously mobilized to improve building quality standards under the umbrella of the Brazilian Programme for Quality and Productivity in the Habitat (PBQP-H). This programme is coordinated by representatives of different construction-related industrial associations and can potentially be used to conduct the discussions and coordinate the implementation of sustainable construction in Brazil.

Balancing building quality with low environmental impact and satisfaction of basic needs of the population

In countries where a significant portion of the population relies on really scarce resources, the social dimension of sustainable building must be heavily weighted, due to the need to confront poverty and satisfy basic needs for large proportions of the population.

Brazil has great needs for housing, infrastructure and sanitation services. In large cities in particular, the public transportation system is far from efficient, encouraging people to use private cars, the collected sewage is not properly treated and contaminates watercourses, significant parts of the streets are not paved, and shanty towns (*favelas*) are a common component of the urban landscape, some of them built in environmentally protected areas.

Sustainability in Argentina increasingly implies and is often in preoccupied with social subsistence and provision of basic living conditions. The growing gap between the rich and poor implies that these two worlds coexist within Argentinean cities, but each has a different criterion for qualifying and measuring requirements for sustainable building [1]. A similar panorama is found in Brazil, which shows one of the highest income disparities in the world. Owing to the vast hydroelectric potential, up to a few years ago the energy scenario in Brazil was based on supply side management, with economic growth based on cheap and abundant electric power. A huge energy crisis was deflagrated in 2002, and severe rationing imposed, including financial penalties for over-consumption and risk of unprecedented blackouts. On one side was the majority of the population, which already consumed less than the target amounts or and could not change their habits and reduce consumption. On the other end, there was a minority of the population that could afford to pay the fines and who simply maintained their extravagant consumption patterns, even though the whole population could be left literally in the dark.

In many Latin American countries, the urgency to solve the housing deficit led to official systems of housing financing and large scale production. This resulted in the introduction of industrialized models, a separation between the end user and the habitat production; a loss of community work, sense of identity and linkage to the residential unit, all essential values for a healthy development of human settlements, which are natural for a low income population. Moreover, such models frequently have enormous environmental impact, causing significant alteration of landscape and local topography, vegetation cover and top soil removal, and erosion, such as dominated official housing production in the 1970s in Brazil.

Mitigation of housing and urban services deficits demands a closer partnership between the construction sector and other interested social agents. The accomplishment of this goal indeed calls for the development of new technological solutions that are challenged by the need to balance low financial cost and low environmental impact. However, some actions are not dependent on further technological development and can be taken immediately, with likely positive effects.

Definition of sustainable building regional parameters

Sustainable building implementation calls for a definition of regional parameters related to local materials, climate, cultural factors and living conditions, that can be significantly different from the definitions found in developed countries. Unlike in colder climates, most Latin American countries have the possibility of using outdoor spaces as valuable living areas. The influence of Portuguese and Spanish colonization on local architecture has resulted in housing projects that contain an open courtyard ('patios') that can be used almost all year round. This is a valuable resource giving thermal comfort at virtually no cost beyond the greater use of land, which, one should notice – is not a scarce resource in many Latin American countries. However, it is barely considered in most environmental assessment methods.

Formal and informal construction add complexity to sustainable building development

Sustainable building development faces difficulties posed by a wide range of variables of formal and informal construction, each one with a different definition of acceptable levels and sustainability, in order to reduce contrasting standards and to improve quality. Both gradual and parallel implementation strategies are needed.

As sustainable building is not perceived as a matter for government intervention beyond the limits of central government legislation, there is a lack of basic environmental regulations related to construction activities. A Brazilian regulation referring to C&D waste destination was recently published and, for the first time, a Brazilian municipality (Salvador, State of Bahia) introduced efficiency concepts into its building code. Anyway, *regulatory action is only able to reach the formal construction sector.*

In most Latin American countries this is not enough. In Brazil, roughly 60% of the total housing production is 'self-help' construction. In Bolivia, this kind of production scheme is responsible for between 45 and 55% of the total urban houses produced every year [6]. Mexico shows similar figures: 60% of the total construction is officially classified as informal, and develops in parallel to urban regulations. Two thirds of the dwellings are acquired with no financial or technical support from the private or public sector, based on self-construction, mutual aid and self-management construction systems that find their roots in local cultural aspects [7]. As a consequence, most basic building materials are sold to small consumers by small-scale sellers.

In low income settlements, there is a trade-off between low quality construction and the social and cultural aspects of communal life, such as participative work, economic solidarity and mutual help. Self-managed construction processes increase local productivity and the community identity. These models manage to recover the cost of participation of the homeowner and have values to be retained and enhanced by specific sustainable building policies and actions. However, they do little to reduce the high life-cycle environmental impact of materials and construction techniques used in self-construction.

Self-builders rarely have any help in drawing up the architecture design, resulting in either inadequate indoor space or in a wastage of materials during further modifications and demolition works. Sustainable building policies and initiatives related to the creation of new services and products to be used by them [7], and to the provision of information and technical assistance to raise the quality of self-construction with participation of the end-users and institutions responsible for social housing production are needed and have great potential.

One example of such good practice is Project TITAM (*transfer of technology innovation to self-builders*), managed by State University of Campinas, Brazil. Based on user information on site characteristics and design needs, a technical team runs software on a CAD platform that generates the plans for approval and legalization, as well as options for further extensions [8]. A national prize was awarded to this initiative in 2001.

The need of education for sustainable building

Sustainable building claims for a market transformation composed of, at one end, the transformation of the nature of the demand (client-side) and, at the other, an adequate supply of sustainable design, products and equipments. This market transformation will only be possible if sustainable building concepts are consistently introduced into professional education.

Currently, sustainable building concepts are scattered among several traditional disciplines in Brazilian universities, and none of them includes a structured core of sustainable construction-related disciplines. Though this is beginning to happen at post-degree level, only a small group of professionals is reached. Action is needed both at undergraduate level and to educate professionals who are already practising.

Environmental assessments require the collection of a high amount of information. The ideal situation is to *integrate and practise environmental evaluation of buildings in the design process*. To achieve this goal, strategies are needed for two stages of professional education: *first*, for integration of sustainable design concepts in the design process at undergraduate level; and *second*, for the provision of education and design aids for professionals in the market. Local professional organizations should collaborate to reinvigorate their agendas to bring sustainable building to their members.

The first step taken in Brazil was to build on previous experiences and develop a building sustainability assessment and rating method assessment and rating tool, fully adapted to national and regional conditions. Though it is recognized that design tools for sustainable design and building management are also urgently needed, the ongoing work in Brazil is aimed at newly completed buildings, while designers can use established targets as guidelines for the design aspects on which they should focus.

All construction industry stakeholders in the country were recently engaged in a public consultation process. The idea is to begin with a small, reasonable and consensual set of assessment parameters that are feasible to measure in the short term, and gradually evolve this towards a more complete assessment system. To amplify the consideration of social and economic aspects, the assessment process is designed from the outset to involve not only the building but also agents in the building process, starting with the building contractors [9].

Conclusions

Latin America is clearly under construction, and more construction is needed to satisfy the basic needs of the population. The relatively low degree of industrialization and the prominence of self-built and unregulated construction schemes increase the latent environmental impact of construction-related activities in Latin America, but proper consideration of social and economic development is of the utmost importance.

The most important barriers to a consistent implementation of sustainable building in the region are the low degree of environmental concern of the citizens, the lack of active government environmental agencies and the urgent need to solve huge social disparities.

Education and institutional integration and international cooperation are therefore essential in instrumenting sustainable building. Sustainable construction requires professionals with better environmental knowledge, demanding involvement and active participation in professional associations, as well as building up a specific body of knowledge to be incorporated and developed in the education of architects and engineers, and to promote sustainability-oriented training and technical information in all fields.

Quite a number of companies and industrial associations in Latin America already have environmental friendly products in their catalogues and/or their own environmental agendas. A number of symposia and conferences have been organized and there are several ongoing sustainable building-related initiatives within the region, but better coordination between the different players is plainly necessary. From this point of view, the *construbusiness* approach adopted by the Brazilian PBQP-Habitat seems promising. Such synergic networks have the potential to offer comprehensive environmentally friendly solutions for buildings and other construction products, so increasing their marketing appeal.

Learning from international experience and exchanging knowledge at regional level are also paramount facilitators. One remarkable initiative in that sense is the so-called *Green Building Challenge* process (GBC), which engages a number of countries in the collective development of a building environmental assessment method [10].

Under strong environmental pressure, the industrialized countries taking part in the first rounds of the GBC process have developed environmental policies and construction-oriented research investment. This solid foundation preceded the work on the environmental assessment of buildings. Argentina, Brazil, Chile and Mexico are now part of this group, but clearly they cannot replicate a method based on the success it had in other regions. Regional sustainable building parameters must be established, closely atuned to local contexts.

The large amount of input required by the assessment tool developed and used in the process, *GBTool*, is certainly not often available in most developing countries, but the GBC process itself can, however, be very helpful as an instrument to raise discussion and introduce sustainable building concepts, as the SB02 outcomes made clear. There is also the undeniable validity of taking the GBC process as a catalyst to stimulate the development of national assessment methods, as they happened and continue to happen in many countries, independently of their industrialization status. In must remain clear, however, that though the *developed–developing countries* model of cooperation is important and very welcome, it has intrinsic limitations and, in many cases, South–South cooperation appears to be more effective.

Keynote speakers at previous sustainable building conferences have emphasized the need to increase the focus on tropical and developing countries. In July 2004, a Latin American sustainable building conference was held in São Paulo, Brazil (see website www.clacs04.org). This was an opportunity to increase awareness, analyse region-delineated critical points and ensure that local knowledge on LA-relevant topics was exchanged and will be appropriately brought up in subsequent global events, particularly the mega conference SB'05, to be held in Tokyo in September 2005.

The resultant knowledge base can be further extended by integration with the SBIS (Sustainable Building Information System), an initiative coordinated by the International Initiative for a Sustainable Built Environment (iiSBE). English, French and German interfaces for SBIS are already available and Portuguese and Spanish interfaces will also be available by the time the Latin American conference occurs.

Acknowledgements

The authors thank the State University of Campinas (UNICAMP); the Federal University of Espírito Santo (UFES); the Studies and Projects Financing Agency (FINEP);

the Brazilian Council for Technological and Scientific Development (CNPq); and the Coordination for Human Resources Improvement at Graduate Level (CAPES) for the support provided.

References

[1] Schiller, S. (2002) Social and living conditions become starting point, *Sustainable Building Magazine*, AENEAS Technical Publishers: Boxtel, NL. Issue 2, 32–35.

[2] Goijberg, N. (2002) No demand, but slowly gaining ground, *Sustainable Building Magazine*, AENEAS Technical Publishers: Boxtel, NL. Issue 2, 22–24.

[3] John, V. M. and Agopyan, V. On the Agenda 21 for Latin American and Caribbean Construbusiness – A perspective from Brazil. (To be published.)

[4] Ocampo, J. A. (1999) Políticas e instituciones para el desarrollo sostenible en América Latina y el Caribbean, *Serie Medio Ambiente e Desarrollo 18*, United Nations, ECLAC, Santiago de Chile, 22pp.

[5] Hill, D. (2000) Latin America: R&D spending jumps in Brazil, Mexico, and Costa Rica, National Science Foundation, Division of Science Resources Studies (NSF 00-316). Arlington, VA, 5pp.

[6] Clichevsky, N. (2000) Informalidad y segregación urbana en América Latina. Una aproximación, *Serie Meio Ambiente y Desarrollo*, United Nations, ECLAC, Santiago de Chile.

[7] Pacheco, P. Apuntes sobre vivienda y sostenibilidad en America Latina, Monterrey. 5pp. *s.d.* (unpublished).

[8] Kowaltowski, D. C., Fávero, E., Borges Filho, F., Labaki, L. C., Rushel, R. C. and Pina, S. A. (2001) Automated Design Assistance for Self-Help Housing in Campinas, Brazil, in: *Cultural and Spatial Diversity in the Urban Environment*, 1st edn, Istanbul: YEM Yayin.

[9] Silva, V. G., Silva, M. G. and Agopyan, V. (2003) Avaliação do desempho ambiental de edificios: estágio atual e perspectivas para desenvolvimento no Brasil, *Revista Ambiente Construído*, **3**(3), 7–18.

[10] Cole, R. J. and Larsson, N. (2000) Green building challenge: lessons learned from GBC'98 and GBC2000, in: *Proceedings Sustainable Buildings 2000*, Maastricht, NOVEM/CIB/GBC, October 22–25 2000, 213–215.

3 Demonstrating New Zealand's future residential buildings

K. M. Bayne[1], C. D. Kane[2], R. J. Burton[3] and G. B. Walford[4]

Summary

The signing of the Kyoto Protocol heralds a number of changes that are likely to occur in the building industry over the next 5–10 years. The New Zealand Government's preferred policy package of strategies assumes that steps will be taken to reduce greenhouse gas emissions to the 1990 levels required in the first monitoring period. Additionally, today's buildings themselves are becoming less able to meet the needs of their residents. Changing demographics and lifestyles mean that the functional needs of today's housing is moving away from that offered by 'traditional kiwi' detached quarter-acre houses.

There is a need to demonstrate how to retrofit existing buildings, and also how to build from scratch, to create 'smart and sustainable' residential buildings, suitable for New Zealanders' future lifestyles, and at an affordable price!

This project has three parts, which will be undertaken between 2003 and 2007:

(1) the creation of a 'Now house' – which establishes a benchmark for further demonstration houses, and demonstrates best use of today's technologies in the pathway to creating these types of buildings
(2) the creation of a 'Then house' – which demonstrates how to retrofit an existing house to turn it into the required and desired building
(3) the creation of a 'Future house' – which is a Now house set in the year 2012–15. This will be built from new systems, technologies, and materials that are currently not yet commercialized.

This chapter introduces the concept and strategy of this vision, and outlines the steps and learning we are taking in creating the first of these buildings, the 'Now' house, due to be completed by late 2005.

---

[1] Built Environment Unit. New Zealand Forest Research Institute Ltd. Rotorua, New Zealand.
[2] Built Environment, Building Research Association of New Zealand. Wellington, New Zealand.
[3] Science Futures. New Zealand Forest Research Institute Ltd. Rotorua, New Zealand.
[4] Built Environment Unit. New Zealand Forest Research Institute Ltd. Rotorua, New Zealand.

Introduction

The built environment encompasses the area in which we live, work and play. This usually implies an urban area, and incorporates buildings, parks, roads and infrastructure. In New Zealand, 85% of the population live in an urban area, and therefore the urban regions account for much of our resource use, waste generation, and carbon dioxide emissions. The construction industry has a challenge from two divergent trends: the growing demand for convenient, affordable homes encouraging standardization of design, and rapid construction techniques often incorporating prefabricated modular systems; *versus* a growing awareness of the wider implications of the sustainability agenda, and increased insistence by regulators and the public alike to deliver buildings that perform to increasingly high standards, but have positive environmental and social benefits. The Kyoto Protocol (already ratified by New Zealand) heralds a number of changes that are likely to occur in the building industry over the next 5 to 10 years. The New Zealand Government's preferred policy package of strategies to address the required changes includes a number of end-goals to enable a better lifestyle by 2012. These assume that steps will be taken to reduce greenhouse gas emissions to the 1990 levels required in the first monitoring period. It is also increasingly apparent that today's buildings are becoming less able to meet the needs of their residents. Housing solutions will have to be creative to keep up with demographic trends – an aging population, later marriages, fewer children, growing divorce and re-marriage rates, single-parent families, and people living alone. New Zealand housing consists mainly of three bedroom, single unit houses on an 800–1000 m^2 section.

Low-density development at one unit per 1000 m^2 was common 10–15 years ago, but today is more likely to be aimed at one unit per 500 m^2 [1]. Despite this aim, there are three times as many lifestyle blocks as there were in 1985 (see Bayleys Real Estate information), and rising proportions of four-, five- and six-bedroom dwellings [2]. There is a trend due to student debt, later marriage and child-rearing, and single-parent families towards renting over home ownership. Additionally, poor housing conditions and overcrowding have led to higher risk of disease, and are linked with poor education and work prospects. These conditions are unfortunately commonplace in areas of South Auckland, particularly amongst immigrant Polynesian and Asian families [3, 4].

Global trends are also indicating a number of important elements in the future housing environment, including: energy efficiency (especially affects water heating); waste (deconstruction and prefabrication); resource use (land, water); materials choice (LCA); affordability; easy-care (low-maintenance and little housework); acoustics; lifestyle (house functional needs, flexibility, amenities).[5] A report by CIB [6] outlined similar criteria for building, and the IEA report *Towards Sustainable Buildings* [7] identified in 1997 the 15 high priority collaborative R&D activities to enable sustainable building development as:

- development of building energy systems
- advanced envelope technologies
- indoor environment quality
- whole building design processes and tools
- protocol for environment rating systems

- strategies for sustainability in existing buildings
- sustainable building market development programs
- standards, codes and project specific requirements relevant to energy aspects and sustainable buildings
- finance schemes and incentives to promote the sustainable building market
- information and dissemination systems
- education
- professional training for sustainable buildings
- product/process data sheets
- a factor 10 city in 30 years – what will it take?
- environmental government policy.

The 'post-Kyoto' vision

In assessing the changing needs of consumers and the building industry, and the building stock of today, there is a mismatch between both what we have available and being built today and the types of buildings that will be required at a national and individual level by 2012 to 2015. Additionally, in discussions with industry colleagues, research organisations and government agencies, there appears to be a unified vision of 'raising the bar' with respect to the level of housing the public should rightly expect. This difference is seen in Figure 3.1 in terms of the 'pre-Kyoto' and 'post-Kyoto' worlds, and the disjoint in the pathway to getting there.

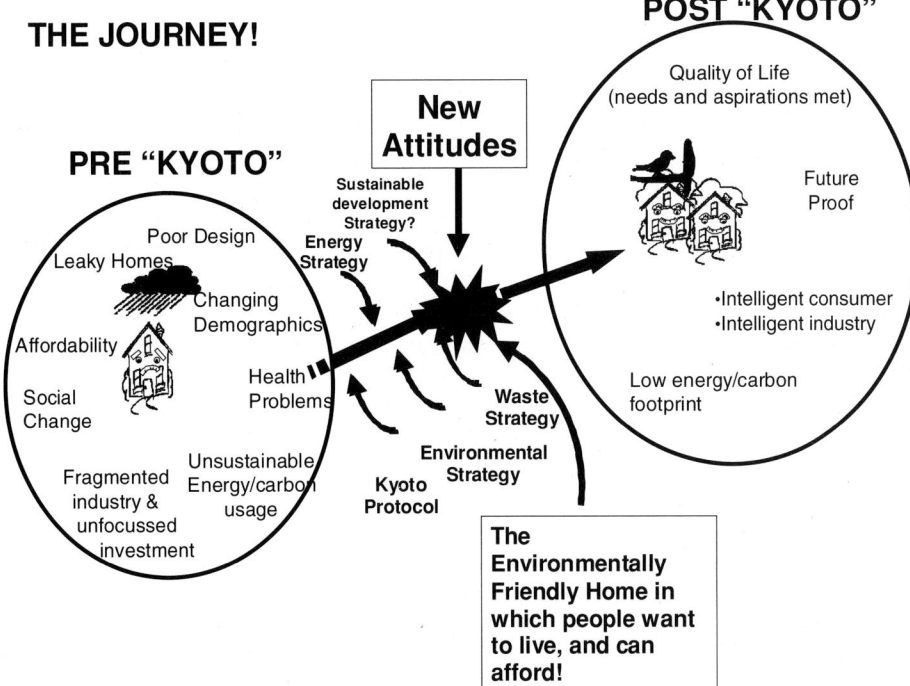

Figure 3.1 The disconnect in buildings today, and the needs of the future decade.

The term 'post-Kyoto' refers to the time period we will be living in after the Government reporting in 2012, in other words, 10–15 years hence. A 'post-Kyoto' building is therefore a building designed for the life and times of the next decade, which meets the likely needs outlined in a recent scenario planning journey [8] and is verified through international Built Environment research, and the recognized needs of central Government for meeting Kyoto reporting [9]. A 'post-Kyoto' building is defined as one *'that enhances the whole of life and quality of life of the inhabitants and the natural environment which nurtures it'*.

Key successes for this vision are:

- a sustainable built environment that people can afford and in which they want to live
- consumer demand for buildings consistent with national climate change, energy and sustainability objectives
- innovations arising from these outcomes that New Zealand industry can exploit for financial gain.

The key to bridging this gap is through provision of three things:

- the changing attitudes to the way people build and use their built environment
- understanding what the 'post-Kyoto' world actually encompasses and educating people about the impacts
- the provision of tools and demonstrable solutions to show people and industry how to move from the 'pre-Kyoto' to the 'post-Kyoto' world.

Although many sustainable demonstration building projects are new-build, in order to make a serious in-road into the housing stock, and deliver buildings that fully meet the needs of the coming decades, addressing what is already here is crucial. This is because only 20–25 000 houses are built per annum in a total stock of 1.4 million. At this rate, it will take 70 years to convert the housing stock fully to meet the needs of the next decade – hardly a successful campaign! Of the future New Zealand housing stock in 2015, 85% is already built [10], 62% of which were built prior to the 1977 insulation requirements and may still have uninsulated walls; many use localized space heating. Most of the newer homes from the last decade are standardized design-build structures, some of which are already experiencing weather-tightness problems. Bringing existing building stock up to new sustainable regulatory requirements, particularly in the case of energy efficiency and durability, will be a major challenge over the next decade. The culmination of this 'post-Kyoto' vision, is the ability to demonstrate how to retrofit an existing building, as well as to build from scratch, to create a 'post-Kyoto' building. This Vision has three parts:

(1) the creation of a 'Now house' – establishes a benchmark for further demonstration houses through best use of today's technologies in the pathway to creating a 'post-Kyoto' building

(2) the creation of a 'Then house' – demonstrates how to retrofit an existing house to turn it into a 'post-Kyoto' building

(3) the creation of a 'Future house' – built from new systems, technologies and materials currently not yet commercialized.

As can be seen, this is a highly ambitious goal, clearly beyond the capability of a single entity or organization's ability to achieve fully. The project has grown in both collaboration and integration of skills and competencies, in order to put together a programme of research to achieve the required end results of success, in order to undertake the practical aspects and fully deliver a first practical step, the 'Now house'.

The 'Now' house project

The 'Now house' project is about a house building approach or concept, for houses in the 'post-Kyoto' environment (2012–2015), but constrained in that it can only use materials/technologies that are currently available/able to be achieved today. The 'Now house' is not a show home, but it physically demonstrates current best knowledge and practice in one possible solution. The 'Now house' is designed with the 'average' New Zealander in mind, rather than as a social housing project, or for the more wealthy customer who would normally gain the expertise of an architect. The 'Now' house, though affordable, is an aspirational 'stretch' target – something that is within reach for the median household income of $NZD48,500 [11], but for which the homeowner would still need to save and work quite hard towards obtaining the 10–20% deposit required for a mortgage.

   The identified likely characteristics of people buying/building new homes from our consumer insight and demographic knowledge has shown that the lifestyle of people living within a 'post-Kyoto' home will no doubt be similar to that of the average society today, although it is quite likely to incorporate persons who work from home, a blended 'family' household (either ethnically, through re-partnering or inter-generational), and children.

   The Forest Research Built Environment team in conjunction with research staff from BRANZ, the Building Research Association of New Zealand, have collaborated with government agencies (EECA, HNZC), local government (Waitakere City Council), industry players through the Fletcher Building GIB Living Solutions partnership programme and sustainable architects in an advisory role throughout the 'Now house' project. These people have been instrumental in deriving the vision and direction of the larger project, and are providing the key skill base to give credibility to a hands-on practical building research project. The partnership of these organizations is leading not only to a strong desire to enact and effect change, but in the desire to see this followed through via active participation in practical research areas.

   Figure 3.2 shows how the Vision aims to be achieved over the next 5+ years, from a rough demonstration of best-practice today through to integrated 'post-Kyoto' houses in a sustainable urban community that meet the needs (and aspirations) of both government, industry and the public, for the coming decades.

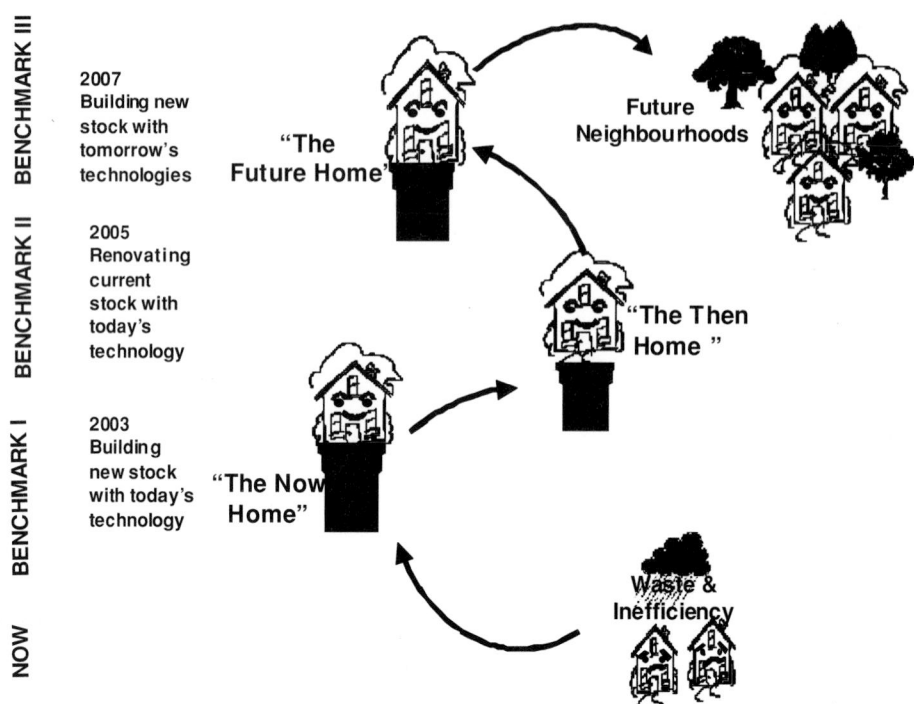

BENCHMARK III — 2007 Building new stock with tomorrow's technologies

BENCHMARK II — 2005 Renovating current stock with today's technology

BENCHMARK I — 2003 Building new stock with today's technology

NOW

"The Future Home"

Future Neighbourhoods

"The Then Home"

"The Now Home"

Waste & Inefficiency

Figure 3.2   Building milestones in the 'post-Kyoto' vision.

The 'Now house' needs to cater to societal and 'green-market' changes, as well as possible, but still have people appeal – demonstrating the potential of future materials and design in a way that people can relate to. The future will be about delivering experiences rather than products, and housing is no exception. Houses are built as a nest, a retreat, a workshop, for entertainment, for storing assets, for wealth accumulation and financial security, and people need to be able to stamp elements of their individual household personality and culture on the home. All these elements are to be provided for (in greater or lesser ways) in the design of a 'Now house'.

The research undertaken during 2003 to underpin the practical demonstration house has encompassed four main stages: Vision; Identifying key features and benefits; Filtering of product and system solutions; and Formulating a design brief. These are shown in Figure 3.3.

Vision

The key mission for the 'Now house' project is in raising awareness of the 'post-Kyoto' issues to the building industry professionals, materials manufacturers, government agencies and the general public. Similar to the scenario planning evaluation [8], successful execution will be seen to be accomplished when the consumer and industry begin to ask questions about the issues they face, and thought-provoking discussions ensue about the future for both the industry and the nation given these needs. In

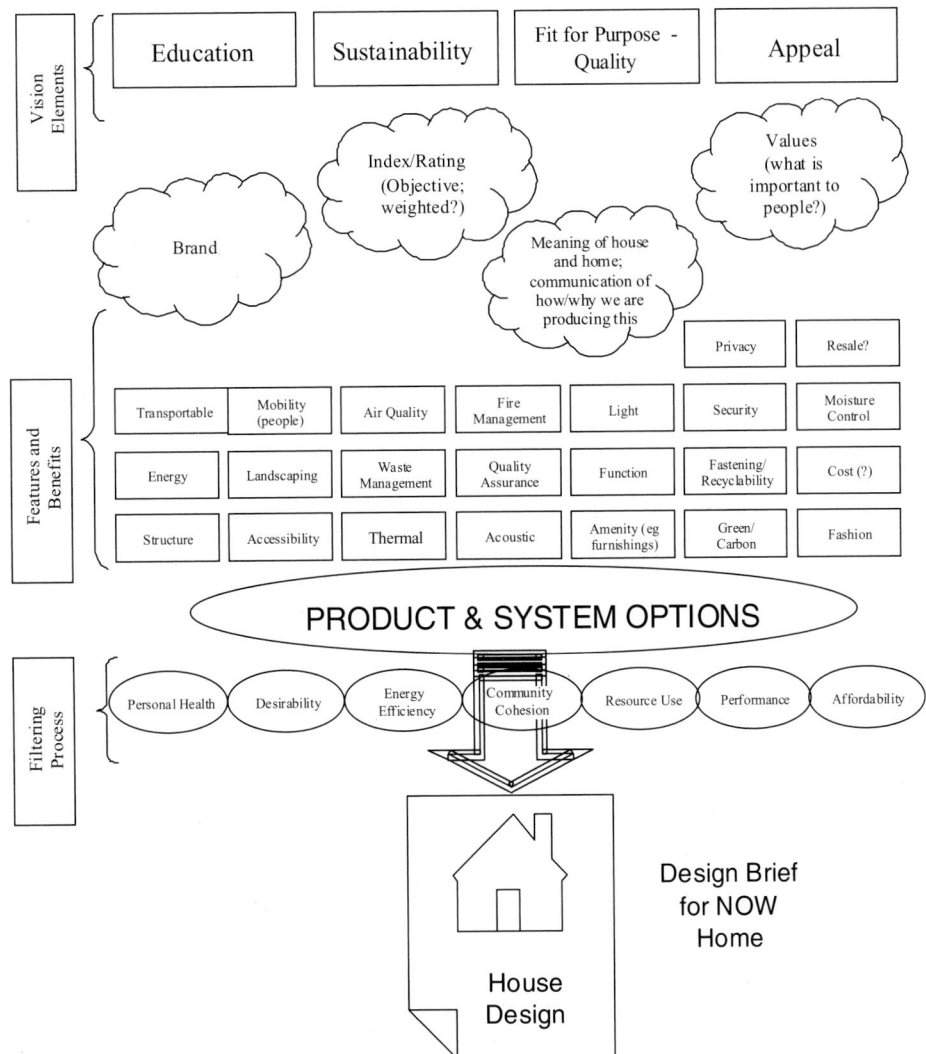

Figure 3.3　The research path for development of the 'Now house' concept.

enabling this to happen, four main success criteria were established:

- Environmental sustainability: Notably, this is just one of the four factors for success-ful 'post-Kyoto' living, however, it was felt that any demonstration house needed to show a 'raising of the bar' with regard to (particularly climate change and energy) issues.
- Quality: Good craftsmanship, code compliance, weathertightness and functionality are key success factors.
- Appeal: To attract interest and stimulate sustainable building as a desirable market factor, rather than for a fringe or 'weird' market sector.

- Education: Through innovation and marketing, to provide and show people smarter ways of tackling the issues.

Linkages

To ensure a good link between these success criteria of the Vision, and the important features and benefits of the house, a series of linking mechanisms were researched to provide a solid basis for filtering products and systems, and ultimately designing the house. These included looking into likely branding opportunities, the values of New Zealand culture, the way houses are used and their inherent meanings, and the index and rating systems available.

Feature and benefits assessment

Over twenty features and benefits were identified as being important for house design, and these were studied in depth to create necessary criteria for a 'Now house', and then transferred to become performance specifications in the design brief.

Design brief

The design brief has been written outlining the core performance specifications for each feature and benefit category. The reason why this has been seen as key to successful implementation of the feature or benefit is spelled out alongside. Usually this relates to best-practice knowledge, research reports or expert advice, so that to the best of our ability all the decisions given priority in the 'Now house' have been justified from a scientific or 'best-practice' position. The brief also has an indication of how the research team aim to measure successful implementation of the performance specifications, and provide a data log for the designers to show tradeoffs between conflicting performance specs, how these were resolved and the justification for the decision.

Products and systems choice

Products and systems are being passed through a seven-point filtering index to choose the elements suitable for the Waitakere demonstration 'Now house'. The filter includes ratings of personal health of the occupants, community health and wellbeing, affordability, desirability, energy efficiency, resource use (meaning land and minerals use, water and waste generation and labour) and future performance.

There has also been incorporated adequate provision for energy, thermal, water and moisture monitoring (wired-house), as well as studies of the construction process in terms of labour and materials use. The project is also to evaluate how far we have moved along the path in terms of provision of the features and benefits listed in order to determine key 'gaps in knowledge' (future research questions) and improve on the provision of these features and benefits in subsequent houses. Adequate evidence of performance, sustainability, appeal and innovation (to educate the public on how to change behaviour/systems and get a 'Now house') will be required to be able to 'brand' the concept, and market it with credibility.

The reasoning for such detailed decision-making comes in the packaging of the information to be useful to other designers and users of the 'Now house' project information. Unlike other demonstration projects, the real value of the work comes not from the physical house demonstration, but from the concept and methodology, which can be used and applied to other sites and situations throughout the country to create further 'Now houses'.

The 'Now house' realized

The project aims also to demonstrate the 'Now house' concept via one possible built solution on a given site. One expression of the 'Now house' concept is presently being designed according to brief, and will be located near Olympic Park, in Waitakere City, Auckland. This version of the 'Now house' concept incorporates a three-bedroom single family home, and is being built with a budget of $NZD150 000. While recognising the limitations of studying a single house in isolation, the project is being built with the needs of a young family in mind, and designed specifically to reflect location and microclimate requirements. The house will also incorporate appropriate landscaping to add value to the site. This building is to be completed in late 2005.

References

[1] Williams, M. (1997) The management of suburban amenity values, Parliamentary Commissioner for the Environment. Wellington.
[2] Statistics New Zealand (2002) www.stats.govt.nz [accessed October 2002].
[3] Statistics New Zealand, SNZ Housing Indicator Project, Stage 1, http://www.stats. govt. nz/domino/external/web/prod_serv.nsf/htmldocs/Housing+Indicators
[4] Joint City Councils Report, Quality of Life in New Zealand's Six Largest Cities 2001, http://www.bigcities.govt.nz
[5] Bayne, K. M., Barnard, T. D. and Maplesden, F. M. (2002): From the shopfront to the street – the future for wooden buildings in a sustainable world, *iiSBE Sustainable Building Conference SB02. CDROM Proceedings*, Oslo, Norway, 22–28 September 2002.
[6] Bourdeau, L., Huovila, P., Lanting, R. and Gilham, A. (1998) *Sustainable Development and the Future of Construction*, CIB Report Publication 225, CIB Working commission W82.
[7] International Energy Agency (1998) Towards sustainable buildings – a workshop on defining collaborative R&D needs, Concept papers and conference report, Hilton Head, South Carolina, USA, 31 August–3 Sept 1998.
[8] Bates, S. L., Bayne, K. M. and Killerby, S. K. (2001) Room for a view. Three visions of the future urban environment in Australasia. *FR Bulletin*, #224. Forest Research 2001.

[9] NZCCP (2002) Climate change: the Government's preferred policy package. *Climate change discussion document*. Department of Prime Minister and Cabinet, April 2002.

[10] Bates, S. L. and Maplesden, F. M. (2001) Futures for wood in the built environment. *Proceedings, Forest Industries Engineering Association Annual Conference*, Nelson, New Zealand. May 2001.

[11] Statistics New Zealand (2002) Thirty-five percent receive investment income, *New Zealand Income Survey June 2002 Quarter*, Media Release, 26 September 2002.

4 Smart and sustainable city – a case study from Hong Kong

Stephen S. Y. Lau[1], J. Wang[2] and R. Giridharan[3]

Summary

According to John Laswick, 'Smart Growth' emerged as a planning concept in the mid-1990s in response to the concern for the proliferation of sprawl development in the USA (Laswick, 2002). He cited the principles of smart growth to include 'Mixing land uses, directing development towards existing communities, preserving farmland and open space, creating walkable neighbourhoods, and providing a range of transportation choices'. The authors of this paper suggest that there is an analogy between the principles of smart growth with those of the multiple and intensive land use (MILU) applications in the compact city of Hong Kong.

In this chapter[4], the authors study the Hong Kong interpretation of smart growth by relating it to the objective of MILU, which is the effective use of land resources within the city through higher residential densities and mixing of other land uses, serviced by a multitude of public transport and infrastructure facilities. The research shows that successful implementation of MILU in Hong Kong depends on mixed land uses. The chapter presents a quick introduction to mainstream mixed development approaches based on MILU, followed by a discussion on the socio and cultural acceptance of a MILU lifestyle.

Hong Kong's challenges

Hong Kong today is a city undergoing phenomenal changes. There is an increasing trend towards land reclamation, there has been insufficient buildable land for the last 150 years, and an incessant population increase due to immigration. Given the hilly territory that presents physical and economic constraints on land supply, the government

---

[1] Center for Architecture and Urban Design for China and Hong Kong, Faculty of Architecture, The University of Hong Kong, HKSAR, People's Republic of China.

[2] Faculty of Architecture, the University of Hong Kong, HKSAR, People's Republic of China.

[3] Faculty of Architecture, the University of Hong Kong, HKSAR, People's Republic of China.

[4] This chapter is developed from a research and serial publication initiated by the HabiForum Research Foundation, the Netherlands, from 2000 to 2002, a project on the applications in New York, London, Hong Kong, Japan and the Neitherlands study of MILU in different countries.

has opted for a high density approach with only 21% of Hong Kong's land area being urbanized. Waves of urban migration from mainland China contribute to a growth in population of one million people per decade. Coupled with the common preference of living in the inner city, where over 50% of its population was recorded, the population density rises to 46 000 people per km^2 in the densest areas (Hong Kong Government, 2002). These habitat and land density figures present a unique debate over a liveable and sustainable urban form.

Multiple use of space – the Hong Kong application

First of all, the high density development is accompanied by mixed-use planning. According to the literature, Rowley (1998) suggests that multiple-use development is, to a large extent, a relative concept, varying in definition from country to country. There are both a horizontal as well as a vertical mixing of uses, but there are no clear guidelines to differentiate between horizontal and vertical developments, or to indicate the number of uses that are necessary to call a building or a complex a multiple-use project. Some planners in the UK argue that any development that is more than 300 m^2, should include a mixture of uses. Similarly, planners in Germany have stipulated that commercial developments should allocate at least 20% of their gross floor area to residential activities (Coupland, 1997). Most researchers agree that properly conceived multiple-use development can bring variety, vitality and viability to an area (Roberts and Lloyd-Jones, 1997; Rowley, 1996). Rowley (1998) argues that it is important to have mixed use development not only within a city block or township but also within buildings both vertically and horizontally. Fine grains of various uses is a precondition for sustainable development.

In Hong Kong, the fine grain pre-condition is viable by using a mixed use approach that offers a close-match to the variety, vitality and viability model discussed. By means of a case study, the authors aim to map the central elements of such a match by studying the following concepts, evolved through a market-driven initiative.

Formulating elements of the compact city

Hong Kong is a classic example of a compact city. The workings of a compact city may be translated into a 'connected' city, owing to observation and analysis of case studies initiated by the authors. The study is based on a selection of different generations of mixed-land use residential developments comprising residential housing, community, recreation and transport facilities, commercial buildings and offices, in some cases.

It can be found that the compact approach to mixing land use is unique because of the high concentration of the facilities mentioned, all of which are within easy reach of one another by means of a web-like system of pedestrian-based connections, that is, sky-walks or pedestrian footbridges, and easy access by public transport. It is observed that vertical development results in a high concentration of residential space in high rise towers and also leads to a close proximity between living, entertainment, landscaping, shopping and transport, exemplifying an important characteristic of a compact city. The efficiency of a transport network relies, to a large extent, on the successful deployment of multi-level transport, whereby pedestrians and vehicles move at separate times and locations from each other, to avoid competition for space.

In this sense, the streets of most developments in Hong Kong occur at different levels. There are streets for people at ground level and at levels much higher than ground zero, as in the podium concept, seen in the case study. There are also roads for cars and service vehicles, again on multi-levels to serve best any development. In a true sense, it is a three-dimension city, where events happen at different levels, above and below ground level.

The degree of proximity has a social-cultural bearing for privacy issues, but in almost every private housing estate, surveys and interviews demonstrated that most residents have either accepted or adapted to the trade-off between privacy and convenience. Besides close proximity, as an integral element of residential developments it is also, in transport terms, a significant contributor to the compact city. As a transport function of a compact development is an efficient and sophisticated city transport system that offers different options of mass transit facilities for all weather circulation to users.

In a compact city, the housing estates are linked and connected with one another to benefit from the 'connectivity' of facilities. Connectivity, in layman's terms, means convenience; its measure can be best represented by the physical distance between home and work, which in turn depends on the physical planning of and efficiency of public transport facilities (Burgess, 2000). In Hong Kong, connectivity or convenience is expressed by the average travel time between home and work, instead of by physical distance such as kilometres. This way of thinking may be explained by the fact that, for locals, time as a measure is more important than physical distance. For Hong Kong, travel time to work (30 minutes) compares comfortably well with that of neighboring cities like Tokyo (45 minutes) and Seoul (40 minutes). For most MILU examples in Hong Kong, the travel time indicator is often employed as an important measure for the social acceptance of high-rise and compact developments.

Vertical living has a close relationship with social acceptance and preferred lifestyle for a particular society. In the Asian cities where there is an abundance of high-rise commercial and residential towers, building height is certainly not a threat to most urban dwellers. Recent experience in public housing in Hong Kong, for instance, has identified an apparent preference of the older residents to choose to live in high-rise towers in the urban centre, in order to enjoy social interaction with neighbors and the community at large. A survey in 2003 in Beijing also revealed that in high-rise towers, senior citizens and residents found the elevators a conductive form of transport, and could go out rather than stay idly at home.

Today, the cities of Hong Kong, Seoul, Shanghai, Guangzhou, Beijing and Singapore are decorated with jungles of high-rise commercial as well as residential towers, to proclaim the way of life of the compact city. Working couples and other walks of life seem to enjoy and take advantage of the compact urban form, in order to benefit from the vitality, viability and variety aspects of mixed land use planning discussed by other researchers above.

A recent survey (Lau, 2001)[5] interviewed residents who lived in private and self-owned apartment buildings, and revealed that the majority of them liked to live on higher floors to enjoy a better view and fresh air, rather than anywhere else, such as low-rise houses in the suburbs. The same survey also allayed concern over the damaging effects on children growing up in high-rise apartments. Mothers were asked about the distancing and separation from the ground of high-rise living that led them to their choice of vertical living – preferences for views, fresh air, and the influence on young children. The majority of their replies were positive with an added explanation that artificial ground, which was a common feature on the roof of most podium decks on which the residential towers sat, offered a welcome solution. The wide adoption of high-density high-rise living is a fact of life to meet the challenge of compact cities and continues to be an incentive for future research. The authors are coordinating a similar opinion poll (2004) for high rise living with the Architectural Society of China on a national scale survey. Preliminary results in cities such as Beijing have indicated a positive preference for high rise living.

Implementation principles for MILU

The following design variables play an influential role in the concept and implementation of MILU concept:

- residential density and development intensity
- land use and amenities
- form and design
- transport modes and pedestrian links
- local property market and environmental quality.

From above analyses, the authors have observed that the success of a MILU planning implementation may be qualified by a number of inter-relating attributes for planning reference. First of all, there is the 'density' of the residents expressed over the built floor area that may be used as a measure of 'intensity' of activities, which in turn depends on 'connectivity' within and with other developments. As already discussed, connectivity is related to the mode of linkage and is 'activity, pedestrian and transport' dependent. In this way, the magic formula for an optimum mix of residential, recreation, amenity, commercial, office, community and transport in any compact development is actually a function of density and connectivity as formulating principles. However, as has already been explained, the success of MILU implementation hinges on social acceptance of it as a preferred lifestyle, in the first place.

---

[5] A 9-month research survey carried out by the author in 2001–2002 on subjective responses of households on various aspects of high-rise living. A total of 102 families in Hong Kong were interviewed about their opinion towards the advantages and disadvantages of high-rise living. Of the 102 surveyed, 98 households enjoyed high-rise living as an acceptable form of urban living.

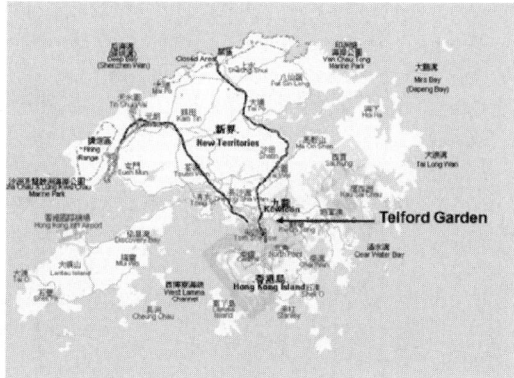

Figure 4.1   Location of case studies.

Case study: Telford Garden, Kowloon Bay MTRC development

Kowloon Bay district has many industrial buildings and open storage areas. The coast-line area has been used for assembling and disassembling cargo in recent years; as more and more industrial activities have been moved to Mainland China, the area is being filled with commercial or industry/office (I/O) buildings. The KMB (public bus) Depot, Sing Tao Press Building and Oriental Press Centre are re-located to Kowloon Bay. There are also many training centres such as the Vocational Training Council, the Clothing Industry Training Centre, the Construction Industry Training Authority, Kowloon Bay Training Centre, etc. to be mixed with a few large public and private residential es-tates such as Richland Gardens, Kai Yip Estate, Kai Tai Court, and the well-known Amoy Garden, where many died during the 2003 SARS outbreak. The general atmo-sphere is not too good for habitation due to heavy traffic and pollution (see Figures 4.1 and 4.2).

The Telford Garden is the largest private estate in Kowloon Bay and is well-used, thanks to its mix-used podium with the station of the Mass Transit Railway

Figure 4.2   The Telford garden.

Figure 4.3   Podium of the Telford garden with all kinds of amenities.

(MTR) involved. The podium, which fully covers the whole site with a four-storey or 15-m tall building, is connected by 24-hour accessible covered walkways from various directions, conveniently linking the MTR with neighbouring developments. On the ground floor is a terminal for long-haul and local commuter buses, maxi-cabs and taxis; while the roof comprises green park, playground, and jogging paths, all for the exclusive use of the residents living in the towers above the podium, as well as visitors. It is interesting to find that these facilities for amenity and transit, at the same time, work as an environment buffer from noise and air pollution (see Figure 4.3).

Being the largest regional shopping and entertainment complex in East Kowloon, Telford Plaza's catchment area is both far and wide. In addition to the 560 000 East Kowloon residents and the 20 000 residents of Telford Gardens, close proximity of the Kowloon Bay MTR station provides a steady flow of commuters averaging 150 000 per day. The rapid development of the Kowloon Bay commercial area also offers a substantial source of shoppers. As mentioned, there are professional and testing institutes such as the City University Campus located in the neighborhood. Telford Gardens consist of 41 residential blocks with a cruciform plan (8 flats per floor, the size of the flat ranges from 473 to 667 ft^2, or 46 to 65 m^2), the number of storeys ranges from 11 to a maximum of 26, making the total number of flats 4992. The total commercial area is 561 569 ft^2 or 55 000 m^2, while government institution and community areas are 9784 ft^2 or 960 m^2. The total number of car parking spaces is 723. There are many different types of stores in the podium: retail and entertainment outlets, cinemas, boutiques, brand name shops, Chinese restaurants, western restaurants, fast food shops, complete grocery markets and all kinds of suppliers and goods. Residents can also visit shops and grocery markets at nearby Amoy Gardens, 10 minutes walk away (Figures 4.4 and 4.5). The main modes of transport in Kowloon Bay are buses, minibuses and the MTR subway. There are bus routes that go to Hung Hom, Sai Kung, etc. Along Kwun Tong Road, there are also airport bus stops. This makes it very convenient for the airport. By MTR it takes 35 minutes to Central, 30 minutes to Tsim Sha Tsui. By shuttle bus,

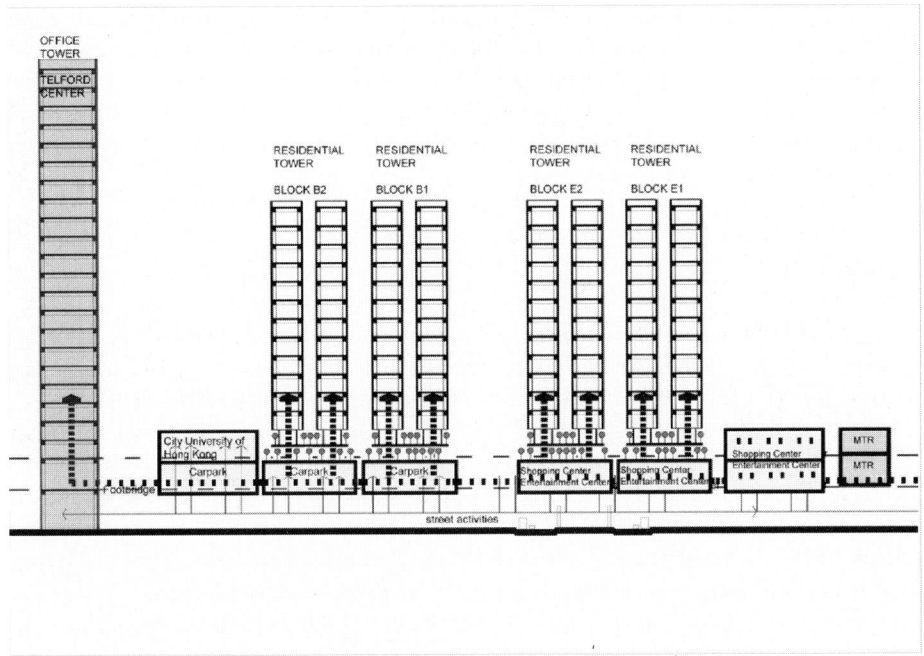

Figure 4.4   Segregation of pedestrians and vehicles.

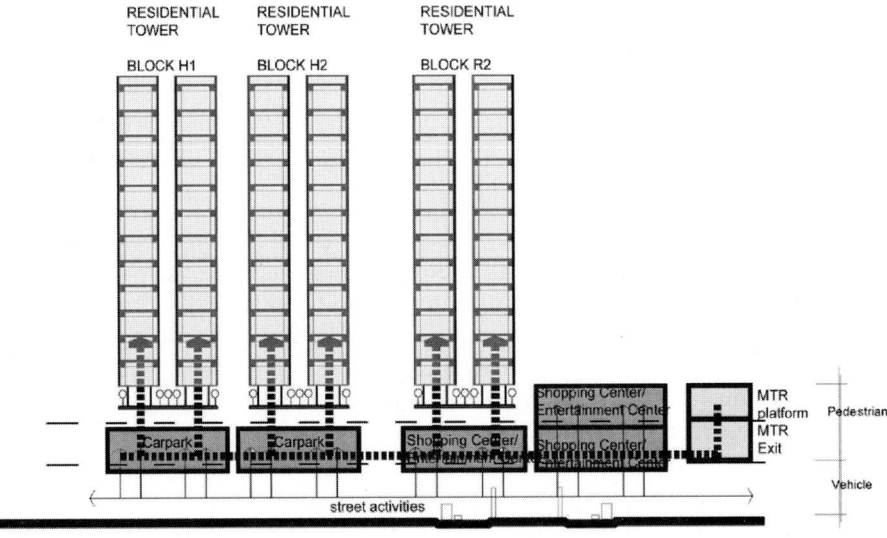

Figure 4.5   Conceptual diagram of a MILU planning in the Telford garden development.

it takes 50 to 60 minutes to reach the airport. There are many sports facilities, such as Kowloon Bay Sports Ground, with football courts, basketball courts, rugby courts, indoor games hall, etc. Nearby is a public park in Kowloon Bay called Hoi Bun Road Park. The Telford Gardens presents the ultimate evolution of MILU in Hong Kong that has come about by a pursuit of high quality combined living and working urban environment.

Conclusions

From the above study, the following is recommended for further research. The case study of Telford's represents a market led approach to achieving optimization of resources that would meet the expectations of the majority of users; in other words, a planning strategy where the physical environment gives a familiar way for human activities and interaction in a constrained-resource situation – given a shortage of land and large population. Over time, private builders have worked out a strategy for designing affordable buildings and built environments that give a high quality of service, and are secure, safe and comfortable to the users. As mentioned, some aspects of living, such as privacy, are compromised by economically deficient home dwellers.

So, the study has highlighted an avenue for further research in the realm of the social acceptance of urban lifestyle as a way of understanding the appropriate urban form for a particular society.

In Hong Kong, the relationship between people and the environment reflects a unique dichotomy between privacy and communal life. In Hong Kong, people sought privacy by avoiding interaction with others on the same floor but interacted with people on other floors (Forrest *et al.*, 2002). Once out of their small flats, people intermingled with people in the neighborhoods and districts without restriction, regularly dining out with their friend and relatives. Living in a constrained environment could be attributed to Chinese history and culture (Zhang, 2000). In the case of present-day Hong Kong, space proximity or the limited distance between buildings is taken as a tolerable rather than an acceptable spatial attribute. This close proximity has given Hong Kong residents both economy of time and space (Forrest *et al.*, 2002).

In Hong Kong the use of urban space for MILU development is closely linked to the culture and life of the people. According to Hughes (1968), the people of Hong Kong enjoy a transient way of life. The cultural aspect of this model of compact multi-functional use of land resource is interesting, as research shows that inhabitants seem to be satisfied with living in a small apartment in a high-rise building as it offers a high standard and quality of life, with a great view of beautiful Hong Kong harbour, where green hills and the sea are always close by. It is also a way of urban life, with proximity of facilities, friends and family, reliable and convenient public transport, work and shops, as well as an acceptable travel time from home to work. For the children there are always schools within walking distance and lots of friends to play with (Hong Kong Government, 2002).

In essence, there is a strong belief among professionals that multiple intensive land use has the potential for achieving sustainable development, but as yet little empirical evidence is available, except that the comparisons of GDP, crime rate, divorce rate, health, etc. provide some confirmation for such a belief.

Since all land in Hong Kong is leased by the Government, lease conditions become an instrument to integrate planned land uses, building densities, achieve desirable building forms and to provide essential urban provisions. Various development controls, in particular the lease conditions and related building development control regulations, have facilitated the evolution of the current multiple intensive land use forms in Hong Kong. Hong Kong's high-rise, high-density and mixed-use developments are enhanced by four design concepts at the primary nodes: space proximity, compactness, verticality and sky city, which together enhance variety, vitality and viability. For Hong Kong, MILU is integrated with high-density, high-rise, high floor-to-area ratio (plot ratio), which continues to offer urban dwellers an exciting and comfortable lifestyle that prospers and, more significantly, influences the present and future prospects of the 600 or so expanding cities of China.

The formulating elements for a livable compact city or development, have been been discussed. The most significant conclusion relates to the complex and dynamic interplay of the formulating elements, which warrant a more systematic approach of case study in future in order to find the magic formula for a successful mix, and intensity of land uses by both qualitative and quantitative methods.

Acknowledgements

Thanks are due to S. Ganesan, H. M. Chan, P. Y. Li for their support.

References

Burgess, R. (2000) The Compact City Debate: A Global Perspective. In *Compact City – Sustainable Urban Forms for Developing Countries*, Jenks, M. and Burgess, R. (eds), Spon Press, London, pp. 9–24.

Coupland, A. (1997) *Reclaiming the City: Mixed-Use Developments*, E&FN Spon.

Forrest, R., Grange, A. La. and Ngai-ming, Y. (2002) Neighbourhood in high-rise and high-density city: Some observations on contemporary Hong Kong. See http://www.neighbourhoodcentre.org.uk

Hong Kong Government (2002) *Hong Kong in Figures*, Department of Census and Statistics, Hong Kong.

Hughes, R. (1968) *Hong Kong Borrowed Place – Borrowed Time*, Andre Deutsch, London.

Laswick, J. (2002) Evolving beyond sprawl in America: smart growth, new urbanism, and sustainable development, paper presentation at the Habi-Forum Expert Workshop on M.I.L.U., May 2002, The Netherlands.

Lau (2001) A survey on residents' response to high-rise living in Hong Kong, Hong Kong.

Planning Department (2002) Hong Kong 2030 Planning Vision and Strategy, Stage 1 Public Consultation Report, Hong Kong: Printing Department, 141 pp.

Roberts, M. and Lloyd-Jones, T. (1997) Mixed uses and urban design'. In Coupland, A. (ed), *Reclaiming the city – Mixed use development*, E&FN Spon, London.

Rowley, A. (1996) Mixed use development: ambiguous concept, simplistic analysis and wishful thinking? *Planning Practice and Research*, **11** (1) 85–98. See http://weblinks2.epnet.com/citation.asp

Rowley, A. (1998) Planning mixed use development: issues and practice. Research report funded by Royal Institute of Chartered Surveyors, UK.

Zaman, Q. M. M., Lau, S. S. Y. and So, H. M. (2000) The compact city of Hong Kong: a sustainable model for Asia? In Compact Cities – Sustainable Urban Forms for Developing Countries, Jenks, M. and Burgess, R. (eds), Spon Press, pp. 255–268.

Zhang, X. Q. (2000) High-rise and high-density urban form. In Compact Cities – Sustainable Urban Forms for Developing Countries, Jenks, M. and Burgess, R. (eds), Spon Press, London.

5 Evaluation of scenarios of a Southern-European intelligent city of the future

Patrizia Lombardi[1] *and Steve Curwell*[2]

Summary

Recently, EU policies and programmes (EC, 2002) emphasized the role of information and communication technologies (ICTs) in creating an integrated city information infrastructure for the future, founded on the four pillars of sustainability (social, economic, environmental and institutional), as well as cultural heritage, participation in governance and quality of life.

More specifically, the EU Information Society Technologies (IST) programme argues for new opportunities of the knowledge society (KS) by a networked 'know how' and 'know why' that can break the vicious circle of poverty and waste.

ICTs can help deliver more sustainable urban decision making through 'virtual' urban spaces, simulation of innovative physical forms and infrastructures and interactive decision support systems. ICTs make a networked knowledge world society possible, facilitating fast and cheap global communications and widening the scope for change.

This chapter illustrates and evaluates a range of alternative scenarios for the city of the future. These scenarios would seek to combine both more sustainable urban forms and infrastructures that facilitate the knowledge society and governance, planning and city management processes using ICTs that will deliver more sustainable cities.

These scenarios have been developed by the INTECITY roadmap project funded by the EU Information Society Technologies (IST) Programme during the first workshop organised in Oslo in September, 2002.

The evaluation makes use of multicriteria analysis, which provides a systematic process for trading off the effects of various alternative, synthesizing individual contributions and values. In this application, the evaluation criteria and the weights and preferences were identified by the INTELCITY Mediterranean stakeholders during a workshop organized in Turin in October, 2002.

The structure of the chapter is as follows: the next section illustrates the alternative scenarios established by the INTELCITY consortium; followed by the application of an MCA to the scenarios; and finally some concluding remarks.

---

[1] City and Housing Department, Polytechnic of Turin, Turin, Italy
[2] School of Construction and Property Management, University of Salford, Salford, UK

The scenarios for the city of the future

INTELCITY (www.scr.salford.ac.uk/intelcity) is a one-year research and technical development (RTD) roadmap project funded by the EU Information Society Technologies (IST) Programme. It aims to explore new opportunities for sustainable development of cities through the intelligent use of Information and Communication Technologies (ICTs). The INTELCITY network developed visions and scenarios for a city of the future with a time horizon of 2030 during the first workshop held in Oslo on 24–25 September 2002. All of the scenarios try to link the use of ICTs with sustainable urban planning and design activities.

The alternative scenarios are:

- the e-democracy city
- the virtual city
- the cultural city
- the environmental city
- the post-catastrophe city.

In the first scenario, the *e-democracy city*, ICT is seen as an enabling mechanism, changing peoples' opinions and behaviour patterns through information provision and empowerment. It provides new ways of decision-making and negotiation through inclusiveness and accessible participation in the decision-making processes that affect the community. This would involve the deployment of a variety of tools such as hard set accessible web based group decision support for visioning, advanced visualisation tools, scenario planning and automatic translation, as well as democratic participation tools such as e-voting and highly devolved decision-making. A peer-to-peer information system architecture is necessary, rather than a client/server relationship, which would seek to provide educational support to enable the understanding of the socio-economic and environmental impacts of the various options under consideration in a way that dissolves boundaries, e.g. between interest groups and between physical and virtual worlds.

The e-governance characteristics include:

- rethinking the role of community education and ways of opinion-forming
- the creation of new platforms for articulation of interests
- decentralised, transparent and open decision making at all levels
- enhanced organizational capacity (structures, knowledge, resources, motivation) to enable an iterative two-way flow of information between the community and decision makers (public and private organisations) that together will lead not only to better ways of making decisions, but also to better, more sustainable, community decisions through a more self-determined, informed consensus and more accountable leadership.

Thus the scenario is one in which self-determinism occurs within a socially cohesive community, respecting the wider community and the environment.

The second scenario, the *virtual city*, represents a knowledge society of networks and flows, where citizens are able to work and live anywhere in the city, supported by intelligent environments that are 'lean, green and smart' (economically efficient and ecologically sound). Ubiquitous computing and telework lie at the centre of work and life, supported by smart transportation and logistics networks. It also depends on the development of intelligent agents to provide personalized, self-tailored information to support a culturally rich, mobile lifestyle.

It is recognized that this will require major infrastructural development in the supply of utilities and that both the economic and environmental transformation that this predicates would also require 'strong' governance. This requirement however, provides the opportunity to use the ICTs and smart technologies both to empower citizens and to make the corporate sector socially responsible (i.e. by taking on economic and environmental issues). New social support groupings are established such as local virtual village halls, as well as more dispersed partnerships (e.g. pressure groups) across the whole city and beyond. Together this offers the potential for a more socially inclusive and, therefore, more progressive form of governance.

If dispersed, virtual cities are to emerge, they need a stronger cultural basis and their socio-economic and environmental implications need to be more clearly articulated.

The third scenario, the *cultural city*, is underpinned by a strong social and environmental ethic. The principles of the cultural city define its social and political organization, emphasizing liberal democracy, equity, multicultural diversity, and cultural and spiritual integrity. It provides the ideal community environment where individual needs and well-being can be balanced with that of society in general.

Advanced technology is present although entirely invisible – it is embedded. This applies to all forms of technology, including mobility and communication technologies. In the cultural city the additional means and channels of communication that ICTs provide assist, imperceptibly and unobtrusively, in all aspects of life, work, education, art, leisure and the democratic processes. It is particularly useful in the negotiation and mediation necessary for agreeing collectively, e.g. for urban re/development action.

The cultural city has a distinct physical form, which is human in scale. This is a high quality environment in which the buildings and landscape are predominant, rather than the mobility systems. The built heritage provides cultural symbols, a strong, sustainable, binding factor between the present and the past. At the heart is a civic entity, space(s) for cultural and artistic exchange and democratic engagement. Thus it is framed and is facilitated by a fundamentally democratic decision-making process. This begins with commitment to stakeholder definition of goals and objectives for a new sustainable future and its delivery according to a democratic expression of rights and responsibilities, individual, corporate and governmental.

The fourth scenario, the *environmental city*, is based on the realistic concept of incremental change in which all the current environmental and social problems are gradually addressed, solved or ameliorated over the period up to 2030. ICTs would play an important part in this enviro-evolution through integration into society in ways that are both enabling and contribute to the quality of life.

In physical and ecological terms the environment city would be very resource efficient achieving factor 4 reductions (at least). An important legal aspect supporting this is the requirement on all manufacturers to recycle/reuse at end of life, thus closing liability loops, which in turn would reduce pollution to very low levels. Resource reduction targets would constrain the outward expansion of cities, with polycentric (internal) growth patterns to reduce mobility requirements and provide greater flexibility and with a better balance in the investment in public and private spaces. Regeneration of the existing fabric would provide greater density of habitation, more effective and efficient public transport and local food production.

In economic terms, full employment (in its wider sense meaning some form of re-warding activity) would be stimulated through the adoption of a service economy where innovative means of extending product life would be sought. The value of 'public goods' would be rediscovered.

In social and cultural terms a multi-cultural, secure and safe society will be achieved through a combination of governance measures and the application of ICTs. This would seek to increase 'self-organizing' capacities through better public information and par-ticipation in urban re/generation so that cities could become lively cultural places, which should tend to decrease current problems with crime and insecurity.

The environment city can be summed up as being of long life, low energy and adaptable. This could be applied equally to the buildings and infrastructure as much as to the citizenry.

The last scenario, the *post-catastrophe city*, is driven by the albeit small possibility of high impact events, such as a natural catastrophe (e.g. a very large volcanic eruption or a meteor strike), environmental catastrophe (e.g. serious global warming causing loss of ice cover at the poles with significant sea level rise of metres) or global war, which may be sufficiently serious to also cause breakdown in the economic system and/or global trade. This would considerably alter the relationship between citizens and the state and impact significantly on the freedom of choice of citizens. This could be exemplified by a crisis in the use of fossil fuel and its adverse impacts on climate and environment and the possible imposition of a carbon tax or very large resource reduction quotas.

It is likely that wealth and quality of life would suffer initially with effective action being seen to be imperative in order to improve matters. Free market conditions would be unlikely to respond quickly enough to events or to respond sufficiently to alleviate the threat(s). Government(s) would have to impose regulations to control behaviour of markets, citizens and industry, assuming, for example, that the cost of carbon misuse is sufficient to make it worth saving. Complexities and uncertainties in this situation are large. Means of encouraging rather than enforcing more sustainable behaviour would be relevant, e.g. trading carbon credits, giving interest on carbon accounts, etc.

Among these uncertainties, information and communication technologies would be highly relevant to such new forms of rationing, i.e., in the equality in allocation of resources. Clearly it can enable real-time monitoring of resource use, carbon in the example, by individual citizens, companies and other organizations.

Evaluation of the scenarios

A multicriteria analysis, named the *Analytic Hierarchy Process* (AHP) is used in this study with the aim of evaluating the five scenarios described in the previous section. AHP is able to determine a list of priorities from a finite series of choice options (the alternative scenarios), which are assessed and compared in relation to identified characteristics of the problem (criteria) when it is appropriately broken down into its fundamental elements (Voogd, 1983).

The selection of a relevant set of evaluation criteria and their weights is crucial for the development of an MCA, as it directly influences the results (Voogd, 1983). There are two possible approaches for selecting a set of criteria in an evaluation context: a deductive approach, starting from the characteristics of the scenarios or, alternatively, an inductive approach, starting from the personal view of decision makers and from their definition of SUD.

The criteria used in this study have been identified by the delegates of the Regional INTELCITY Workshop held in Turin on 30th October 2002, by adopting an inductive approach.

In term of skills employed, the group of delegates included: planning authorities, civil and traffic engineers, architects, planners, economists, technical services, academics and post-graduate students. In addition to the above group of experts, the workshop included a number of under-graduate students from the School of Architecture of the Polytechnic of Turin, and a group of non-European post-graduate students (architects and engineers) from developing countries (Mexico, Columbia, Africa, Vietnam).

The delegates were asked the following questions.

What are the attributes (issues or criteria) that better illustrate your city of the future (year 2030)? What are the indicators or parameters for specifying them? What are their priorities or weights?

This exercise required the delegates to have an explicit individual vision and definition of sustainable urban development and to discuss (eventually negotiate) it with the group. The final discussion emphasised the variety of issues characterising the city of the future in a Southern-Mediterranean knowledge society. However, it was possible to recognize some common issues among the participants, and particularly the strong emphasis placed on the social and cultural issues of sustainability and quality of life, and the limited role assigned to ICTs aspects, seen simply as a transparent means of innovation in the infrastructure and institutional sectors rather than a crucial factor of change for the urban scenario (e.g. the virtual city).

From a more detailed analysis of the individual sheets (31 sheets in total) filled out by the workshop delegates, the following set of common (more frequent) issues emerged.

(1)	Ecological issues	This includes environmental quality, greenery, no pollution, etc.
(2)	Infrastructure efficiency	This refers to transport systems, ICT, accessibility to services.
(3)	Social interaction	This includes social integration, multi-ethnical environment.
(4)	Quality of life	This refers to architectural renewal, cultural life, urban security.
(5)	Institutional efficiency	This refers to governance, participation, democracy.
(6)	Economical development	This includes growth (occupation, GNP) and equity (no poverty).
(7)	Spatial issues	This deals with population and the need to limit the current urban sprawl.

Table 5.1   Frequency anaiysis of the issues.

Issues	Experts	Students	Non-EU	EU	Total
Ecological issue	11	4	4	15	19
Infrastructure eficiency	15	6	4	21	25
Social interaction	9			9	9
Quality of life	13	4	1	17	18
Institutional efficiency	9	5	3	14	17
Economical development	7	1	3	8	11
Spatial issue	3	1	3	4	7

Table 5.1 shows the number of time the above issues were indicated by the respondents in their individual criteria-identification sheets.

It also shows that:

- Infrastructure efficiency is placed first by all three groups of respondents.
- Quality of life is more important than ecological issues only for the EU delegates in term of frequency in which the issue recurs (but it is less important than ecological issues in terms of priority assigned by respondents, especially for the non-EU respondents).
- Institutional issues are less important than ecological issues for everyone.
- Quality of life is more important than social issues for everyone.
- Social issues are more important than economic issues only for the EU respondents.
- Spatial issues are the least less important issues for the EU delegates.

A final vector of criteria weights W1 (Table 5.2) was adopted in the multicriteria evaluation, which is based on a normalization of the numbers included in Table 5.1 – column EU (group experts and students).

The following step is a pairwise comparison of the five alternative options: (a) the e-democracy city, (b) the virtual city, (c) the cultural city, (d) the environmental city and (e) the post-catastrophe city, in relation to the selected evaluation criteria.

Using the AHP method ('Expert choice' package, 2000), seven matrices have been developed; each matrix contains the result of a technical evaluation based on pairwise comparison of the scenarios with regard to each individual criterion, answering the following question:

'Which is the more important between the two alternatives (a) and (b), with respect to this criterion? And by how much?'

These comparisons are based on the nine-point measurement scale (Saaty's scale). In addition, the consistency ratio (R.C.) was checked for each matrix, as a value lower than 0.1.

Table 5.2   Weights vector, W1.

Criterion	1-Ecologic	2-Efficiency	3-Social	4-Quality	5-Institution	6-Economic	7-Spatial
Weight	17	24	10	19	16	9	5

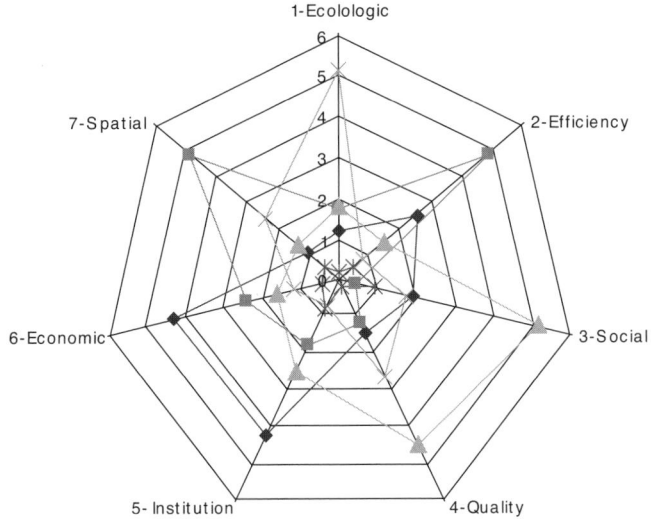

1-Ecolologic

7-Spatial

2-Efficiency

6-Economic

3-Social

5- Institution

4-Quality

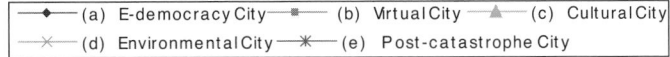

| ──◆── | (a) E-democracy City | ──■── | (b) Virtual City | ──▲── | (c) Cultural City |
| ──✕── | (d) Environmental City | ──✳── | (e) Post-catastrophe City | | |

Figure 5.1 Radial graph of the scenarios' performance.

Finally, the eigenvector (w_{alt}) from each matrix is automatically obtained; this provides the ranking of the alternative scenarios in relation to each criterion, as illustrated in Table 5.3.

Figure 5.1 provides a visualization of the performance of the scenarios illustrated in Table 5.3 on a radial model.

The final vector of composite weights for the alternative scenarios that was obtained in the last step of the AHP application from the criterion-weights of W1 is shown in Table 5.4. These final weights show that the *Cultural city* is winning, followed by the e-democracy city, the virtual city and environmental city. The first three scenarios are quite close to each others in terms of final weights. This is mainly because of the high weight given to all the criteria of infrastructure efficiency, quality of life and institution capacity, by the Mediterranean respondents.

In the following, an additional exercise is developed in order to control what in literature is defined 'the uncertainty of the method' (Voogd, 1983), i.e. the dependency of the final results on the MCA method used.

In this analysis, a different MCA method is adopted, namely 'Regime analysis', developed by Hinloopen *et al.* in the Netherlands (Hinloopen *et al.*, 1983) (MCA package version 4.4, developed by Nijkamp and Albers, 1991).

The regime method is a qualitative multiple analysis, developed in the area of soft econometrics. It is based on a combination of Kendal's paired comparison method for ordinal data and logit analysis. The method does not require cardinal data, but only

Table 5.3 Relative weights of the five scenarios with regard to each criterion.

Issues	1-Ecologic	2-Efficiency	3-Institution	4-Social	5-Quality	6-Economic	7-Spatial
(a) E-democracy city	1.2	2.5	4.2	1.9	1.5	4.3	1.0
(b) Virtual city	1.7	4.8	1.8	0.4	1.2	2.4	4.8
(c) Cultural city	1.8	1.4	2.5	5.1	4.5	1.6	1.3
(d) Environmental city	5.1	0.8	0.7	1.7	2.6	1.2	2.4
(e) Post-catastrophe city	0.2	0.5	0.8	0.9	0.2	0.4	0.5

Table 5.4   Final weights.

Alt c	Cultural city	2.6
Alt a	E-democracy city	2.5
Alt b	Virtual city	2.4
Alt d	Environmental city	2.0
Alt e	Post-catastrophe city	0.5

a classification of the criteria in increasing or decreasing order of importance (Fusco *et al.*, 1993).

The results from this application of regime analysis provide the following final ranking of the alternative scenarios:

- First position: Cultural city
- Second position: E-democracy city
- Third position: Virtual city
- Fourth position: Environmental city
- Fifth position: Post-catastrophe city.

This ranking is stable and does not change among the four criterion-weights vectors.

Concluding remarks

In this study, an AHP application was developed in order to select the preferred scenarios for a city of the future from a Southern-European perspective. The evaluation criteria adopted in this study are those selected by the INTELCITY Mediterranean Platform users. This application has recognized that Southern-European people:

- enjoy traditional cultural environments
- require more efficiency in infrastructure, services and institutions
- see the role of ICTs as a transparent means for innovation in the institutional sector (governance) rather than a crucial factor of change in the urban scenario.

The above results have been checked against the problem of 'uncertainty of method used' and have been confirmed by the application of the regime analysis developed afterwards.

Acknowledgements

Although this work is a joint effort of the two authors, Patrizia Lombardi is responsible for the 1st, 3rd and 4th sections, while Steve Curwell is responsible for the 2nd section. The authors of this paper wish to thank all the INTELCITY team, Think Tank and the Mediterranean End-Users.

References

Brandon, P., Lombardi, P. and Bentivegna, V. (eds) *Evaluation in the Built Environment for Sustainability*, Chapman & Hall, London (in press).

Brandon, P. and Lombardi, P. (2004) *Evaluating Sustainable Development*, Blackwell Publishing, Oxford.

Curwell, S. and Lombardi, P. L. (1999) Riqualificazione urbana sostenibile, *Urbanistica* No. 112, June, 96–103 (in English, 114–115).

Deakin, M., Mitchell, G. and Lombardi, P. (2002) Valutazione della sostenibilità: una verifica delle tecniche disponili. In *Urbanistica* No. 118, 28–34 (in English, 50–53).

European Commission (2002) *Visions and Roadmaps for Sustainable Development in a Networked Knowledge Society*, EC report, February 2002.

Fusco, Girard L., Nijkamp, P. and Voogd, H. (1993) *Conservazione e sviluppo*, Franco Angeli, Milano.

Fusco, Girard L. and Nijkamp, P. (1997) *Le valutazioni per lo sviluppo sostenibile*, Franco Angeli, Milano.

Hinloopen, E., Nijkamp, P. and Rietveld, R. (1983) Quantitative discrete multiple criteria choice models in regional planning. In *Regional Science and Urban Economics*, **13**, 77–110.

Lombardi, P. (1995) Non-market and multicriteria evaluation methods for public goods and urban plans. In *Financial Management of Property and Construction*, Meban A. G., Shaw S. W., McCluskey W. J. and Hanna I. C. (eds), University of Ulster, Northern Ireland, pp. 231–245.

Lombardi, P. L. (1997) Decision making problems concerning urban regeneration plans. In *Engineering Construction and Architectural Management*, **4**(2), 127–142.

Lombardi, P. L. (2001) Responsibilities toward the coming generation forming a new creed, *Urban Design Studies*, **7**, 89–102.

Lombardi, P. (2004) Analytical hierarchy/network process. In *Sustainable Urban Development: The Environmental Assessment Methods*, Deakin, M., Mitchell, G., Nijkamp, P. and Vrekeer, R. (eds.), Vol. 2, E&FN Spon, London.

Nijkamp, P. and Perrels, A. (1994) *Sustainable Cities in Europe*, Earthscan, London.

Roscelli, R. (ed.) (1990) *Misurare nell'incertezza*, Celid, Torino.

Roy, B. (1985) *Mèthodologie, multicritére d'aide á la dècision*, Economica, Paris.

Saaty, T. L. (1995) *Decision Making for Leaders*, RWS Publications, 4922 Ellsworth Ave., Pittsburgh, PA 15213. Vol. II, AHP Series.

Saaty, T. L. and Vargas, L. G. (1991) *Prediction, Projection and Forecasting*, Kluwer Accademic, Boston.

Saaty, T. L. (1996) *The Analytic Network Process*, RWS publications, Pittsburgh, USA.

Stone, P. A. (1989) *Development and Planning Economy*, E&F Spon, London.

Voogd, H. (1983) *Multicriteria Evaluation for Urban and Regional Planning*, Pion, London.

Part 2
Emerging technologies and tools

6 Building as power plant – BAPP

V. Hartkopf[1], *D. Archer*[2] *and V. Loftness*[1]

Summary

The Building as Power Plant (BAPP) initiative seeks to integrate advanced energy-effective building technologies (ascending strategies) with innovative distributed energy generation systems (cascading strategies), such that most or all of a building's energy needs for heating, cooling, ventilating, and lighting are met on-site, under the premise of fulfilling all requirements concerning user comfort and control (visual, thermal, acoustic, spatial, and air quality), organizational flexibility and technological adaptability. This will be pursued by integrating a 'passive approach' with the use of renewable energies. The project has progressed through preliminary architectural design and engineering and five workshops (Ascending Energy Strategies, Floor-by-Floor Infrastructures, Interior Systems, HVAC systems, and Cascading Energy Strategies). BAPP is designed as a six-storey building, located in Pittsburgh (a cold climate with a moderate solar potential) and has a total area of about 6000 m², which houses classrooms, studios, laboratories and administrative offices. At present, the combined cooling, heating, and power generation option that is being considered for the demonstration building is a Siemens Westinghouse 250-kW Solid Oxide Fuel Cell (SOFC).

In this chapter, a preliminary engineering concept of the SOFC based energy supply system will be described. The purpose of this preliminary engineering is to determine an energy supply system configuration (flow diagram), equipment selection, and mode of operation that will effectively, efficiently, and economically meet the energy needs of the building occupants. This work will provide guidance for detailed engineering and will serve as a pattern for similar efforts to plan effective overall energy supply systems for buildings.

---

[1] School of Architecture, and the Center for Building Performance and Diagnostics, Carnegie Mellon University (CMU).

[2] Department of Mechanical Engineering, CMU. Dr David H. Archer is a member of U.S. National Academy of Engineering for his leadership in electric power and energy systems engineering.

Introduction

A national need

Almost 40% of the energy in the USA is being consumed to heat, light, ventilate and cool buildings (EIA, 1995). Adding to this figure, the energy required to fabricate, transport and assemble the materials, components and systems of buildings, conservatively estimated, results in an additional 10% of the US national energy budget.

Substandard building performance, such as buildings that sicken their inhabitants (sick building syndrome), can lead to a reduction of as much as 20% of the productivity of the workforce (Loftness, 2002). The Environmental Protection Agency has estimated the cost to the US economy to amount to about $60 billion annually. During 1993 $508 billion for new construction and $339 billion for the renovation of existing facilities was spent in the USA. This total of $847 billion amounted to 12.5% of the US GNP. Considered long-term, 5/8 of the nation's reproducible wealth is invested in constructed facilities. Collectively the US construction industry only expends 0.5% of sales on R&D. The industrial average for the US is 3.5% (Construction Industry Whitepaper, 1994). In summary, commercial buildings in the US require significant resources to be constructed, operated and adapted and are judged by the occupants to fail principal tests. Research and development expenditures are inadequate.

The Robert L. Preger Intelligent Workplace: the living laboratory

The Robert L. Preger Intelligent Workplace™ (IW) (Figure 6.1) is the result of an unprecedented collaboration between the Center for Building Performance and

Figure 6.1   The Intelligent Workplace™.

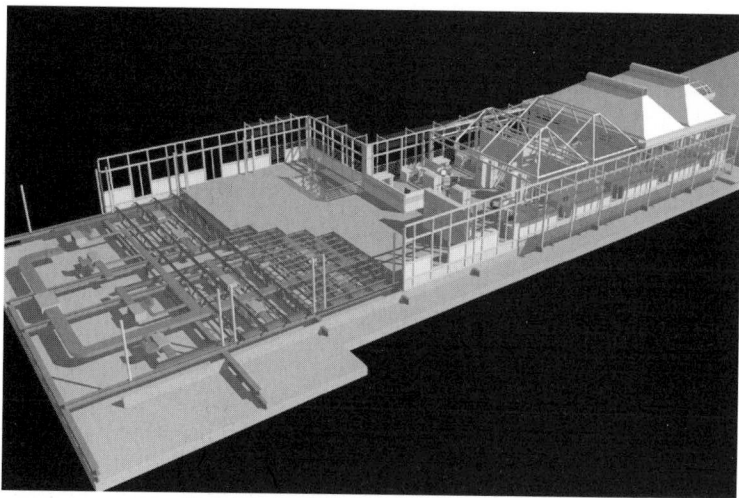

Figure 6.2   Peel-away view of the Intelligent Workplace™ illustrates the integration of all the building systems.

Diagnostics, a National Science Foundation Industry/University Cooperative Research Center, and its supporting industry and governmental members, organized in the Advanced Building Systems Integration Consortium (ABSIC). The 7000-ft² IW is a living laboratory of office environments and innovations. Completed and occupied in 1997, the IW is a rooftop extension of Margaret Morrison Carnegie Hall on the Carnegie Mellon campus. The project provides a test bed for organizational innovations for the advanced workplace.

As a 'lived-in' occupied office, research, and educational environment, the IW provides a testing ground to assess the performance of new products in an integrated, occupied setting (Figure 6.2).

Goals of the Robert L. Preger Intelligent Workplace™

(1) *Individual productivity and comfort*   Both interior system and engineering infrastructures are 'plug and play' to ensure that furniture and space reconfigurations for individual productivity and creativity are immediately matched by technology and environment reconfigurations.

(2) *Organizational flexibility*   The community of workplaces are reconfigurable on both annual and daily levels to ensure 'organizational re-engineering' for collaboration supporting regrouping and sharing for organizational productivity, creativity and innovation.

(3) *Technological adaptability*   Vertical and horizontal pathways for connectivity are accessible and both interior systems and engineering infrastructures support changing technological demands for horizontal and vertical work surfaces, lighting, acoustics, thermal conditioning and ergonomics.

(4) *Environmental sustainability*   Both energy and materials are used effectively over a building's life cycle. System efficacy, user controls, micro-zoning for flex-time, just-in-time delivery of infrastructures, environmentally sustainable and healthy materials, and natural conditioning are demonstrated.

The IW is not envisioned as a one-time 'show-and-tell' demonstration project, but rather as a dynamic environment for the teaching and evaluation of how integrated building components, systems, and assemblies affect building performance. In-house post-occupancy research is critical to validating predicted performance through simulation and to assessing the performance in the integrated setting. As a test bed of new ideas, and a demonstration centre for successful innovations, combined with innovative officing concepts and portable diagnostics, the IW is a unique living laboratory of office environments.

The IW is conceived as a modular system, the units of which can be stacked or reconfigured to adapt to the needs of multiple office settings. The inherent rules of this system – enabling decisions affecting such aspects as building configurations, size of work neighborhoods, cabling and wiring schemes, ratio of shared services and collaborative workspaces to workstations – makes feasible its application on a wider scale.

BAPP – Building as Power Plant

Building on the concepts of and experiences with the Intelligent Workplace™, a living (always adapted and updated) and lived-in laboratory at Carnegie Mellon University (Hartkopf and Loftness, 1999; Napoli, 1998), a research, development and demonstration effort is directed at the 'Building as Power Plant – (BAPP)'. This project seeks to integrate advanced energy-effective enclosure, heating, ventilation, and air-conditioning (HVAC) and lighting technologies with innovative distributed energy generation systems, such that most or all of the building's energy needs for heating, cooling, ventilating and lighting are met on-site, maximizing the use of renewable energies. Figure 6.3 schematically illustrates this idea.

BAPP is designed as a six-storey extension of the existing Margaret Morrison Carnegie Hall Building with total area of about 6000 m^2 that houses classrooms, studios, laboratories and administrative offices for the College of Fine Arts. It is our intention to develop a building that will be equipped with a decentralized energy generation system in the form of a combined heat and power plant. This will include a 250-kW Siemens Westinghouse Solid Oxide Fuel Cell (SOFC) and absorption chiller/boiler technologies. In addition, advanced photovoltaic, solar thermal, and geo-thermal systems are being considered for integration.

Figure 6.4 illustrates a conceptual scheme for an 'ascending–descending energy scheme' that integrates energy generation and building HVAC and lighting technologies. In an 'ascending strategy', fenestration, shading and building mass will be configured to minimize the lighting, cooling and heating loads and maximize the number of months for which no cooling or heating will be needed. Then, passive strategies such as cross-ventilation, stack ventilation, fan-assisted ventilation and night ventilation would be introduced. Passive cooling would be followed by desiccant cooling when humidity levels exceed the effective comfort zone. Geothermal energy will be

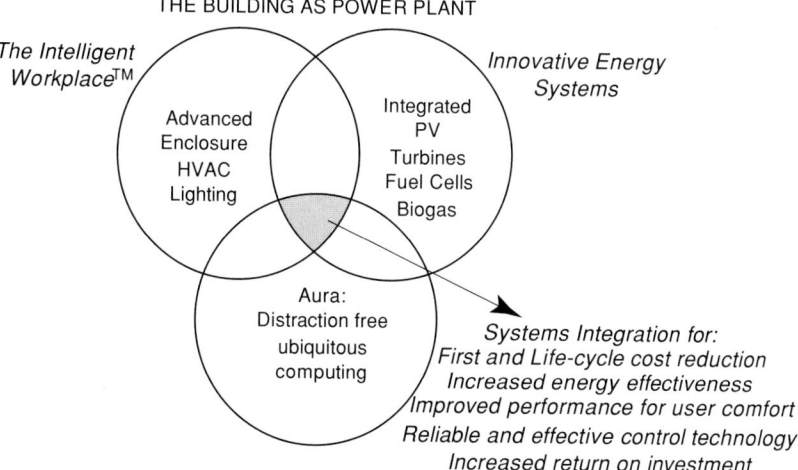

Figure 6.3   The Building as Power Plant concept.

used to activate the building mass for cooling and heating. As outdoor temperatures or indoor heat loads exceed the capability of these systems, then absorption and finally refrigerant cooling will be introduced, first at a task comfort level. Only the last stage of this ascending conditioning system will be a task-ambient central-system refrigerant cooling system.

Complementing these 'ascending' energy strategies is a 'cascading' energy strategy designed to make maximum use of limited natural resources. In a cascading system, a fuel cell and photovoltaic panels might be bundled for the building's power generation;

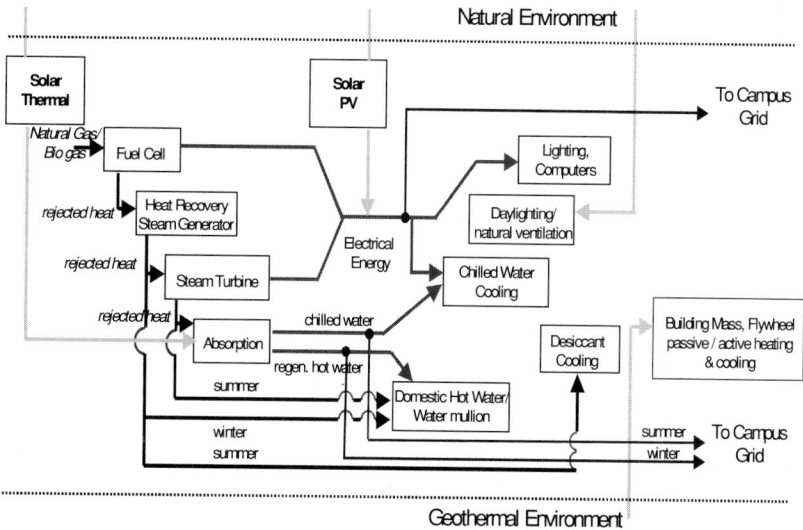

Figure 6.4   Conceptual scheme for a building-integrated 'ascending-descending' energy system.

Table 6.1   Building information.

Building dimensions	
Building length	49.8 m
Building width	16.80 m
No. of floors	6
Area per floor	836.64 m^2
Total area	5019.84 m^2
Floor-to-floor height	4.65 m – in order to connect appropriately with the existing building
Floor-to-ceiling height	3.15 m
'Interstitial space' (underside ceiling to surface of raised floor)	1.50 m
Total building height	27.9 m above grade (plus roof)
Atrium dimensions	
Length	32.00 m
Width	22.00 m
Height	3 storeys
Area	704 m^2

reject heat can be converted into steam, which can be used to first drive desiccant, absorption and refrigerant systems; and finally the resulting reject heat can be used for space heating and hot water.

Figures 6.5–6.8 give an idea of the building design and Table 6.1 summarizes the dimensional information of the building.

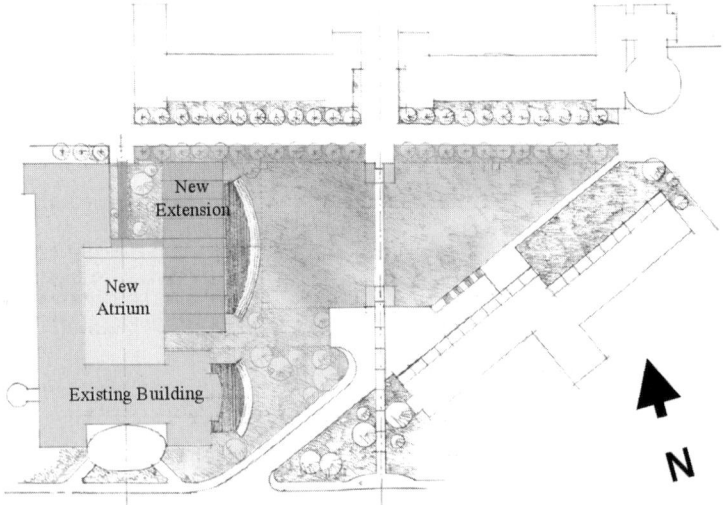

Figure 6.5   Site plan showing the new extension planned for the BAPP project.

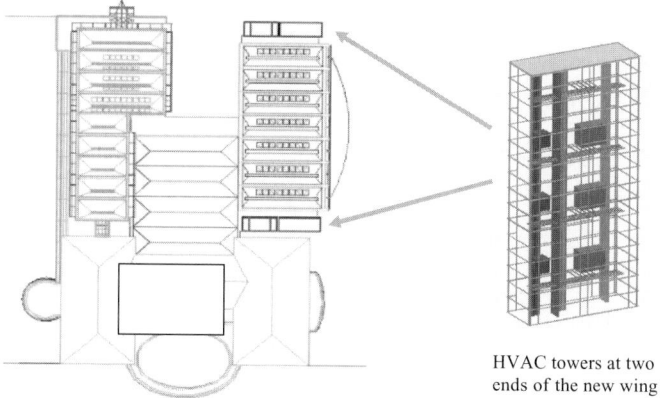

HVAC towers at two
ends of the new wing

Figure 6.6   Building roof plan and a schematic diagram of the plug-and-play service towers.

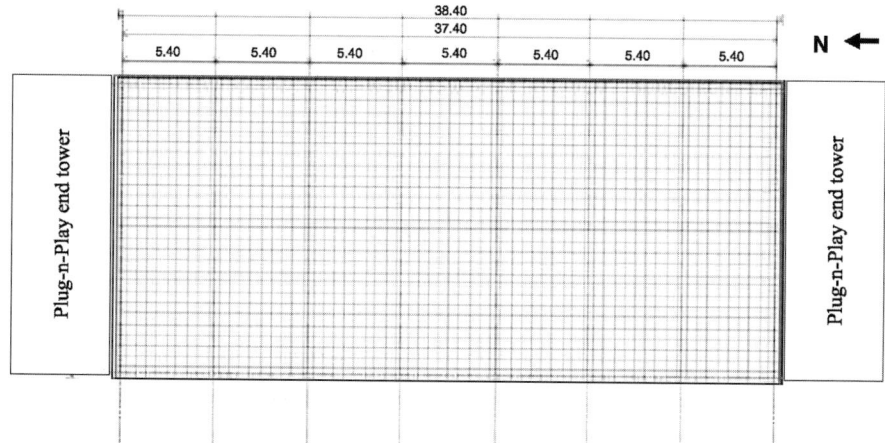

Figure 6.7   Plan for the new wing (dimensions in metres).

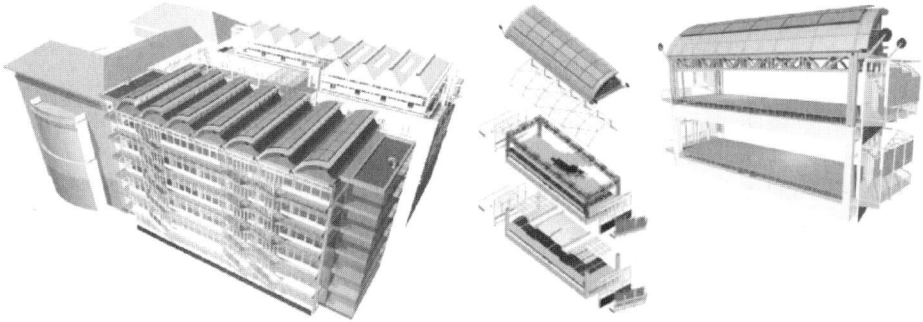

Figure 6.8   Building as renewable asset and energy generator.

BAPP workshops

Based on a preliminary design described above, BAPP project concepts were further developed with the help of the five workshops mentioned below.

(1) Ascending Strategies Workshop, December 4th, 2001, Pittsburgh, PA, USA. This workshop analysed major functions and requirements for the building's facade and roof.
(2) Floor-By-Floor Systems Workshop, January 31st, 2002, Ottawa, Canada. The objective of the workshop was to develop a set of preliminary strategies for BAPP concerning HVAC, Lighting, and Connectivity.
(3) Interior Systems Workshop, April 2nd, 2002, Pittsburgh, PA, USA. This workshop focused on developing project concepts and generating new ideas in the area of Interior Systems – 'Collab Kits'.
(4) HVAC Workshop, April 4th, 2002, Pittsburgh, PA, USA. The objectives were to obtain feedback from experts about HVAC approaches that are being considered for the BAPP in relation to cascading energy systems.
(5) Cascading Energy Workshop, May 30th, 2002, Pittsburgh, PA, USA. The objective of the Cascading Energy Workshop was to obtain feedback from experts in the field about the Power Generation and Primary Energy Generation system approaches that are being considered.

Intelligent workplace systems and performance analysis – lessons learned

Reduced waste in the construction of the IW

The IW project exemplifies how the design, engineering, and material selection, can result in 70–90% reduction of emissions and waste during production of the materials used for the exterior wall, floor and roof, compared with a conventional building. There was no on-site waste during most of the construction phase because of the IW's modular design and its off-site fabrication with complete recycling capacity of all by-products.

Reduced waste in operation

The IW is conditioned for six or more months through 'natural' energies alone during daylight hours. In addition to the resource savings of operating a building, there is significant potential to reduce material waste through the management of material and subsystem obsolescence. Demonstrated in the IW, the reconfigurable/ relocatable interior systems, with modular interfaces to the envelope, HVAC, lighting, communication, structure, power systems, enable organizational change on demand, as well as technological change on demand.

The integrated, modular and demountable systems reflect the fact that buildings are made from components that have different life cycles. The envelope as a system should have a life of 50–100 years, with a possibility of exchanging glazing materials, photovoltaic elements and other components, when superior performance becomes economically feasible. The structural system should have a life of 100 years, and when becoming obsolete at a particular site should become redeployable elsewhere (a column is a column, a truss is a truss). Whereas interior systems have considerably less 'life expectancy', down to computing systems that might have a useful life of 2–3 years.

New design approaches to absorb change and avoid obsolescence: flexible grid – flexible density – flexible closure

These are a constellation of building subsystems that permit each individual to set the location and density of HVAC, lighting, telecommunications, and furniture, and the level of workspace enclosure. These services can be provided by separate ambient and task systems where users set the task requirement and the central system responds with the appropriate ambient conditions.

The concept of grids and nodes: ensuring seven basic needs for each individual

Access to all of the basic needs for a healthy, productive workplace – air, temperature control, daylight and view, electric light control, privacy and working quiet, network access and ergonomic furniture – can only be provided by a shift away from blanket and centrally controlled infrastructures to the concept of grids and nodes (Figure 6.9). The 'grids' establish the overall level of capacitance available to support the working group or neighborhood (fresh air, cooling, power and network capacitance). Then, the 'nodes' or user interfaces must be flexible in terms of location, density, and type of service offered.

Flexible infrastructures begin with accessible and expandable vertical service

There should be a significant shift towards distributed systems to support local control by organizational units with differing equipment and occupant densities, or with

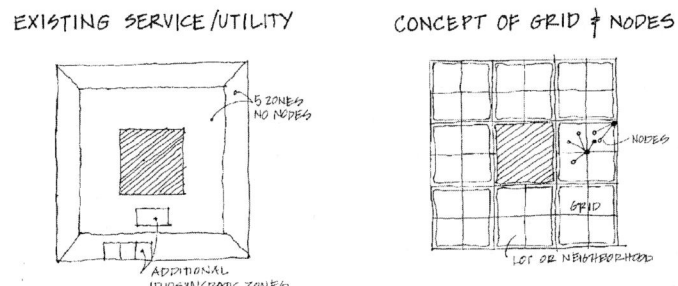

Figure 6.9 Conventional large zone approaches to thermal conditioning and lighting are incapable of delivering adequate environmental quality to accommodate the dynamics of technology, workstation density and teaming concepts.

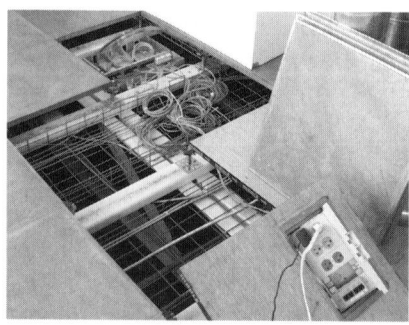

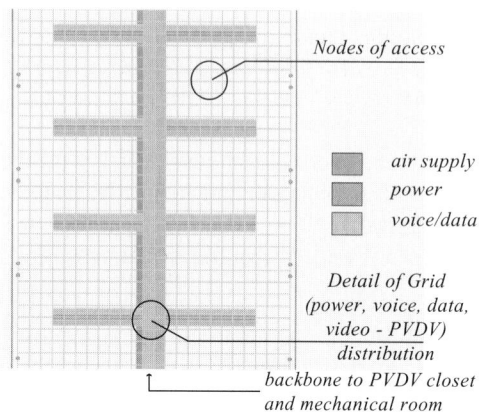

Nodes of access

air supply
power
voice/data

*Detail of Grid
(power, voice, data,
video - PVDV)
distribution*
backbone to PVDV closet
and mechanical room

Nodes of access in the IW

*Detail of Grid (power, voice, data,
video) distribution in the IW*

Figure 6.10 Rational underfloor infrastructure distribution allows for just-in-time additions and changes to support the dynamic workplace.

different work schedules, ensuring appropriate technical and environmental service without excessive costs.

Flexible infrastructures require collaborative horizontal plenum design and relocatable 'nodes' of service

Advanced buildings today demonstrate that floor-based servicing may more effectively support the dynamic workplace (Figure 6.10). Since networking, ventilation and thermal conditioning needs to be delivered to each workstation, services at floor level or at desktop offer a greater ease of reconfiguration than ceiling-based systems. In addition, floor-based systems such as electrical and telecommunication cabling and outlet terminal units can be continuously updated to meet changing needs.

Flexible infrastructures can support reconfigurable workstations and workgroups

It is critical to design the furniture/wall system to support rapid changes between open and closed layouts, between individual and team spaces, as well as rapid changes in occupant density, equipment density, and infrastructure/service to match these configurations.

Smart interior systems

Interactive multimedia and web-based technologies create the possibility of working within ever changing teams, both locally and globally. This requires that built environments must be responsive to ever changing organizational and rapidly evolving technological circumstances.

Intelligent workplace energy systems analysis

We analyzed the energy usage and performance of the intelligent workplace and its building control systems. The lessons learned during this study can be used to better design and operate the BAPP. The analysis focused on:

(1) data acquisition system
(2) building control systems
(3) building design.

The IW uses several energy systems to provide heating, cooling, ventilation, dehumidification, and lighting (Figure 6.11 shows the HVAC systems). Heating is provided by warm water mullions on the facade. The cooling is provided through multi-modal strategies consisting of radiant panels, COOLWAVES by LTG, Johnson Control Personal Environment Modules (PEMs), a make-up air unit to supply the PEMs and floor vents, and a SEMCO air handling unit. The SEMCO unit, which is controlled by an Automated Logic system, is a 100% outdoor air system with enthalpy wheel for dehumidification. A JCI Metasys system controls the rest of the HVAC system. The lighting system is controlled by a Zumtobel-Staff LUXMATE system.

The IW uses three different systems to record energy data. Energy Sentry (72 data points) records electrical energy consumption, Metasys (160 data points) is used for HVAC related data and Weather-Station (8 data points) records outdoor environmental

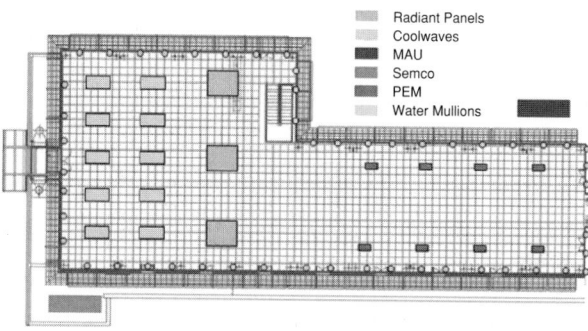

Figure 6.11   IW HVAC systems.

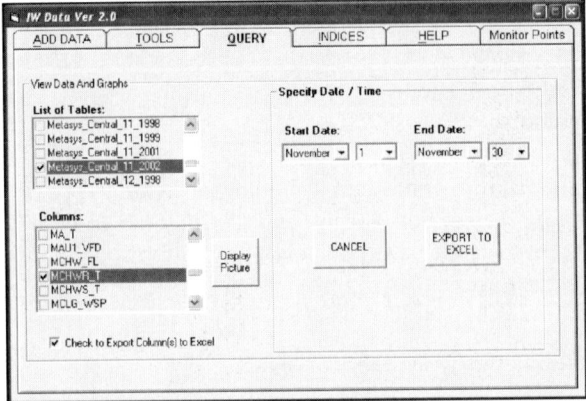

Figure 6.12   CBPD data acquisition and analysis tool.

data. These three systems record data in different formats in different locations and within the systems, each sensor records data in a different file. To expedite and facilitate the analysis process, it is necessary to bring the data into the same format. As a result, we developed a tool (Figure 6.12) to collect data from the different systems (that is in different formats and in various locations) and organize it into one common easily usable database. This tool has several features that allow easy analysis of the building data.

The data collected from sensors in the building was analyzed to determine energy usage and trends (Figure 6.13). It was found that the data contained had missing values. Reasons for this were incomplete documentation of the file storage structure, IP address problems and the system going offline for various reasons. Statistical techniques and simulation were used to fill in missing data to make the calculations more accurate. The existing DOE 2.1E simulation model of the IW was calibrated to match

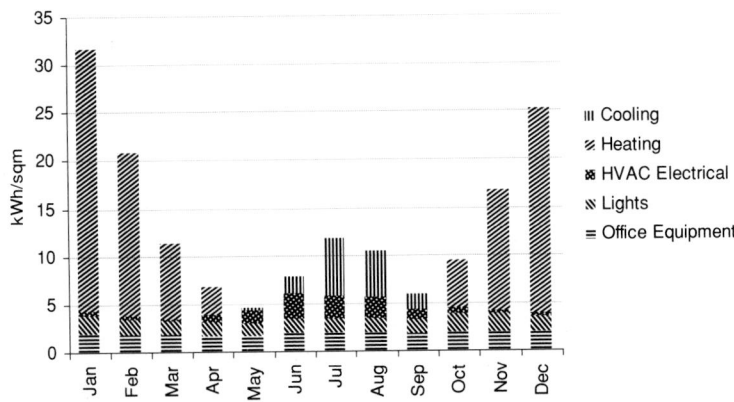

Figure 6.13   Summary of monthly energy consumption in the IW building.

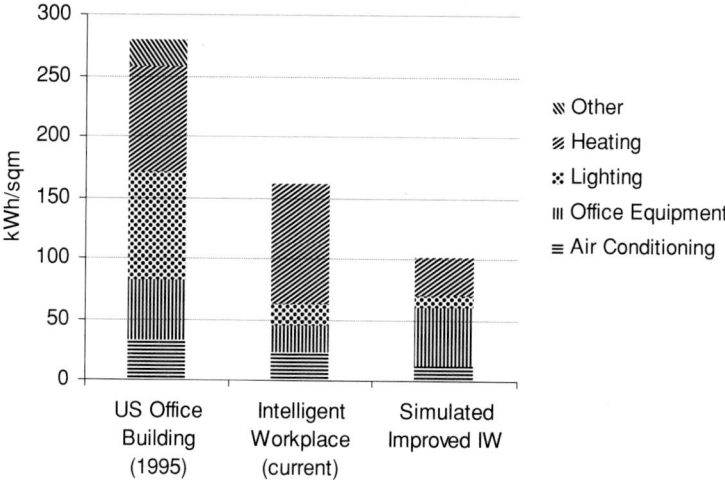

Figure 6.14 Comparison of IW energy with typical US office.

the current measured conditions in the IW. This model was then used to predict energy consumption under different scenarios.

It was found that although energy usage was less in the IW than in standard US office buildings (Figure 6.14) there were still areas where the energy consumption could be reduced further. Several hypotheses were suggested to explain the results obtained from the analysis. These were based on the design of the building and its mechanical systems:

- The high air infiltration and one un-insulated area in the IW caused the heating load to increase.
- The heating setpoint in the IW is higher than that of a standard office building, which further increased the load.
- The under-floor plenum of the IW contains steam pipes for the floors below that are not measured by the IW systems, this caused a small reduction in the heating load.
- The higher than typical amount of glass and exposed surface area caused an increase in cooling energy.
- This cooling load was increased by the un-controlled operation of windows.
- The unconditioned underfloor plenum was measured to have an average temperature of 31 °C during the summer. Since it is not insulated from the IW living space, it also increased the cooling load.
- The lighting in the IW is through fluorescent fixtures that use dimmable ballasts. At the time of construction these were available only for 220 V therefore transformers were used. It was found that of the total annual lighting load of 18.92 kWh/m^2 (about 1/5th of good US practices), 10.12 kWh/m^2 was transformer losses (the 'parasitic' load therefore is 50% of this reduced load).

Building control systems

Based on our experience of the control systems in the IW, the goals for the future control system should be:

- to meet the needs of the building users. For the occupants it should be intuitive to use, facility managers should be able to maintain it with their level of knowledge, energy managers should be able to use it to analyze energy consumption, organizational managers should be able to modify it easily, and researchers should be able to extract and analyze the data easily
- to allow for easy expansion to integrate new technologies such as wireless sensors and controllers. The control system should allow equipment to be plug and play installable by untrained installers
- to allow intelligent monitoring and decision support with the ability for continuous process improvement and economic analysis. It should allow for conflict resolution;
- to allow for preventive maintenance with the ability to predict future system and component behavior
- to have control processes that are interoperable and integrated to allow for management of the whole building and its constituent parts in terms of energy efficiency;
- to allow control strategies to be simulated and checked before they are implemented.

BAPP building loads

Based on the preliminary architectural design of the building and its proposed mode of operation, both annual operating energy consumption and the related emissions have been simulated using EnergyPlus V1.1 (EnergyPlusV1.1.0 2003). In addition to the high office equipment load (due to the high density of computers on two floors housing the Entertainment Technology Center and power tools in the architecture and design shops), typical office equipment load is also being simulated. The energy data for a typical load, as opposed to high load, office building in the USA and Europe has been used as the baseline for the energy performance analysis. In order to see the effect of ascending strategies, operating strategies and descending strategies, the simulation is carried out step by step, from the case based on ASHRAE Standard 90.1-1999 to the case with distributed power generation. The simulation specifications for the ASHRAE case are shown in Table 6.2.

Table 6.2 Simulation parameters of ASHRAE case.

Parameter	Value	Parameter	Value
R-value: Roof	$3.58 \, \mathrm{m^2 \, K/W}$	Visible Transmittance: Window	0.18
R-value: Above grade wall	$2.89 \, \mathrm{m^2 \, K/W}$	Electric lighting load	$15 \, \mathrm{W/m^2}$
U-value: Window	$3.15 \, \mathrm{W/m^2 \, K}$	Office equipment load	$8.9 \, \mathrm{W/m^2}$
SHGC: Window	0.35	Infiltration rate	0.2 ACH

Table 6.3   Energy performance of BAPP with typical office equipment load.

	Site energy (kWh/m^2 yr)	Primary energy (kWh/m^2 yr)	Operating cost ($)	Carbon equivalent (kg)
ASHRAE	122.2 (100%)	381.4 (100%)	$50 945 (100%)	202 480 (100%)
Ascending strategies	96 (79%)	225.2 (59%)	$27 841 (55%)	112 503 (56%)
Operating strategies	79.3 (65%)	199.5 (52%)	$25 304 (50%)	101 295 (50%)
GSHP	56.1 (46%)	196.8 (51%)	$35 505 (70%)	106 551 (53%)
GSHP + PV	45.1 (37%)	158.2 (41%)	$31 140 (61%)	85 700 (42%)
GSHP + PV + SOFC	**80.4 (66%)**	**80.4 (21%)**	**−$7413 (−15%)**	**21 270 (11%)**
Electric chiller and heat Exchanger + PV + SOFC	90.3 (74%)	90.3 (24%)	−$5462 (−11%)	23 889 (12%)

Then ascending strategies, such as:

- high performance building envelope (R-value of roof: 7.06 m^{2}K/W, U-value: 1.37 W/m^2 K, SHGC: 0.27, and visible transmittance of window: 0.56, with an infiltration rate of 0.1 ACH)
- daylighting based dimming (lighting setpoint: 500 lux)
- high performance electric lighting (lighting load: 5.4 W/m^2) (Campbell, 2002)

are applied. In the case with better operating strategies, natural cooling (natural ventilation and night ventilation), and demand controlled ventilation (the ventilation volume is decided according to the number of occupants in the space) are deployed to reduce the annual energy consumption and provide improved thermal comfort as well as indoor air quality.

In addition, a ground source heat pump (GSHP) is also considered as an alternative to a typical chiller–boiler plant configuration. Finally, photovoltaic (PV) cells on south facing roofs and a 250 kW solid oxide fuel cell (SOFC) are included as distributed power generation strategies.

In Table 6.3, the percent in parentheses shows the percentage of site energy, primary energy, operating cost and carbon equivalent compared with those of the ASHRAE case. The minus sign in operating cost shows that instead of paying more for energy, the construction of BAPP could reduce the energy bill of the campus by $7413 per year. The comparison of BAPP with other typical office buildings is shown in Figure 6.15, which indicates that the primary energy consumption of BAPP is predicted to be only 11% of a typical US office building.

Elements of the energy supply system

SOFC power generator

Power generation for the building energy supply system design has focused on a SOFC power system (Figure 6.16) being developed and commercialized by Siemens

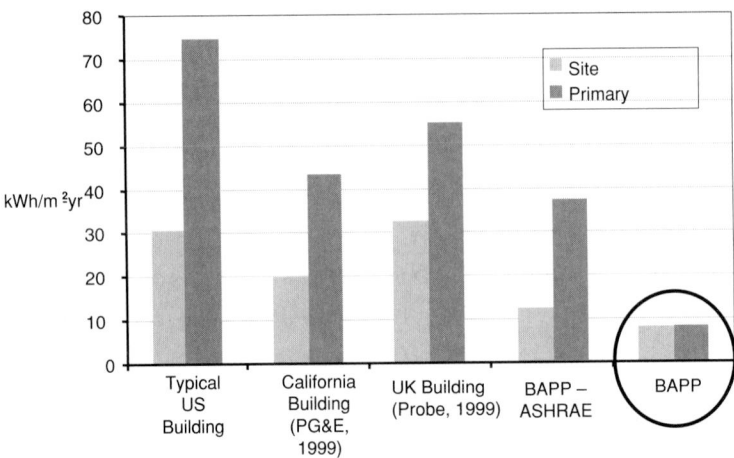

Figure 6.15   BAPP compared to other office buildings.

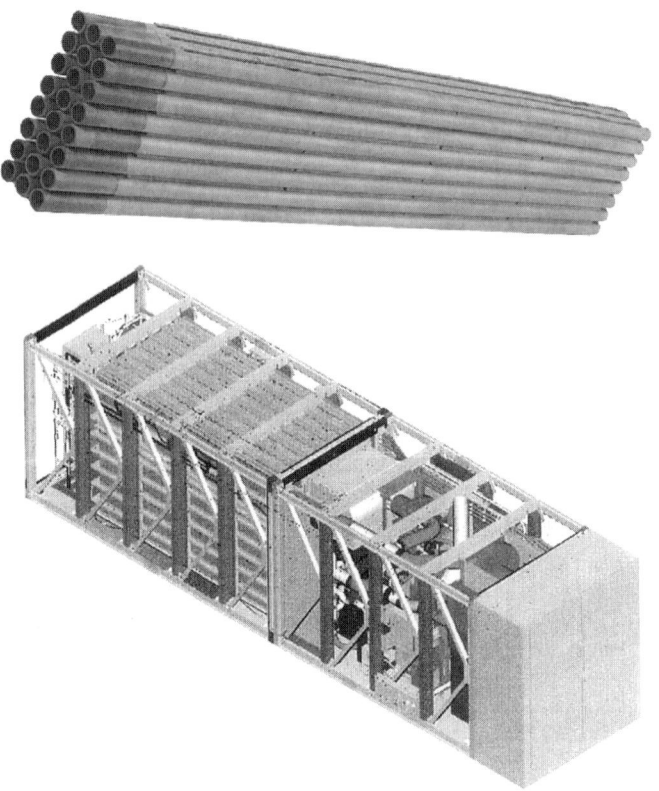

Figure 6.16   Siemens Westinghouse solid oxide fuel cell, cell bundle and assembled system.

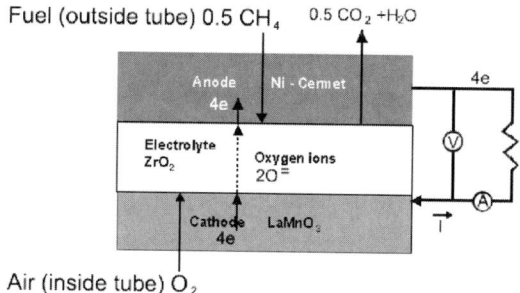

Fuel (outside tube) 0.5 CH_4 0.5 CO_2 +H_2O

Air (inside tube) O_2

Figure 6.17   Siemens Westinghouse solid oxide fuel cell cross-section.

Westinghouse Power Corporation, Stationary Fuel Cells Division. The capacity of the commercial system currently envisioned for production in 2006 is 250 kW, approximately the maximum power required by the building. Its electrical efficiency operating on natural gas fuel is 50%. The reject heat from the system, 40% of the fuel energy input, is potentially available from the system as steam and hot water at the conditions and quantity required by this building.

The SOFC power generation and heat recovery system

A flow diagram for the natural gas fired SOFC power generation and heat recovery system that serves as the basis of the overall building energy supply system is illustrated in Figure 6.17. The symbolic representation of the generator shows three cells with hemispherical closed ends. Each of these cells has an internal tube that feeds preheated air at its bottom. The air flows upward inside and fuel outside each cell. As current is drawn and power produced, oxygen is transferred through the cells from the air stream into the fuel stream. At the top of the cells a portion of the largely oxidized, combusted, fuel stream is drawn out of the generator by means of the ejector and mixed with the desulphurized, natural gas fuel feed. The H_2O in this oxidized fuel gas reacts with and reforms the methane, CH_4, in the natural gas fuel producing hydrogen and carbon monoxide. The reformed fuel, primarily H_2 and CO mixed with combustion products H_2O and CO_2, flows from the reforming passages into the region at the base of the cells and then upward around the cells. The portion of the oxidized fuel gas at the top of the cells not returned and mixed with the fuel feed mixes with the spent air stream leaving the top cells, and the combustion process is completed.

The combustion product gases then are passed to the heat recovery steam generator, the HRSG, comprising sections for superheating, boiling and feed water heating. The possibility of auxiliary firing with natural gas fuel provides additional flexibility in operation of the heat recovery system, enhancing steam production or continuing steam production in the case of fuel cell outage.

Absorption chiller, air dehumidification systems

Both the absorption chiller and the air dehumidifier systems are based on liquid desiccants, an aqueous solution of an inorganic salt such as lithium bromide (LiBr) or lithium chloride (LiCl). In both systems a concentrated desiccant absorbs water vapor, and the resulting heat release is removed by cooling water. The desiccant is subsequently re-concentrated (regenerated) by heating it with steam and driving off the water vapor. Liquid desiccant dehumidifiers, single- and multi-staged, are under development both in the USA and abroad[3]. Their advantage in dehumidification over condensing air coolers is the reduction of steam and cooling water usage by a factor of about two.

Cooling water supply system

Cooling water is needed for both the absorption chiller and the air dehumidifier systems. Two possible sources of cooling water are a conventional cooling tower and a geothermal cooling water supply and return system.

Solar energy utilization

The capture of solar energy by photovoltaic and/or thermal energy panels mounted on the facade and roof of the building is also planned. The incremental energy supplied by such systems will be integrated with the building's overall energy supply system, and the incremental cost will be evaluated.

Mode of operation for the building overall energy supply system

In the initial phase of operation of the system it is planned to:

- operate the SOFC power system continually at its 250 kW design level, interconnected with the campus power grid. The grid will provide for differences between the building power needs and the fuel cell output
- operate the heat recovery steam generator, HRSG, producing steam at the design pressure, temperature, and flow and providing hot water for heating and for domestic use in the building. The HRSG will be interconnected with the campus steam grid and will interchange steam with the grid as needed at the campus charge rate for steam
- operate the absorption chiller system when air conditioning is required, producing chilled water at the design flow and temperature conditions. The chiller will be interconnected with the campus chilled water supply and return grid and will interchange with this grid dependent on differences between the building cooling requirement and the chiller output

---

[3] Dr Andy Lowenstein and his AIL research group represent a single stage absorber and regenerator, Dr. Jiang Yi and his research group in Tsinghua University, China, are demonstrating a multi-stage absorber and regenerator

- operate the liquid desiccant dehumidifier system and the water evaporation humidifier as needed. The fresh air dehumidification system will remove sufficient moisture to provide comfort for the building occupants, to prevent condensation in the air cooling equipment of the building and fresh air systems, and to compensate for the humidity, latent heat, added by the building occupants.

Building energy supply system alternatives, for evaluation

A number of additional alternatives and extensions were considered. These are to:

- replace the SOFC power generator by a molten carbonate fuel cell, MCFC, generator or by a gas turbine generator
- add a small backpressure or condensing steam turbine generator for additional power from the system during spring and autumn, in which steam-driven absorption chiller and liquid desiccant dehumidifier are not running or are in partial operation
- use exhaust air from the building for exchange of heat and humidity with the incoming fresh, ventilation air (currently it is planned to exhaust air directly to an atrium adjoining the building)
- Consider both single-stage and multi-stage processing and regeneration of liquid desiccant in both the absorption chiller and the air dehumidification systems.

Conclusion

The work at the Center for Building Performance and Diagnostics, supported by the Advanced Building Systems Integration Consortium, has established and demonstrated the economic and technical feasibility, as well as social/political desirability of creating commercial buildings which consume substantially less energy compared with best U.S. practices, while offering the occupants dramatically increased user satisfaction, providing for organizational flexibility, and technological adaptability.

The Building as Power Plant Project demonstration aims to show that the building can be a net exporter of energy. The preliminary analysis of the cascading system shows that the fuel cell produces more power and thermal energy than needed by the BAPP building, which can be exported to the CMU campus. The next steps in this project include the analysis of multi-modal conditioning systems. The process will consist of establishing partnerships with industry and studying the performance and systems integration issues for each of the strategies. A complete design of the building and the systems will be finalized in consultation with project architects and engineers. Preliminary engineering of a novel, solid oxide fuel cell based building energy supply system for a multi-purpose building of advanced design at Carnegie Mellon University is now underway. The purpose of this work is to provide economic justification, guidance for detailed engineering, and a basis for the evaluation of new technology.

Their ideas have attracted the attention of the U.S. Congress and are therefore contained in the 2004 Energy Bill, creating the framework for the BAPP to function as a National Test Bed.

References

Campbell, S. (2002) Can intelligent office lighting systems really save 80%? *Energy User News*, April.

Construction Industry Whitepaper (1994) National Science and Technology Council; Committee on Civil Industrial Technology. Subcommittee on Construction and Building. Civil Engineering Research Foundation, April 28, 1994.

EG&G Technical Services, Inc. (2002) *Fuel Cell Handbook*, 6th edn, Chapter 7, Science Applications International Corporation, 195 pp.

EIA (1995) EIA: Energy Consumption in Commercial Buildings in 1995. Energy Information Administration, U.S. Department of Energy, 1995.

EnergyPlus V1.1.0 (2003) http://www.energyplus.gov

EnergyPlus Engineering Documents April 2003, pp.262–267.

Hartkopf, V. and Loftness, V. (1999) Global relevance of total building performance, *Automation in Construction*, **8**, pp. 377–393.

Komor, P. (1997) Space cooling demands from office plug loads, *ASHRAE Journal*, December.

Loftness, V. (2002) Building Investment Decision Support, U.S. Green Building Council's International Green Building Conference and Expo, Austin, Texas, November 13–15, 2002.

Napoli, L. (1998) Where every worker is ruler of the thermostat, *The New York Times*, Money and Business, Section 3, Sunday February 15.

Pacific Gas and Electric Company (1999) Commercial Building Survey Report, 1999.

PROBE Strategic Review (1999) Final Report: Technical Review, August.

Wang, S. K. (2000) *Handbook of Air Conditioning and Refrigeration*, 2nd edn, Chapters 14, 28, McGraw-Hill, New York.

Siemens Westinghouse Power Corporation, Stationary Fuel Cells (2002) Solid oxide fuel cells – the new generation of power, The CHP250 SOFC Power System.

Streitz, N. A., Geißler J. and Holmer, T. (1998) *Roomware for Cooperative Buildings: Integrated Design of Architectural Spaces and Information Spaces*, pp. 4–21, Springer, Heidelberg.

7 Development of a self-contained micro-infrastructure appliance

U. Staschik[1], H. Hinz[2], Do Hart[3], G. Morrison[4] and C. Ives[5]

Summary

This chapter provides an overview of the ongoing research and development of a stand-alone micro-infrastructure 'appliance'. The self-contained micro-utility was developed as an alternative to large-scale grid-based infrastructure systems. It provides safe potable water, reliable heat and electricity, and environmentally benign sewage disposal to single dwellings or dwelling clusters. The synergistic use of all resources available allows the unit to become cost-effective and reduces greenhouse gas emissions. The micro-utility appliance deliberately de-couples individual infrastructure from communal infrastructure. The user of the micro-infrastructure becomes at the same time the provider of the infrastructure. This fosters increased technology awareness and allows the user to participate more actively in the development of novel integration and conservation strategies – directed by individual awareness and site-specific requirements rather than large-scale external infrastructure developments. The chapter also touches on product acceptance and conclusions derived from in situ testing.

Introduction

For several years, Canada Mortgage and Housing Corporation has worked on the advancement of site-specific micro-infrastructure solutions. In 1995, the 'Toronto Healthy House' was constructed in the heart of urban Toronto, showcasing a single-family dwelling unplugged from the conventional utility grid [1].

The present project builds on the 'Toronto Healthy House' experience and continues the goal of seamless integration of on-site micro-utility, with an energy and resource efficient building envelope and a healthy living space. As a major distinguishing feature,

---

[1] Architectural and Community Planning Inc., Winnipeg, Canada.
[2] Canadian EcoConcepts, Niagara-on-the-Lake, Canada.
[3] Watershed Technologies Inc., Toronto, Canada.
[4] Canada Mortgage and Housing Corporation, Thunder Bay, Canada.
[5] Canada Mortgage and Housing Corporation, Research Division, Ottawa, Canada.

Figure 7.1   The first prototype unit was installed in August 2000. The micro-utility equipment housed in the combined mechanical utilities container (centre) supplies all infrastructure and utilities for a 107 m² three-bedroom single-family dwelling. The left-hand dwelling is not connected at all to the central utility grid.

the micro-utility appliance is specifically developed for rural, remote or northern applications.

To be successful, the integration of self-contained micro-infrastructure technology has to be effortless, transparent and convenient for the user/supplier. The consumer must be able to use the self-contained utility appliance in exactly the same way as the conventional grid-based infrastructure. In an effort to develop a factory-assembled micro-infrastructure appliance, a novel, eco-appropriate solution, the EcoNomad™ was designed and prototyped and is presently undergoing in situ testing.

Figure 7.1 shows the first prototype unit being installed in August 2000. The micro-utility equipment housed in the combined mechanical utilities container supplies all infrastructure and utilities for the attached 107 m² three-bedroom single-family dwelling. The dwelling is not connected at all to the central utility grid.

An important motivation in the development of alternatives to conventional utilities is the very high cost of providing grid-based infrastructure in the northern communities of rural Canada.

The Canadian Ministry of Indian and Northern Affairs [2] reports typical costs of a central community grid-based power connection, including potable water and sewage, to be in the range of $CAD 80 000 per single-family dwelling. In some cases, costs have been reported as high as $CAD 180 000 per serviced dwelling. In most cases, the cost of providing grid-based utility servicing from central supply to the properties is equal to or higher than the cost of the dwelling construction itself.

The Business Plan of the Nunavut Housing Corporation [3] outlines that, in Canada's remote and northern Territory, the operating and maintenance cost for a two-bedroom single-family dwelling (80 to 90 m²) can be as high as $CDN 48 per day, or an annual cost of $CDN 17 484.

To enhance the functional optimization further, the micro-utility appliance not only provides the utilities, but also the full mechanical system for any host building.

Micro-infrastructure

Description of existing technology

It has been common practice for housing and subdivision developers to handle the installation of all required infrastructure such as power generation and distribution, central wastewater management, on-site septic tanks, potable water distribution and mechanical systems in an uncoordinated way – finding a solution to each requirement independently.

The individual components are all physically present, but the equipment has all been selected, and delivered independently – each item had to be installed individually and tested on site. Although the equipment can be expected to operate and have a reasonable service life, the overall utility system, is a 'make-the-best-of-it' aggregate of individual mechanical units, rather than a planned, purposeful integration of suitable equipment.

The combined mechanical utilities container

The 'combined mechanical utilities container', by contrast, permits important synergies between the various components and, in this way, promotes operational efficiency and resource conservation.

The physical layout of the components was researched and planned to minimize interference, to create functional synergies and to maximize beneficial interactions. All equipment is factory assembled and, before delivery, the whole system is factory tested (and benchmarked) as an integrated micro-utility appliance.

The completely assembled module is shipped and installed in the as-tested format. On site only seven piping connections, one electrical and one communication connection are required to tie the utilities module functionally to the host building. If the host dwelling and infrastructure appliance are designed as integrated units, the host building can be fully operational less than 4 hours after delivery of the utility appliance.

The design of the equipment and components are not new in themselves. The distinctiveness lies in the fact that all parts are assembled and secured in one singular utility appliance. In addition, the system allows the host dwelling to operate independently from any communal service connections, making it invulnerable to problems and outages that might plague a centralized system.

Figure 7.2 shows a conceptual system diagram and the main component assemblies. To allow for maximum system flexibility and ongoing opportunities for upgrades the EcoNomad™ uses a proprietary 'platform and module' approach.

The module builds on proven technologies and combines the following systems into one singular unit:

- potable water storage and a three-barrier disinfection system
- biological wastewater management and treatment

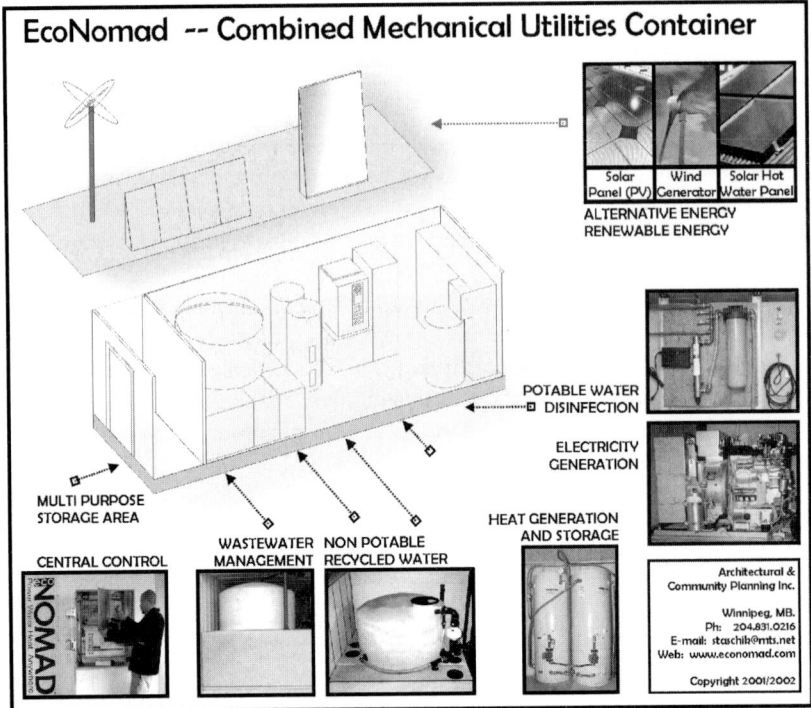

Figure 7.2   Conceptual layout of the mechanical equipment housed in the utility appliance. The system diagram shows the main component assemblies. To allow for maximum system flexibility and ongoing opportunities for upgrades, the EcoNomad™ uses a proprietary 'platform and module' approach.

- storage and management of reclaimed non-potable water (used for non-contact applications such as toilet flushing, process water, gardening, etc.)
- heat and power generation (primarily) by means of a custom designed micro co-generation engine (fueled by diesel, biodiesel, natural gas or propane), power conversion and storage;
  - ○ presently, an internal combustion engine is used but, in the future, integration of Stirling engine, micro-turbine or fuelcell is planned
- integrated opportunistic use of alternative energy components (micro wind-turbine, solar PV, micro-hydro, solar hot water panels)
- heat storage and thermal transfer for space and domestic water heating
- stand alone fire suppression system for utilities module and host dwelling
- air conditioning, air purification and heat recovery ventilator for host dwelling
- central control and alarm function
- PLC based data logging and remote dial-in monitoring (optional, site-specific).

Individually, all these are technologies usually found as separate mechanical systems or external utilities in conventional buildings.

Equipment and technical implementation

The use of existing and proven technologies was a key aspect of the research project. This approach reduces system development cost and engineering time and it ensures the availability, serviceability and reliability of the system components. The concise design of the system and inter-connection of the components allows for factory assembly, commissioning and maintenance of the system by laborers with basic skills and training – reducing costs and making the technology accessible and understandable to the end-user.

The methodology applied in the research and prototype development was based on the establishment of performance criteria for each subcomponent. Contrary to common engineering practice, none of the components was optimized as an individual stand-alone piece of equipment – rather, an in-depth experimental optimization procedure was applied to the entire system. This research methodology led to the development of subcomponent 'system interconnection sequences'.

Commissioning of first prototype

In 1997, Canada Mortgage and Housing Corporation supported the construction of a 'Northern Healthy House' (located 400 km east of Winnipeg, Manitoba, Canada) to showcase the beneficial interaction between a single-family dwelling and stand-alone infrastructure module.

Computer modelling indicated that if an energy-conserving family were to occupy the house, the monthly cost of operating all utilities would be between $CAD 60 and $CAD 100. Data logging since November 2001 shows that the actual operating costs were approximately 20% higher than anticipated, but substantially below conventional systems.

In the three coldest months of the year (December to February) the average monthly cost of fuel was less than $CAD 140, this being the only input required to produce all electricity, space heating, water heating and all related pumping, system monitoring, and household appliance needs. A comparable three-bedroom single family dwelling in the same community would typically have a monthly utility cost of $CAD 400.

Synergistic interaction of infrastructure components

Major benefits in micro-utility development result from the synergy created by optimized interaction of the various mechanical infrastructure components:

- A micro co-generation unit provides electricity as well as space and water heating with a useful fuel conversion factor of 80 to 90%, much higher than conventional utilities. The renewable energies – solar and wind – are used in an opportunistic manner when available. The multiple power sources allow for a significant reduction in battery storage capacity, and at the same time encourage the use of renewable energy systems in areas that are usually regarded as 'marginal'. By sharing common components (thermal and electrical storage, pumps and system controls) synergies are generated that reduce costs and improve performance.

- The integration of the biological wastewater treatment and management system into the infrastructure module allows for a controlled generation of non-potable reclaimed water. As discussed by Jowett [4] this water can safely be used for non-contact applications such as toilet flushing, washing machines, process water or lawn watering. In residential applications a reduction of water requirements of up to 40% (commercially up to 85%) can be achieved. Further stages can be added to the wastewater management/reclaimed water cycle. Several test projects are ongoing and Paloheimo *et al.* [5] state that the final product meets aesthetic and approval guidelines for re-use in contact applications (e.g. the shower).

- The container houses a reservoir for potable water and a three-barrier disinfection system. The cascading of potable water uses allows for the use of primary treated water for bath, shower, kitchen and cooking use, less treated water for washing machines, and non-potable, reclaimed water for toilet flushing. Cascading of uses reduces the demand for treated potable water – directly resulting in water and energy savings.

- In many rural or remote locations, firefighting capabilities are below acceptable standards. The utilities container can provide a complete fire-suppression means inside the mechanical module as well as to the host dwelling.

In addition to the synergy benefits outlined above, the platform and module approach to the manufacturing process provides further advantages.

Each sub-component is constructed as a stand-alone unit. With the development of new and more environmentally benign technologies, individual components can be exchanged without affecting the system as a whole.

For example, the entire spatial concept of the utilities module is ready made to accept a fuel cell, micro-turbine or Stirling engine, once pricing and consumer demand make this form of combined power and heat production acceptable to the public at large. No change in the overall mechanical or electrical interconnection is required. The novel concept of 'plug and play technology' for infrastructure and housing is emerging.

The micro-utility offers economies over centralized grid systems: it eliminates transmission losses and recaptures heat that would otherwise be lost at source. A widespread system of micro-utilities is also inherently more stable and resistant to breakdowns, natural disasters or terrorism that could interrupt central facilities.

Application in emerging economies

Initially, the infrastructure appliance was developed for remote applications in Canada. But in view of the socio-economic change that follows new infrastructure developments, the deployment of pre-fabricated, ready-to-use infrastructure will be a tremendous tool in rural and urban economic development, especially in emerging economies.

One of the benefits from the deployment of stand-alone micro-infrastructure units (or clusters) is the speed with which one can install a significant supply of infrastructure modules. The 'off-the-shelf' modularity allows almost immediate shipment of the combined utility module with additional units following on a scheduled delivery basis. With proper site preparation, small units can be installed and operational in less than one day. The utility user, government, relief organization or infrastructure developer can deploy 'infrastructure convenience' to targetted sites selected for maximum disaster relief, community development or socio-economic impact.

The mobile and modular characteristics allow deployment for short-term, temporary and permanent use with minimal stranded cost risk. A political administration can take credit for measurable results while still in office and perhaps even enjoy some of the benefits from the resulting rural, remote or urban development.

Slattery [6] describes, that many developing countries regard rural development as a significant catalyst for socio-economic change. Following a project announcement, the quick delivery of stand-alone micro-utility modules to the customer's site will demonstrate political responsiveness and potentially enhances project and government stability. Effective deployment can stabilize the infrastructure and utility convenience for both rural and urban facilities, as well as address local (employment) needs.

Affordability and site design

The technology has the benefit of increasing energy efficiency, decreasing environmental impact and lowering total infrastructure costs. It makes utilities economically viable particularly in northern, remote and rural locations as well as in developing countries without a well-established utility generation and delivery grid. The micro-utilities modules address all aspects of conventional infrastructure, with the following improvements:

- the economic integration of renewable energy sources, using shared components
- a reduction of fossil fuel requirements for space heating purposes
- conservation of potable water sources and processing
- controlled wastewater management and improved effluent quality.

To reduce the cost of servicing further, preference is given to the development of cluster applications. Figure 7.3 outlines some of the site-specific configurations that can be developed.

- Individual homes or businesses can be developed completely off-grid – without need for any grid based utility connection.
- Individual power and heat modules can be attached directly to host dwellings – individual houses, duplexes or row housing units. Potable water, reclaimed water and wastewater management are supplied by a secondary communal micro-utility module or by conventional grid based water and sewage systems.
- Alternatively, potable water, reclaimed water and wastewater management units can be attached to the individual host dwellings. Power is supplied by a conventional grid connection. Space and water heating is supplied by conventional furnace or boiler equipment.

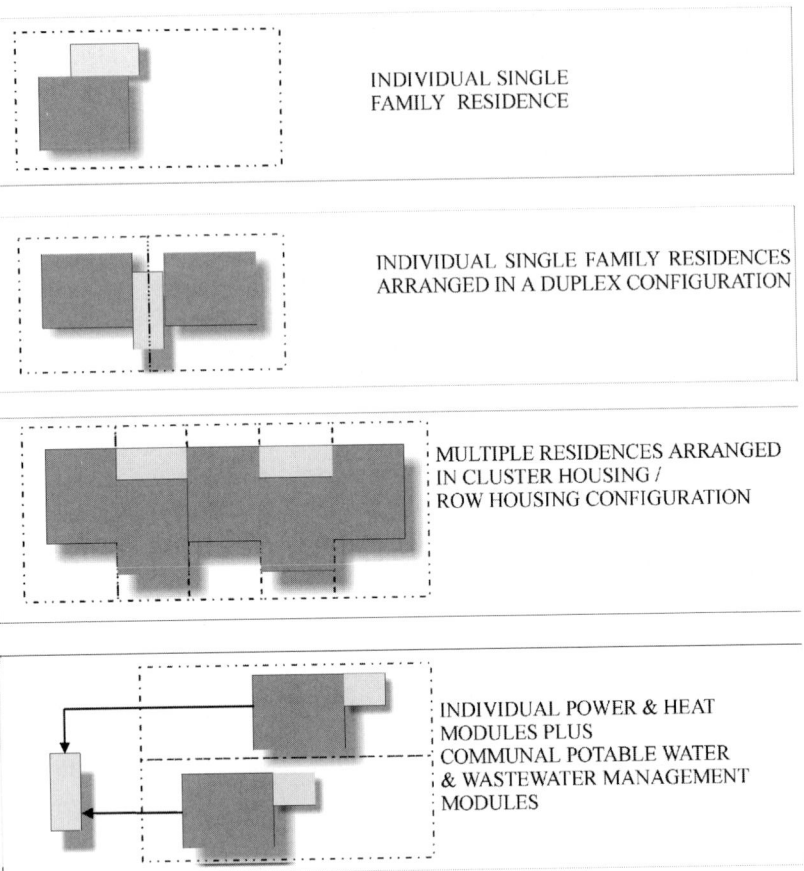

INDIVIDUAL SINGLE
FAMILY RESIDENCE

INDIVIDUAL SINGLE FAMILY RESIDENCES
ARRANGED IN A DUPLEX CONFIGURATION

MULTIPLE RESIDENCES ARRANGED
IN CLUSTER HOUSING /
ROW HOUSING CONFIGURATION

INDIVIDUAL POWER & HEAT
MODULES PLUS
COMMUNAL POTABLE WATER
& WASTEWATER MANAGEMENT
MODULES

Figure 7.3   Conceptual layout to illustrate the connection options between the infrastructure appliance and host dwellings. Cluster arrangements for host buildings allow for an optimized sizing of the micro-infrastructure appliance.

On a larger scale, entire community infrastructure concepts can be developed to make the best use of dispersed or 'embedded' infrastructure, combining the benefits of the conventional grid based infrastructure with the distributed generation. Micro infrastructure modules are strategically installed to allow for site-specific load growth, load matching, grid stability and reduction of (power) transmission lines, (water) distribution and (sewage) collection systems.

Benefits of micro-infrastructure development

Dunn [7] best describes the 'Eight Hidden Benefits of Micropower' in his Worldwatch Institute study *Micropower, The Next Electrical Era*. Unfortunately, the term 'micropower' is very limiting and refers only to electricity generation and distribution.

Based on our research, the scope of the technology development needs to be broadened to include all site-specific micro-utilities and it is offering the benefits of:

- modularity
- short lead time
- fuel diversity
- load growth insurance and load matching
- reliability and resilience
- avoided plant and grid construction and losses
- local and community choice and control
- avoided emissions and other environmental impacts.

Conclusions

The continued use of the 'combined mechanical utilities container' has validated the research objectives. The prototype has been in operation since August 2000, providing utility conveniences to a single-family dwelling located in Northern Canada.

The introduction of the micro-utility appliance has opened new avenues to provide infrastructure convenience. However, the micro-utility concept will only gain widespread acceptance once governments and local utility providers are clearly in support of this new type of infrastructure and make the necessary policy decisions [8]. Canada's remote, northern and arctic communities would benefit most from the use of micro-utility appliances. These communities are heavily dependent on government subsidies to provide housing and adequate infrastructure convenience.

By extension, remote and presently underserviced regions in countries with emerging economies stand to gain the most from introducing the micro-utility concept alongside conventional grid-based utility systems. The micro-utility concept can be seen as a 'leapfrog technology' to allow certain areas of economic viability (for example remote eco-resorts, remote mining operations, geographically isolated communities) to provide 21st century infrastructure convenience without having to wait for the establishment of a national or regional utility supply grid [9].

The concept of dispersed or embedded infrastructure will provide the transition from macro-grids to micro-infrastructure, allowing a coexistence of both technologies.

References

[1] Townshend, A. R., Jowett, E. C., LeCraw, R. A., Waller, D. H., Paloheimo, R., Ives, C., Russell, P. and Liefhebber, M. Potable water treatment and reuse of domestic wastewater in the CMHC Toronto 'Healthy House'. In *Site Characterization and Design of On-Site Septic Systems, ASTM STP 1324*, M. S. Bedinger, J. S. Fleming, and A. I. Johnson (eds), American Society for Testing and Materials, West Conshohoken, PA 1997, pp. 176–187.
[2] Hlady, G. and Durante, R. (2000) Indian and Northern Affairs Canada/Public Works Division Estimated Water/Sewer/Hydro Infrastructure Costs Per Equivalent Housing

Unit For Remote Special Access Communities, Thunder Bay and Sioux Lookout, Ontario, Canada.

[3] Nunavut Housing Corporation (2000) Business Plan 2001/2002, Figure 3, p. 7 (NHC LHO Summary of operating cost 1995/96 to 1999/2000) source: NWTHC/NHC Audited Financial Statements.

[4] Jowett, E. C. (1999) Immediate re-use of treated wastewater for household and irrigation purposes, presented at 10th Northwest On-site Wastewater Conference, University of Washington, Seattle, United States.

[5] Paloheimo, R. and Le Craw, B. (1996) Reusing treated wastewater in domestic housing: the Toronto Healthy House project, presented at Disposal Trenches, Pre-treatment and Re-Use of Wastewater Conference, Waterloo Centre for Groundwater Research and the University of Waterloo, Canada or www.healthyhousesystem.com

[6] Slattery, J. A. (2001) Using new infrastructural technology for sustainable rural and urban development, White Paper, Powercell Corporation, Tulsa, Oklahoma, United States.

[7] Dunn, S. (2000) *Micropower: The Next Electrical Era*, Worldwatch Paper 151, Worldwatch Institute, Washington, DC [based on: Lovins, A. and Lehmann, A. (2002) *Small is Profitable: The Hidden Benefits of Making Electrical Resources the Right Size*, Rocky Mountain Institute, Boulder, Colorado, United States].

[8] Casten, T. R. (19 June 2003) World Alliance for Decentralized Energy 'Dinosaur Rules', speech given at the Canadian District Energy Association's 8th Annual Conference and Exhibition – 'Delivering Energy Solutions'.

[9] Mc Fall, K. (December 2002) *Fuel cells: A Leapfrog Technology for Rural Latin America?* Platts Energy Business and Technology/McGraw Hill Companies.

8 Concept for a DC–low voltage house

M. Friedeman[1], A. van Timmeren[2], E. Boelman[2], and J. Schoonman[3]

Summary

The general expectation is that electricity will increasingly be generated decentrally, for example using solar cells. Solar cells provide direct current (DC) electricity at low voltages. Before this energy can be used in buildings, it needs to be converted to alternating current (AC). However, many AC-powered household appliances work internally on DC. This implies the need for a transformer within the appliance, in order to convert the AC current of 230 V to (low voltage) DC current. Using photovoltaics (PV) this method involves two energy conversions with inherent energy losses. It is reasonable, therefore, to assume that introducing a DC (low-voltage) grid can mitigate these losses. The feasibility of this idea has been studied by ECN (Energy Centre the Netherlands), by comparing a grid-connected house with a stand-alone PV-powered house provided solely with DC electricity. The conclusion was that introducing a DC grid would not lead to a significant reduction of energy losses. The present study reveals that further analysis of the energy losses requires the inclusion of different conditions and circumstances. The main obstacles to energy loss reduction appeared to be that not every household appliance is suited for DC-low voltage electricity and that the use of a normal distribution system for low-voltage currents involves significant losses during current transport. The focus of this research is to find a solution for these problems and develop a concept with these solutions for a DC low-voltage system that does reduce energy losses. The main points of interest studied are:

- alternative ways of transporting DC low-voltage electricity in a single family house with minimal energy losses
- to devise the most efficient spatial lay-out and furnishing of a house to minimize the distance between generation and end-use of DC electricity
- to use a hybrid system consisting of an AC and DC grid connected to the various household appliances, depending on their suitability for either form
- to set out an adequate scale for DC power storage and generation.

---

[1] Shell Renewables, Amsterdam, The Netherlands.
[2] Delft University of Technology, Faculty of Architecture, MTO/BT, The Netherlands.
[3] Delft Institute for Sustainable Energy, Faculty of Chemical Technology & Bioprocess Technology, Delft University of Technology, The Netherlands.

The result is a conceptual design of a house incorporating these points of interest. The design will be based on zero-energy houses, now being planned for a neighbourhood in the city of Zoetermeer, the Netherlands. The design aims at mitigating energy-losses in the house and neighbourhood without loss of comfort for the inhabitants.

Notation

A	Diameter of cable
AC	Alternating current
DC	Direct current
I	Current
l	Length of cable
LV	Low voltage
P_{loss}	Power losses
R	Resistance
ρ	Specific resistance

Introduction

As a result of increasing environmental concern, the impact of electricity generation on the environment is being minimized and efforts are being made to generate electricity from renewable sources. Although widespread use of renewable energy sources in electricity distribution networks is still far away, high penetration levels may be reached in the near future at a local level in new suburban areas, where much attention is paid to renewable energy. One of the sources to be used is the sun. Emergency phone systems, yachts, and caravans increasingly use solar panels to generate the energy they require to operate. The Dutch government offers generous grant schemes for solar panels on houses and offices. Solar energy is the power of the future. There is however a slight drawback, since photovoltaic cells produce direct current (DC) at low voltages, usually 24 V. For normal European domestic use, this would have to be converted into a 230-V alternating current (AC). On the other hand, many appliances these days use low voltage DC power provided by a plug-in adapter that reduces the 230-V AC mains power to a lower DC voltage. Besides that, due to the intermittent nature of solar energy supply, stand-alone PV systems require electricity storage. In locations where a public electricity grid is available, 'storage' by grid connection is usually considered to be more convenient than privately-owned electricity storage. Again, grid connection requires a conversion from direct current (DC) to alternating current (AC). This conversion also enables the use of standard AC supplied household equipment. Using PV with an AC grid thus involves two energy conversions, with inherent energy losses. It is reasonable, therefore, to assume that introducing a DC (low-voltage) grid can mitigate these losses.

Another important and actual aspect related to the use of electricity in buildings is the dissatisfaction of an increasing number of users with the ever increasing jumble of cables, leads, plugs, power points, and boxes with transformers and rectifiers, some of which are larger than the appliances they power. A domestic DC network would shortcut the last conversion stage so all those power adapters could be 'consigned to the waste bin'. In addition, new domestic current distribution systems could drastically reduce the need for power leads. One example is the use of flat power strips in spacecraft, which might well be suitable for adaptation in a domestic environment. The strips could simply be stuck on to the walls, and moved to a different position as the interior design of a room is changed. Appliances can be connected to the strips at any point. Possible innovations like this fit in well with current ideas of flexible, reusable and sustainable building.

There is no simple answer to the question whether a domestic 24 V DC power network, augmented by a limited AC mains system, would be a real improvement over the current pure AC mains setup. It cuts out the power losses that occur during the conversion from sustainable DC power to AC power and vice versa, as well as the losses resulting from the 230 V to 24 V transformation, and it does compensate for the power losses that occur as the electricity is transported at a low voltage.

Research conducted in the Netherlands by the research centre of KEMA in 1997 advocated the introduction of a national DC power grid to provide a more efficient way of distributing electricity.

The feasibility of the idea of a domestic DC grid has been studied by the ECN (Energy Centre the Netherlands) [1], by comparing a stand-alone PV-powered AC house to a stand-alone PV-powered house provided solely with DC electricity. A conventional lay-out was used for both situations, with up to 40 metres of cable length. The ECN research, established in 1998, turned out in favour of the alternating current system. According to the ECN the DC system resulted in energy losses of as much as 11%. However, the study was based on the conventional layout of both houses and distribution networks. An additional re-design of the (integration inside) houses with accompanying adaptation to optimum passive energy related aspects might drastically change the outcome.

Objectives and scope

The aim of this research is to investigate to what extent the energy losses in a PV-powered house can be mitigated by introducing a DC low voltage distribution system. The present study focuses mainly on two issues that have been given relatively little attention in previous research:

(1) What if both AC and DC are used, so that higher power demands are met with AC, and lower power demands with DC electricity?
(2) Is it possible to reduce distribution losses by adapting the design of a standard house, so that the distances in the indoor distribution lay-out are limited? At present the power losses in the DC low-voltage distribution system are high due to large distances in standard indoor grid lay-outs.

Differences in losses between DC/DC and AC/DC converters were also examined in the present study, but to a lesser extent. Previous research by ECN did not take into account the losses at the supply side of household appliances.

Analyzing the power efficiency of an entire house is easier said than done. Which types of house should one look at, and what kind of appliances do they contain? The final choice for the basis of the conceptual design of the DC system was a preliminary design of energy-neutral houses, currently being planned for the 'Groene Kreek', a neighbourhood within a so-called Vinex extension of the city Zoetermeer, in the Netherlands. The reference house is being redesigned into a concept house, which is optimized to incorporate a DC distribution system, photovoltaic electricity generation (PV) and storage. That there is no loss of comfort for the inhabitants is a 'sine qua non condition'.

Research outline

In order to meet the above-mentioned objectives, the following items are studied. First, different types of solar panels are investigated and a choice of solar panel type is made. The amount of generated electricity is calculated based on the availability of sunlight in the western part of The Netherlands and on an assumed efficiency of 10% for the chosen panel. A study is also done on the suitability of household appliances to work on DC low-voltage current, in order to estimate the electricity demand (DC load) and the number of solar panels necessary to feed the selected DC appliances. Combining these data for supply (PV panels) and demand (DC loads) enables a rough estimate of the storage capacity, and a tentative choice of an electricity storage system.

The choice of storage system also takes into account the possibility of spatially integrating electricity storage into the house.

The distribution system is looked at in more detail. This study focuses on whether clustered DC appliances and a different lay-out of the house can lead to smaller distances in the indoor distribution grid, and hence to smaller electricity transport losses. A design is made for a distribution system in a concept DC house, and the transport losses are calculated for the optimized lay-out. Finally the transport and conversion losses in the DC house are compared with the losses in a standard AC house, and conclusions about the feasibility of the DC distribution system are drawn.

Solar-energy conversion and electricity demand in the DC–low voltage house (in 'De Groene Kreek', Zoetermeer)

Solar energy can be converted into electrical energy and into heat. Photovoltaic cells (PV) are used for the conversion of solar energy into electricity. In the concept house, poly-crystalline silicon cells with an efficiency of 10% are used. This technique is the best developed at the moment, and is affordable for consumers. The panels are placed on the flat roof of a building, facing south at an angle of 45° from the horizontal plane (see Figure 8.1). It is estimated that 24 m^2 of solar panel can meet the demand and provide a small overcapacity, to ensure that electricity demand can also be met during bad summers.

For the electricity output, data [2] are used concerning the average daily sun radiation over a year, based on yearly, monthly and daily averages. With these numbers the electricity output for the 24 m^2 of panels with an efficiency of 10% is estimated. The results are shown in the next section.

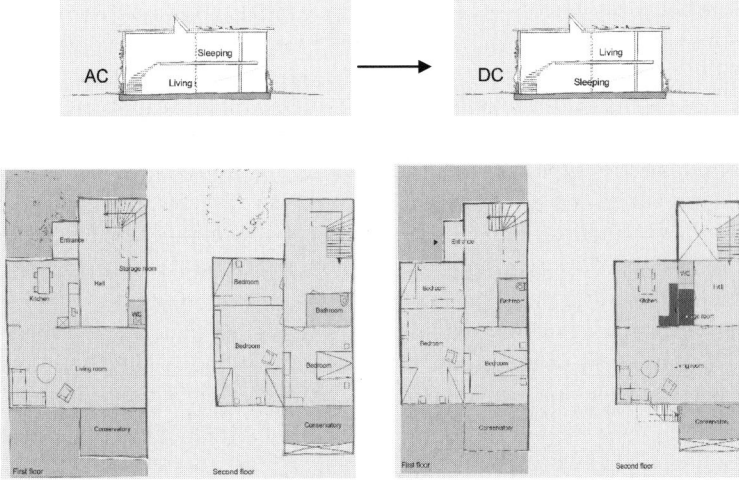

Figure 8.1 Section of the design of the Solar house to be built at the new 'Groene Kreek' development in the commuter town of Zoetermeer. The neutral-energy house, designed by Bear Architects from Gouda, will provide the basis for the design of a house with the DC system [4, 6].

To estimate the DC and AC loads in the concept house, first the total electricity demand is assessed [3]. Average annual electricity consumption in the 'Groene Kreek' is estimated to be slightly higher than the average of 3500 kWh per year in the Netherlands. Based on the income levels of future inhabitants, average yearly consumption per household in the 'Groene Kreek' is estimated to be about 4000 kWh.

This is the total output for both AC and DC electric loads, regardless of which appliances can be supplied with low-voltage DC, and of how much they consume. The suitability for DC supply strongly depends on how electricity is converted. In principle, every appliance can be made suitable, but some groups of appliances are easier to adapt to DC electricity supply, or already function on an internal DC supply. The suitability for low voltages depends on the power demand of the appliance: only low power demands (<150 W) are suited to low-voltage, or else the losses in the distribution system will be too high. Table 8.1 shows the appliances that are suitable for DC low-voltage supply.

Table 8.1 Overview of appliance suitability for DC low-voltage supply.

Kind of energy conversion		Examples	Suitability for DC supply	Suitability for low-voltage supply
Electric	Heat	Boiler, Coffeemaker	++	+/−
Electric	Rational/ Mechanical	Washing Machine, Fan	+/−	+/−
Electric	Sound/ Vision	TV, radio, CD player	++	+
Electric	Light	Fluorescent lamp, Light bulb	+/−	+

The suitability of DC low-voltage supply is investigated for four different groups [4]: TV and video appliances, lighting, small kitchen appliances that do not generate heat, and appliances that use batteries (e.g. electric toothbrushes and shavers). The total electricity demand is found to be 2400 kWh for DC, and 1500 kWh for AC. This DC demand is used for the calculation of the required number of solar panels.

Whether DC electricity supply can actually lead to a reduction in energy conversion losses depends on the efficiency of DC/DC and AC/DC conversion. If the chosen voltage level (24 V) matches the voltage level needed by the DC appliance, no voltage conversion is needed. However, several household appliances work on different voltage levels, which implies that there is also a need for voltage conversion with a DC supply. A possible reduction of energy losses thus depends on the losses in both AC/DC and DC/DC converters. In this study, tentative measurements have been made of the AC/DC losses in household appliances. Significant electricity conversion losses were detected but the evidence is not yet conclusive. For the losses in DC/DC converters, information was obtained and compared from various manufacturers. The various data indicated an average DC/DC converter loss of about 10%.

The DC distribution system

In the concept house proposed in this research, DC electricity is transported by means of a low-voltage distribution system. This has to be done in a safe way, with as little losses as possible. The voltage chosen for the DC grid is 24 V, due to compatibility with most PV panels and storage systems. To transport the same power load as normal at this voltage, a higher current is needed. As a result, power losses are higher because of a quadratic increase of losses with current.

$$P_{\text{loss}} = I^2 \times R$$

These losses can be reduced by lowering the resistance, which can be reduced by shortening the length of the distribution cables or enlarging the diameter.

$$R = \rho \times (l/A)$$

In this study, we investigate the possibility of reducing transport losses by shortening the length of electric distribution cables. The possibility of enlarging the diameter, perhaps even by a totally new electricity distribution concept, is not considered here. In this study, the spatial layout of the entire house is scrutinized and adapted so as to enable shorter lengths of indoor electricity distribution cables to be used.

Architectonic integration

The selected house – two floors under a large, flat roof – was designed to use solar panels. The spatial layout of the concept house is adapted in this study, in order to reduce cable length and enable better use of direct current and passive energy aspects, without loss of personal comfort. Because the DC electricity source (solar panels) are

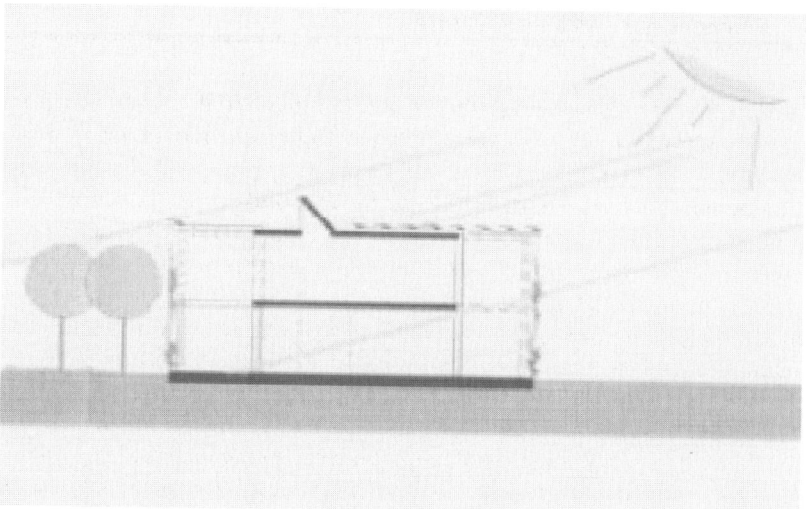

Figure 8.2 Conceptual reversion of the functional typology of the house due to a needed closer situation to the electricity generation and the use of a Direct Current (DC) grid. The intake of DC electricity therefore is situated on the second floor, in the centre of the house.

Figure 8.3 Three dimensional visualisation of the DC-test house, built by ECN (Energy Centrum Netherlands).

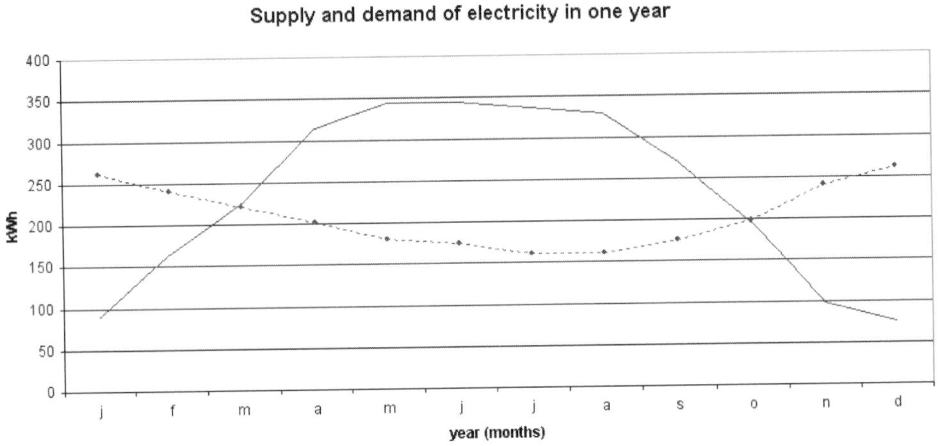

Supply and demand of electricity in one year

Figure 8.4   Estimated yearly electricity demand and supply in the concept house [4].

situated on the roof, the rooms in the house are arranged in such a way that the living room and kitchen (where most DC electricity is used) are placed closest to the source (Figure 8.2). This implies placing the living areas on the upper floor and the bedrooms on the ground floor, which is the opposite layout to that of a typical Dutch house. The intake point of the cables from the solar panels is placed in the centre of the house. With this arrangement, and due to the placement of the electricity intake point in the middle of the house, cable lengths are reduced to a maximum of 12 m, from the normal 40 m in a standard AC house.

The types of appliance inside the house were determined from the popularity of the electric household appliances in Dutch households. If a certain type of appliance was used by more than 40% of Dutch households, it made it into the reference house. Furthermore, all the AC appliances (mostly in the kitchen and bathroom) are clustered in one single zone, so that an extensive AC grid is not necessary. Additionally, electrical installation plans are made for the DC concept house and for the reference AC house, to enable calculation and comparison of electricity losses in the distribution system (see Figure 8.3).

Electricity storage

Since a stand-alone PV house is assumed, there is need for electricity storage. Storage systems that are suitable for use in homes are flywheels, batteries and, if a good storage solution can be found, possibly hydrogen in the future. Currently, lead–acid batteries seem to be the best option. In the future, lithium–ion batteries will form a cheaper and safer alternative. The required electricity storage system capacity is estimated in this study by comparing electricity demand from DC appliances and supply from the PV panels.

A global estimate for a year-round storage system size can be made from Figure 8.4. The storage system needs to cover about 600 kWh. This implies several cubic metres of

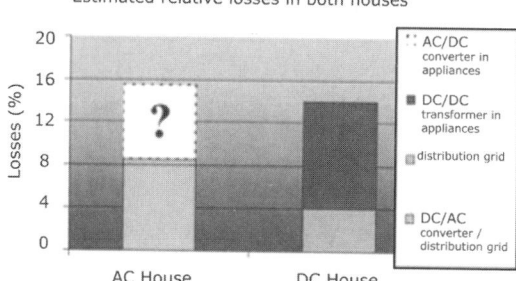

Figure 8.5 Estimated relative losses in the whole distribution system.

battery storage per house. Because of these large dimensions, another option is to cover only night time demand during the summer months (March until October), and use alternative power-generating devices during the remaining months. This topic needs further research, but is not addressed in the present study.

Estimated overall system losses

The relative losses in the whole system, from generation until end-use, are estimated for the DC low-voltage concept house and for the AC reference house [6]. See Figure 8.5.

Conclusions and recommendations for further research

In this study, a mainstream reference house design is optimized in order to accommodate DC electric power distribution and generation with photovoltaic cells. Investigation into electricity demand by an average household, PV supply, DC distribution and storage lead to the following conclusions:

- The introduction of DC-based distribution has the potential to reduce conversion losses, by eliminating the need for conversion between PV and distribution (DC/AC), and within appliances (AC/DC).
- An essential condition for the realization of this potential is building design optimization; in this study an improved house lay-out led to a significant reduction in electricity distribution losses.
- Another essential condition (not verified in this study) is that the relative losses in AC/DC converters in household appliances should prove to be more than about 6%. In this case, the total power losses in the AC house would exceed the total power losses in the DC concept house, so that the introduction of DC based distribution would mean a reduction in the total losses of electricity conversion and transport.

Further research is needed on the following topics:

- losses in AC/DC inverters in household appliances and in DC/DC inverters
- possibilities for new distribution networks, with larger cross-sections and improved comfort (accessibility, flexibility, safety) for residents.

References

[1] Pellis, J. (1998) *The DC Low-Voltage House*, Research by ECN (Energy Centre, The Netherlands.

[2] KEMA (1996) *Electricity Consumption in New Houses in 'Vinex' Extensions in the Netherlands, Based on Enquiries*, December 1996, The Netherlands.

[3] Energiened (1998) *Basisonderzoek Electriciteits-verbruik Kleinverbruikers (BEK 1998)*, Energiened, Arnhem, The Netherlands.

[4] Friedeman, M. M. (2002) The DC low voltage house, graduation project, Delft University of Technology, Faculty of Architecture, The Netherlands.

[5] Graaf, A. van der (2003) Life under the source. Sustainable housing using low-voltage DC power, *Delft Outlook 2003–1*, Delft University of Technology, The Netherlands.

[6] BEAR Architecten (2002) *Voorlopig ontwerp energie-0 woningen 'De Groene Kreek'*, Vinex locatie Oosterheem, Zoetermeer, BEAR, Gouda, The Netherlands.

9 Development of phase change thermal storage wallcoverings in buildings

K. Ip[1], A. Miller[2], T. Corner[3] and D. L. Dyball[4]

Summary

Energy saving wallcoverings that incorporate phase change materials (PCMs) as the thermal storage media are currently under joint development by the University of Brighton and OMNOVA Wallcovering (UK) Ltd. The PCM in the wallcovering functions to reduce the heating and cooling requirements by shifting the peak cooling/heating demand. This will result not only in the reduction of annual overall energy consumption, but also in smaller installed heating/cooling plant capacity. Initial evaluation indicated that energy saving of over 25% on a design day (the day during which the latent heat is fully utilised) for a typical air conditioning office in the UK is possible. Dynamic computer simulation was used to provide some preliminary guidance on the optimum quantities of phase change materials to be embedded in the wallcoverings in different perimeter zones of a model building. A number of development issues have been identified and these need to be addressed before the production of PCM wallcoverings can be introduced commercially.

Introduction

A phase change material (PCM) stores the sensible heat as the temperature increases but, more importantly, it can store or release large quantities of latent heat during the phase transition. This can be illustrated by using a common building material such as concrete, which has a sensible heat capacity of approximately $1.0\,kJ/kg\,K$ whereas a phase change material such as calcium chloride hexahydrate can store/release $193\,kJ/kg$ of heat on phase transition. The potential for peak load shifting, reduced space temperature fluctuations and small volume make PCMs very attractive as an energy saving material. There are numerous applications of PCMs in buildings

---

[1] University of Brighton, School of the Environment, UK.
[2] University of Brighton, School of the Environment, UK.
[3] Omnova Wallcovering (UK) Ltd, UK.
[4] Omnova Wallcovering (UK) Ltd, UK.

including PCMs embedded in concrete blocks and boards [1], PCM modules for underfloor heating [2] and 'CoolDeck' for night cooling [3].

A novel developing application of latent thermal storage described in this paper is the incorporation of phase change materials in wallcoverings. Wallcoverings are widely used in offices, hospitals, hotels and domestic buildings. With an estimated demand in new wallcovering area of three billion square metres per year in North America alone [4], the PCM wallcovering can have significant impact on the energy consumptions and installed capacities of the heating and cooling plants in buildings. A three-year research programme, supported under the UK government's Teaching Company Scheme [5], was set up in 2002 between The University of Brighton and Omnova Wallcovering (UK) Ltd to develop wallcoverings embedded with phase change materials. The research programme involves the evaluation of the potential applications for PCM wallcoverings, the testing and development of prototypes and the marketing of the product.

This chapter reports on some of the initial outcomes of the research. The concept of incorporating PCMs in wallcoverings and a simple evaluation of the energy saving potentials are summarised. An eight-zone computer model representing a typical office building has been developed and adapted in a computer simulation program to assess the effective latent heat storage capacities in each zone. The development team has identified a number of development issues that ought to be addressed before the commercial production of the PCM wallcoverings.

Wallcoverings with PCM

PVC (vinyl) wallcoverings are usually made by one of two manufacturing processes. Perhaps the most commonly used is the 'plastisol' route. The term plastisol is used to describe a suspension of finely divided PVC polymer made by emulsion or micro-suspension polymerisation, in a plasticiser. Typically, plastisols are modified by the addition of stabilisers, inert fillers, pigments and rheology control agents prior to coating a continuous fabric or paper substrate with the resultant paste. Coating can be achieved by various methods but direct coating of the substrate with a knife coater is most common. Once the continuous substrate is coated to the desired thickness, usually between 100 and 500 μm, the plastisol is transformed into a solid substance by passing the coated substrate through an oven which is heated to an elevated temperature for several minutes. As the temperature of the plastisol is raised, the plasticizer penetrates the PVC particles, which then swell. The plasticized particles eventually coalesce to give homogeneous plasticised PVC. Whilst still hot, the softened PVC can be embossed. On cooling, the material solidifies and it can then be printed to give rise to the familiar PVC wallcovering.

The second most common manufacturing route to vinyl wallcoverings is the calendering route. Here, porous PVC granules made by suspension polymerisation are mixed with plasticiser, stabilisers, fillers, etc. in a blender to produce pre-compounded PVC in powder form. The resultant powder is usually dry because the plasticiser is absorbed within the pores of the PVC granules. Next, the PVC compound is fed into a melt-compounder or extruder to convert the powder into a plastic material that can then be fed into a set of heated rollers, the calender. As the plastic compound is fed through successive heated sets of rollers, it is converted into a hot, homogeneous sheet of plasticized PVC. This hot molten sheet can be used to coat continuous substrates such as paper and fabric, or it can be laminated with other sheets of materials by hot rolling. The resultant materials can be embossed and/or decorated as described above. In general, the thicknesses of PVC sheets made by calendering lie in the range 100–800 μm, although sheets up to 1.5 mm thick can be produced.

A number of organic and inorganic PCMs are being evaluated for suitability for use in wallcoverings. Organic PCMs are mainly products of petroleum refinement such as paraffin wax, although it is possible to use fatty acids from renewable sources [6]. Inorganic PCMs are mainly salts and salt hydrates such as sodium sulphate decahydrate. Most inorganic PCMs have high volumetric energy storage capacities [7] but smaller changes in volume in comparison to organic PCMs [8].

Omnova Wallcovering (UK) Ltd filed a UK patent application in April 2002 and a PCT application in April 2003 describing wallcoverings containing phase change materials. The wallcoverings can be used to reduce the amount of energy required to heat and cool buildings and to produce a more comfortable environment by minimizing temperature fluctuations.

Energy-saving potential

A simple steady state thermal model representing a typical office was established to evaluate the energy-saving potential of the wallcovering due the latent heat of PCM embedded in it. The model office has the dimensions of 5 m by 5 m by 2.5 m high with a south facing window of 10 m^2. Taking into account the furniture and other furnishing of the room, the net wall area available for the storage of latent heat is approximately 31 m^2. The wallcovering selected in this analysis had a thickness of 4 mm comprising 35% by volume of inorganic PCM, which has a latent heat capacity of 150 kJ/kg and phase change temperature of 21°C. The PCM wallcovering in this model would have an average latent thermal storage capacity of 641 kJ/m^2 or a total of 19 800 kJ. The room is assumed to be occupied for 250 days a year, split between 166 winter days and 84 summer days.

For the summer cycle, the model assumes that for each day that the air conditioning is used, the PCM releases its heat by night cooling and the wallcovering absorbs its 'full quota' of heat during the hours of occupancy. The total energy saving due to the latent heat of the phase change process alone is the product of the area of wallcovering available for the storage of latent heat, the average thermal storage capacity and the number of days on which air conditioning is used.

For the winter cycle, the low angle Sun allows deep penetration of direct solar radiation on the wallcovering. As the direct solar gains far exceeds the total latent heat capacity of the PCM wallcovering, the model assumes that the air temperature swing would allow the wallcovering to absorb its 'full quota' of heat in the early part of the day and the PCM releases this heat into the room in the late afternoon. The energy saving due to the latent heat transfer of the PCM alone is the product of the area of

Table 9.1 Energy reduction on design days due to PCM wallcovering.

PCM wallcovering applied to:	Energy storage capacity (kJ)	Heating only (Percentage reduction)	Cooling only (Percentage reduction)
1 m^2	641	0.81	0.67
21 m^2	13 712	17	14
31 m^2	19 674	25	20
38 m^2	24 480	31	25

wallcovering available for latent heat storage, the average thermal storage capacity and the number of days when useful solar gain is available.

Using the solar data information from the CIBSE Guide A [9], the reduction in maximum energy consumption for the model room due to the latent heat of the PCM on the design heating and cooling days are summarised in Table 9.1. The energy consumption for a typical air-conditioned office, based on the energy consumption data from the UK Energy Efficiency Best Practice Programme [10], is 79 704 kJ in winter and 96 444 kJ in the summer. Higher energy savings are expected if specific heat transfers are included in the calculations. The results show that savings, depending on the PCM wallcovering area, are between 17% to 31% for heating and 14% to 25% for cooling.

Preliminary dynamic thermal analysis

A validated computer thermal model representing the interaction between the PCM wallcoverings, the thermal behaviour of the room, the mechanical heating and cooling systems being used and the external climate has yet to be developed. The computer thermal model developed at this stage aims to provide an initial assessment of the optimum quantity of PCM to be used in different parts of a building. The optimum quantity of PCM allows the best use of latent heat on a daily cycle basis.

The program used in this assessment is a dynamic thermal simulation software Virtual Environment (VE) produced by Integrated Environmental Solutions Ltd. Virtual Environment is a simulation program that can perform dynamic thermal analyses and energy evaluation of buildings [11]. The program allows users to specify the physical environmental parameters of each room, the external weather climate and the operation of the heating and cooling systems. As the thermal simulation of a room with PCM paper is not available, a special PCM room component was adapted to emulate the thermal performance of the PCM wallcovering.

The simulation model consists of an intermediate floor of an office building, which is divided into eight perimeter zones and one internal zone. Each zone is 5 m × 5 m × 2.5 m high. The corner zones have two 4 m × 1.25 m low-E double glazed windows while the middle zone in each facade has only one 4 m × 1.25 m low-E double glazed window. The construction materials used for the floor, roof, external walls and windows of the building are those that meet the UK Building Regulations 2002. The wallcovering is assumed to have a thermal resistance of 0.177 m^2 K/W and conductivity of 0.5 W/mK. An organic and an inorganic PCM were used in the simulation.

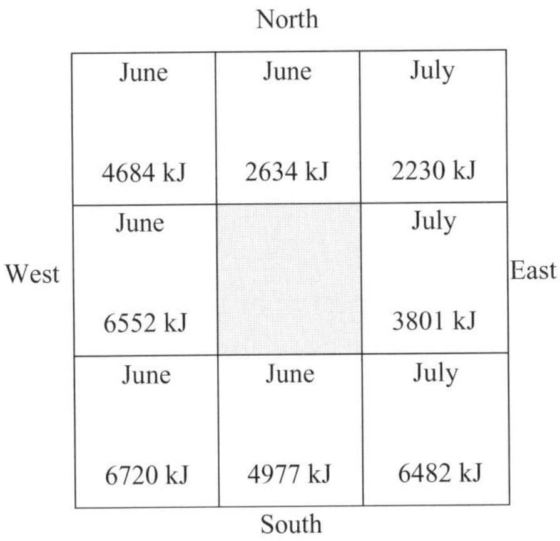

North

June	June	July
4684 kJ	2634 kJ	2230 kJ
June		July
6552 kJ		3801 kJ
June	June	July
6720 kJ	4977 kJ	6482 kJ

West ... East

South

Figure 9.1   Maximum PCM latent heat transfer.

Simulation process

The computer simulation to find the optimum quantity of PCM is based on the principle that the daily cycle of heat gain during the charging period is balanced by the heat loss during the discharging period.

To simulate the latent heat gain of the phase change material, a cooling cycle would be specified in the PCM room that emulated the PCM wallcovering. The phase change material starts to melt once the room reaches phase change temperature and begins to absorb heat. The simulation software would determine the amount of energy that is required to keep the room at that temperature; this represents the amount of energy absorbed by the phase change material.

To simulate the phase change material releasing the latent heat, a heating cycle would be specified. This means that heating is required should the temperature drop below the phase change temperature. The amount of energy required to keep the room 'warm' shows the amount of heat that could be released by a phase change material.

The simulations were repeated, with adjustments to the charging and discharging periods of the melting and solidification processes, until the heat gain by the PCM was balanced by the heat it released. Further simulations were carried out for each month in a year, for every zone in the model building.

Results

The results representing the maximum amount of energy that can be used by phase change materials in the perimeter offices on four sides of the square model building are shown in Figures 9.1 to 9.3. The results indicated that the PCM performs best in the summer months of June and July. Each box in Figure 9.1, representing a building zone, shows the maximum energy involved in the latent heat transfer process.

Figures 9.2 and 9.3 show the amount of inorganic and organic phase change materials that are required in each zone to absorb and release the latent energy indicated in Figure 9.1.

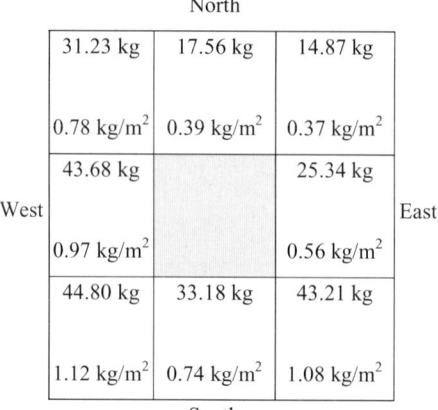

Figure 9.2   Optimum quantity of organic PCM (latent heat storage capacity of 150 kJ/kg).

The amount of organic phase change material required in the model building ranges from $0.37\,kg/m^2$ to $1.12\,kg/m^2$ whereas $0.31\,kg/m^2$ to $0.93\,kg/m^2$ would be required using inorganic phase change materials. Owing to the current limitations of the simulation model, such as disregard of specific heat transfer and direct solar gains, the results can only be considered as useful indications of the effective quantity of PCM to be used.

Development issues

OMNOVA's patent application describes how wallcoverings containing phase change materials can be made. The effectiveness of such wallcoverings in absorbing and then releasing heat is to be studied in a test room environment and later in 'thermally

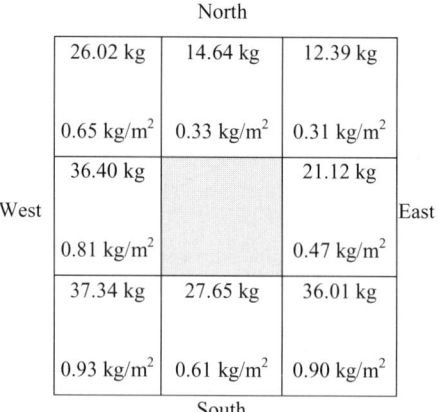

Figure 9.3   Optimum quantity of inorganic PCM (latent heat storage capacity of 180 kJ/kg).

matched' houses. There are, however, many technical issues that need to be addressed prior to a commercial launch of wallcoverings containing PCMs. These include the following:

- Will the microencapsulated PCMs retain their recycling capabilities with respect to melting and freezing after experiencing the processes involved during the manufacture of wallcoverings?
- Can high loadings of PCMs in wallcoverings be obtained so that high thermal mass is achieved at modest and acceptable thicknesses?
- Can heat in a room be readily transferred to and from the PCM containing wallcoverings?
- Will, for example, the fire resistance properties of PCM containing wallcoverings be good enough for wide application within both commercial and domestic buildings?

Other issues that are being addressed include:

- Marketing   How should wallcoverings containing PCMs be marketed and sold?
- Sustainability   There are various environmental issues to be addressed:
  - Recycling of PVC   This is possible with some forms of PVC and recycling plants are being built in Europe using the 'Vinyl Loop' process developed by Solvay Industries [12]. Would PVC wallcoverings containing PCMs be recyclable?
  - Life cycle energy analysis   Would the manufacture of wallcoverings containing PCMs consume more or less energy than they would save through reduced heating and cooling energy?
  - CO_2 emissions   Air-conditioning is becoming more prevalent and it generates large amounts of CO_2. What role could wallcovering containing PCMs play in mitigating the proliferation of air conditioning systems and thence the reduction of CO_2 emissions?
  - Healthy buildings   What is the value of providing a 'more comfortable' environment by reducing the difference between the upper and lower temperatures attained in a room?

Conclusions

Novel energy saving wallcoverings that use the latent thermal storage technique are currently being developed jointly by the University of Brighton and OMNOVA Wallcovering (UK) Ltd. Initial 'snapshot' analyses indicate there is potential for energy saving. Further dynamic analyses to assess the annual energy savings and plant size reduction are necessary to provide more detailed and accurate data. A number of development issues have already been identified; these need to be addressed before the PCM wallcovering can be manufactured commercially.

With the energy saving feature and environmental sustainability principles embedded in their production, the PCM wallcoverings can help to minimize the environmental impact of existing and new buildings.

References

[1] Zalba, B., Marin, J. M., Cabeza, L. F. and Mehling, H. (2003) Review on thermal energy storage with phase change: materials, heat transfer analysis and applications, *Applied Thermal Engineering*, **23**(3), 251–283.

[2] Ip, K. C. W. (1998) Solar thermal storage with phase change materials in domestic buildings, *CIB World Congress*, Gavle, Sweden, 1265–1272.

[3] Maunsell, F. (2003) *Stevenage Borough Council Case Study*, Faber Maunsell, available from http://www.fabermaunsell.com [accessed 2 May 2001].

[4] Dyball, D. L. (2003) Plasterboard market report, OMNOVA Wallcovering (UK) Ltd, UK.

[5] TCS (2003) Teaching Company Scheme Homepage, available from http://www.tcsonline.org.uk/ [accessed 2 June 2003].

[6] Feldman, D. Banu, D. and Hawes, D. (1995) Low chain esters of stearic acid as phase change materials for thermal energy storage in buildings, *Solar Energy Materials and Solar Cells*, 36, 311–322.

[7] Dincer, I. (1999) Evaluation and selection of energy storage systems for solar thermal applications, *International Journal of Energy Research*, **23**, 1017–1028.

[8] Hawes, D. W., Banu, D. and Feldman, D. (1992) The stability of phase change materials in concrete, *Solar Energy Materials and Solar Cells*, 27, 103–118.

[9] CIBSE (1986) *Guide A2 Weather and Solar Data*, The Chartered Institution of Building Services Engineers, London.

[10] BRESU (2000) Energy consumption guide 19: Energy consumption in offices, BRE, London, UK.

[11] IES (2003) APACHE Virtual Environment Software, Integrated Environmental Solutions Ltd, Available from http://www.ies4d.com/ (accessed 20th May).

[12] Solvay.com (2003) Plastics Sector Strategy, available from http://www.solvay.com/strategy/sectorsapproach/0,7,78-2-0,00.htm (accessed 10th June 2003).

10 Investigating the energy efficiency of double-skin glass facades for an office building in Istanbul

I. Cetiner[1] and M. Aygun[1]

Summary

Currently, reducing energy consumption, preventing environmental pollution and using recyclable materials are being intensively considered in facade design. These factors are especially important in multi-storey glass facades because of the high-energy loads resulting from heating, cooling and ventilating systems. Reduction of these loads will be possible as a result of attention to climatic facade design. Such a facade will be able to protect the building by controlling external effects. Double-skin glass facades have been developed as the result of such an approach. They are formed by installing solar control devices in the cavity between two glass skins. In such facades, heat losses are relatively small. The system allows windows to be opened, even on the top storey of the building. Hence the ventilation is provided by fresh air, resulting in the need for less energy to run the mechanical system.

These systems have not yet been applied in Turkey, but some firms consider manufacturing them in the near future. However, there is no available information on the extent to which these systems are effective in local climates. This chapter aims to determine the effectiveness of double skin facades, compared with conventional single skin glass facades in terms of energy consumption.

Notation

CEC	Cooling energy consumption, kWh/m^2
HEC	Heating energy consumption, kWh/m^2
ME	Mass effect
SHGC	Solar heat gain coefficient
TEC	Total energy consumption, kWh/m^2

---

[1] Istanbul Technical University, Faculty of Architecture, Istanbul, Turkey.

U	Heat transmission coefficient, W/m^2 K
WB	With blind
WOB	Without blind
WS	Windows' solar
WT	Windows' transmission

Introduction

Natural life on earth is under serious threat due to the gradual depletion of fossil fuels and rising environmental pollution. These problems necessitate energy efficient, environmentally friendly and ecological approaches in all areas, for improving the quality of life, and contributing to sustainable development. In the building industry, in particular, it is important to consider the amount of energy consumed and the harmful substances that can be produced in the processes of material manufacture, building construction and operation. There are also impacts on facade design. According to this new approach, a sustainable facade should be designed to reduce the energy consumption of buildings further by using natural energy sources, such as wind and sun.

The double-skin facade is one such facade, designed to decrease energy consumption and improve user comfort. It consists of an external skin, an intermediate cavity and an internal skin. The outer skin provides protection against the weather and improved acoustic insulation against any external noise. A sun-shading element is usually installed into the cavity to protect the internal rooms from high cooling loads caused by insolation [1]. The cavity is connected with the outside air so that the windows of the interior facade can be opened, even on high-rise buildings that are subject to wind pressure. This provides natural ventilation and thus results in decreased energy consumption and a lower running cost for the air-conditioning system [2]. Opening the windows in the internal skin also enables night-time cooling of the building [3]. Therefore, double skin facades are especially relevant to multi-storey buildings with high energy consumptions. In these buildings, the amount of energy needed to run the mechanical systems is very high. User illnesses emanating from the continuous operation of the mechanical systems pose a substantial health issue. Double-skin facades help to reduce energy consumption and achieve optimum conditions for building users. These facades have not yet been used in Turkey, but some applications are expected in the near future. This chapter, therefore, aims to determine to what extent such facades are efficient compared with conventional single-skin glass facades in terms of energy consumption in local climates.

Method

The energy consumptions of the skin configurations generated for both single- and double-skin glass facades are determined by computer simulation. ENER-WIN (Energy Calculations – Windows version) simulation program is used to compute the energy loads for the configurations. Some data needed to run this program, such as heat transmission coefficients (U-values) and solar heat gain coefficients (SHGC-values) pertaining to the configurations, are determined with the WIS (Advanced Windows Information System) simulation program. ENER-WIN has been developed by Texas A&M University, with financial support from the DOE [4]. WIS has been developed by the teams cooperating in the European research project (the project WIS), under the coordination of TNO Building and Construction Research [5]. The configuration with the lowest energy consumption is selected.

Simulation

The simulation was carried out for the facades of a hypothetical 30-storey office building located in Istanbul. The building is enclosed on all facades by a glass double skin with a 900-mm cavity. The steps of the simulation process are as follows.

Arranging the data

The inputs needed for the simulation are: the data describing the hypothetical building, the window system and the weather in Istanbul. The data pertaining to the building and window system can be seen in Table 10.1. Weather data are obtained from the field measurements at Goztepe, which is a weather station in Istanbul. The contents of all data are explained below:

- The data describing the hypothetical building are the properties of the building (for instance, building type, building size, ceiling height, total facade area, etc.) and its mechanical system (for instance, the type of heating and cooling systems, ventilation rate, etc.) as in Table 10.1. The mechanical system data are selected from the library of ENER-WIN.
- The window system comprises transparent components, frame and spacer. The values related to the transparent components are taken from a glass manufacturer in Turkey. The frame and spacer types are selected from the library of WIS.
- Weather data include the monthly average of dry bulb temperatures, dew point temperatures, direct normal solar radiation, wind speed, daily maximum temperatures and standard deviations of these monthly average values.

Generating the skin configurations

The skin configurations are generated by changing the type of the glass, the use of the sun shading element and the position of the glass. For this simulation, the clear and low-E glasses are selected as the glass types, and the configurations are formed with the consideration that both of these glasses can be used either on the external or internal skin and can be applied as single- or double-glass units. In addition, the case whether or not there is a sun-shading element in the intermediate cavity is also considered.

Table 10.1 Data needed for the simulation.

<table>
<tr><td rowspan="12" style="writing-mode: vertical-rl">The data related to the building</td><td colspan="4" align="center">**Building properties**</td></tr>
<tr><td colspan="2">Building type</td><td colspan="2">Office</td></tr>
<tr><td colspan="2">Building size</td><td colspan="2">36×36 m</td></tr>
<tr><td colspan="2">Number of floors</td><td colspan="2">30</td></tr>
<tr><td colspan="2">Ceiling height</td><td colspan="2">3.30 m</td></tr>
<tr><td colspan="2">Total facade area</td><td colspan="2">13 896 m^2</td></tr>
<tr><td colspan="2">Total floor area</td><td colspan="2">38 880 m^2</td></tr>
<tr><td colspan="4" align="center">**HVAC system**</td></tr>
<tr><td colspan="2">Air-conditioning system*</td><td colspan="2">Fan coil units</td></tr>
<tr><td colspan="2">Heating system</td><td colspan="2">Gas</td></tr>
<tr><td colspan="2">Infiltration rate</td><td colspan="2">0.4 ach**</td></tr>
<tr><td colspan="2">Ventilation rate</td><td colspan="2">20 l/s</td></tr>
</table>

** Air changes per hour

<table>
<tr><td colspan="4" align="center">**Transparent system**</td></tr>
<tr><td>Pane type</td><td></td><td>Clear glass</td><td>Low-E glass</td></tr>
<tr><td>Thickness</td><td></td><td>6 mm</td><td>6 mm</td></tr>
<tr><td>Emissivity</td><td>Outdoor</td><td>0.840</td><td>0.100</td></tr>
<tr><td></td><td>Indoor</td><td>0.840</td><td>0.840</td></tr>
<tr><td>Solar reflectance</td><td>Outdoor</td><td>0.070</td><td>0.218</td></tr>
<tr><td></td><td>Indoor</td><td>0.070</td><td>0.144</td></tr>
<tr><td>Visual reflectance</td><td>Outdoor</td><td>0.080</td><td>0.041</td></tr>
<tr><td></td><td>Indoor</td><td>0.080</td><td>0.055</td></tr>
<tr><td>Solar transmittance</td><td></td><td>0.770</td><td>0.574</td></tr>
<tr><td>Visual transmittance</td><td></td><td>0.880</td><td>0.825</td></tr>
<tr><td>Cavity width</td><td colspan="3" align="center">900 mm</td></tr>
<tr><td>Gas type in the cavity</td><td colspan="3" align="center">Air</td></tr>
<tr><td>Sun shading element</td><td colspan="3" align="center">Venetian blind</td></tr>
</table>

The data related to the window system

Frame and spacer

Frame type: Metal with 16 mm thermal break
Spacer type: Aluminium spacer

The generated alternatives for both single- and double-skin facades can be seen in Table 10.2 where the groups define the type of glazing in the skin configuration. For instance, the alternatives with single glazing on the external skin and double glazing on the internal skin are in group III, whereas double glazing is used on both skins in group V. The codes cover all the alternatives that were generated as a result of the combination of five groups with both glass types and cases with or without a blind. Code 7.2, for instance, is the case of a double low-E glass on the external skin, single clear glass on the internal skin and a blind in the intermediate cavity. The schematic sections of the skin configurations can be seen in Table 10.2.

Table 10.2   Skin configurations.

		Single-skin glass facade			
Group	**Code**	**Single skin**	**Blind**	. . . . . .	**Skin configurations**
	1.1	DG/CG	WOB	. . . . . .	
I	1.2	DG/CG	WB	. . . . . .	
	2.1	DG/LG	WOB	. . . . . .	
	2.2	DG/LG	WB	. . . . . .	

		Double-skin glass facade			
Group	**Code**	**External skin**	**Blind**	**Internal skin**	**Skin configurations**
II	3.1	SG/CG	WOB	SG/CG	
	3.2	SG/CG	WB	SG/CG	
III	4.1	SG/CG	WOB	DG/CG	
	4.2	SG/CG	WB	DG/CG	
	5.1	SG/CG	WOB	DG/LG	
	5.2	SG/CG	WB	DG/LG	
IV	6.1	DG/CG	WOB	SG/CG	
	6.2	DG/CG	WB	SG/CG	
	7.1	DG/LG	WOB	SG/CG	
	7.2	DG/LG	WB	SG/CG	
V	8.1	DG/CG	WOB	DG/CG	
	8.2	DG/CG	WB	DG/CG	
	9.1	DG/CG	WOB	DG/LG	
	9.2	DG/CG	WB	DG/LG	
	10.1	DG/LG	WOB	DG/CG	
	10.2	DG/LG	WB	DG/CG	
	11.1	DG/LG	WOB	DG/LG	
	11.2	DG/LG	WB	DG/LG	

SG: Single glass, DG: Double glass, CG: Clear glass, LG: Low-E-glass, WB: With blind, WOB: Without blind.
| Single glass, || Double glass, ⦙ Blind

Computing the U-, SHGC-values and annual heating and cooling energy consumption of the skin configurations

The U- and SHGC-values of the generated alternatives are computed with WIS. For this simulation, the position, size and material properties of the window components (external skin, the cavity, sun-shading element and internal skin) are defined in the program. The results pertaining to the U- and SHGC-values of the alternatives can be seen in Table 10.3. The annual heating and cooling energy loads of the skin configurations are computed with ENER-WIN according to the variation of these values. The loads are taken as the total of the heat gain/losses resulting from the windows'

Table 10.3 The simulation and calculation results.

Group	Code	U (W/m² K)	SHGC	Annual heating (kWh/m²)				Annual cooling (kWh/m²)				TEC (kWh/m²)
				WT	WS	ME	HEC	WT	WS	ME	CEC	
I	1.1	2.95	0.55	33.67	20.38	41.72	95.77	0.09	20.11	13.85	34.05	129.82
	1.2	2.70	0.24	31.38	10.48	38.87	80.73	0.72	9.53	13.31	23.56	104.28
	2.1	2.34	0.44	27.35	16.03	36.59	79.97	0.29	13.34	14.74	28.36	108.33
	2.2	2.21	0.19	26.66	7.41	35.05	69.12	0.63	5.81	14.24	20.68	89.80
II	3.1	2.90	0.55	32.99	20.19	40.99	94.17	0.04	19.96	14.05	34.06	128.23
	3.2	2.64	0.22	30.95	9.78	8.57	79.31	0.74	8.70	13.51	22.95	102.26
III	4.1	2.21	0.47	24.74	16.45	32.69	73.88	0.08	17.08	13.30	30.45	104.33
	4.2	2.09	0.17	24.72	7.41	32.94	65.07	0.53	6.52	13.57	20.63	85.70
	5.1	1.93	0.39	21.85	13.30	27.81	62.95	0.22	11.62	10.99	22.83	85.79
	5.2	1.86	0.14	22.05	5.23	27.93	55.21	0.48	4.26	11.32	16.06	71.27
IV	6.1	2.20	0.47	24.53	16.34	32.41	73.29	0.08	17.12	13.11	30.31	103.59
	6.2	2.09	0.27	24.24	10.78	31.93	66.95	0.34	10.34	13.19	23.87	90.82
	7.1	1.82	0.38	20.41	12.70	26.17	59.27	0.22	11.30	10.61	22.13	81.40
	7.2	2.02	0.25	23.93	9.22	32.07	65.22	0.42	7.69	13.41	21.51	86.74
V	8.1	1.86	0.42	20.39	14.12	25.07	59.59	0.08	15.32	9.57	24.97	84.56
	8.2	1.79	0.22	20.41	8.52	25.55	54.48	0.33	8.44	10.10	18.87	73.34
	9.1	1.70	0.35	18.91	11.46	24.76	55.13	0.24	10.33	10.11	20.68	75.80
	9.2	1.66	0.18	19.09	6.27	24.38	49.74	0.35	5.46	10.18	16.00	65.74
	10.1	1.62	0.34	17.85	10.95	23.82	52.63	0.21	10.03	9.53	19.77	72.40
	10.2	1.76	0.21	20.37	7.41	26.09	53.88	0.36	6.38	10.62	17.36	71.24
	11.1	1.50	0.30	16.41	9.61	22.41	48.44	0.19	8.75	9.05	17.99	66.43
	11.2	1.61	0.17	18.47	5.87	23.87	48.21	0.35	5.14	10.12	15.61	63.82

U: Heat Transmission Coefficient, SHGC: Solar Heat Gain Coefficient, WT: Windows Transmission, WS: Windows Solar, ME: Mass Effect, HEC: Heating Energy Consumption, CEC: Cooling Energy Consumption, TEC: Total Energy Consumption.

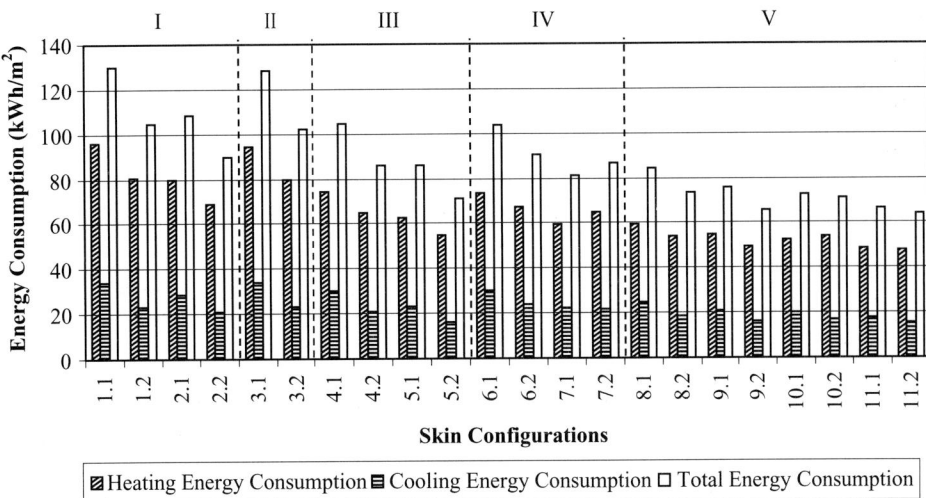

Figure 10.1 The energy consumption of the skin configurations.

transmission (WT), the windows' solar (WS – the heat gain/losses resulting from solar radiation) and the mass effect (ME) of the facade. Then the annual energy consumption values are calculated depending on the efficiency of the energy system to be used. Table 10.3 shows the U- and SHGC-values, the annual heating (HEC), cooling (CEC) and total energy consumption (TEC) values per square metre of floor area resulting from WT, WS and ME for all alternatives. The change in the annual energy consumption of all alternatives can also be seen more clearly in Figure 10.1.

Evaluating the simulation and calculation results

The following deductions can be made from the numerical results in Table 10.3 and Figure 10.1:

- The U- and SHGC-values of single-skin configurations are greater than those of double-skin configurations. The values of the configurations with Venetian blinds are lower than those without for both skin types. In the double-skin configurations, however, there is a noticeable increase in the U-values of the configurations with blinds when low-E glass is used on the external skin. The best results are achieved in the configurations with low-E glasses on both skins. Use of Venetian blinds causes a fall in the SHGC value of the skin types.
- For both single- and double-skin glass facades, the values resulting from WT constitute a large part of the heating energy consumption, but only a small part of the cooling energy consumption, while WS and ME significantly affect both types of consumption.
- In both single- and double-skin facades, the configurations with Venetian blinds (WB) consume less energy per square metre of floor area than the configurations without Venetian blinds (WOB), because the blinds greatly reduce the loads resulting from WS.

- It can easily be seen that the best configurations in terms of energy consumption are formed with the use of the low-E glass.
- The single-skin facade configurations consume more energy than the double-skin facade configurations because of the higher U and SHGC values. The double-skin configuration with the least energy consumption (Code 11.2) of all the alternative combinations is about 61% more efficient than the single-skin facade configuration with the highest energy consumption (Code 1.2).

Conclusion

This chapter aimed at determining to what extent double-skin glass facades are energy-efficient compared with conventional single-skin glass facades. For this purpose, the position and type of glass in conjunction with or without Venetian blinds were evaluated. As expected, the study revealed that double-skin facades would be more energy efficient than single-skin facades for an office building in Istanbul. According to this result, it is possible to infer that the use of mechanical systems can significantly be reduced. This also means that the energy costs of a building with a double-skin facade will be lower than that for one with a single-skin facade.

References

[1] Oesterle, E., Lieb, R-D., Lutz, M. and Heusler, W. (2001) *Double Skin Facades*, Prestel Verlag, Munich, Germany, p.12.
[2] Compagno, A. (2002) *Intelligent Glass Facades*, Birkhauser, Berlin, Germany, p.135.
[3] Wigginton, M. (1998) Glas und Architektur, *Detail-Bauen mit Glass*, April/Mai (3), Deutschland, p.309.
[4] Van Dijk, D. and Goulding, J. (1996) *WIS Reference Manual*, TNO Building and Construction Research, Delft, Netherlands.
[5] Degelman, L. (1999) *ENER-WIN User's Manual*, Texas A&M University, Texas, USA.

11 How to reduce energy-related greenhouse gas emissions in homes and offices in New Zealand

R. M. E. Hargreaves[1]

Summary

The Earth's climate is changing; the expected changes pose a serious challenge to the built environment. This includes our cities, our rural and coastal settlements, our homes and businesses, our infrastructure and ultimately our health and safety. As a response, the New Zealand Government has signed the instrument of ratification to the Kyoto Protocol. Once in force, this will mean that in the first commitment period of 2008–12, New Zealanders will be subject to a number of domestic and international policy measures to reduce their greenhouse gas emissions.

To assist homeowners and small- to medium-sized office owners/managers cope with these requirements, a booklet called the *Easy Guide to Being a Climate-Friendly Kiwi* was developed. The key objective was to encourage individuals to think about the connections between their actions and impacts on the climate, and to enable people to take action to reduce their energy-related greenhouse gases.

Energy-related emissions were chosen, as energy use is the area in people's lives over which they have the most control. As well as providing a range of mitigation options, the guide provides a 'carbon calculator' for readers to calculate how much CO_2 they are responsible for. This paper discusses how this guide was developed and why the information included is an important issue for the building sector.

Introduction

Since 1997, BRANZ Ltd has researched the direct and indirect implications of climate change on homes and offices for the building industry in New Zealand. This has involved studies into the technical and social aspects of climate change and resulted in a number of technology-transfer activities, e.g. reports, articles, seminars and other

---

[1] BRANZ Ltd (wholly owned by the Building Research Association of New Zealand Incorporated), Private Bag 50908, Porirua City, New Zealand.

111

presentations. Concurrent with these activities is the need to provide easy-to-use information that demonstrates practical solutions for individuals to reduce their personal greenhouse gas emissions at home and at work.

A paper-based publication has been developed to achieve this. Called the *Easy Guide to Being a Climate-Friendly Kiwi*, the key objectives are to encourage individuals to think about the connections between their actions and impacts on the climate, and to enable people to take action to reduce their emissions on a month-by-month, or yearly basis. This paper discusses the background of the issue from New Zealand's perspective and describes the development and content of this mitigation tool, along with its future directions.

Climate change and New Zealand

The climate is in a constant state of change. We know that Earth's climate has naturally varied for a variety of reasons [1]. Climate change itself is not the concern; it is the current rate of change and the causative factors that are creating threats and opportunities for all communities around the world. Strong evidence from the scientific community suggests that the relatively rapid climate changes observed over the last 100 years are at least in part due to increasing greenhouse gas levels in the atmosphere. (Greenhouse gases include carbon dioxide, methane, nitrous oxide, sulphur hexafluoride, perfluro-carbons and hydroflurocarbons.) Indeed, most of the warming observed during the last 50 years is attributable to human activities, in particular, the burning of fossil fuels [2].

New Zealand is experiencing changes consistent with global warming trends. In the last 50–100 years, temperatures have risen, on average, 0.7°C, the sea has risen 10–25 cm, and our glaciers are 40% shorter and cover 25% less area. In addition, there have been changes in rainfall averages and extremes, with shorter period rainfall fluctuations due to the counteracting, or reinforcing effects of the 'El Nino' Southern Oscillation, and the Interdecadal Pacific Oscillation [3]. Projections for New Zealand to the year 2080 include [4]:

- changes in average rainfall patterns, with very heavy rain more frequent
- a 1–2°C temperature increase
- fewer frosts
- sea level rise of 30–50 cm
- UV and net solar radiation changes
- enhanced westerlies
- potentially more tropical cyclones
- snow line rise and glacier shrinkage.

Within these projections there are regional variations. For example, the west of the country will become wetter, the east drier [3], and temperatures in the North Island will increase faster than in the South Island [5]. For the built environment, the most significant risk is related to an increased incidence of heavy rainfall and subsequent flooding [4]. How is New Zealand's built environment positioned to minimise these future climatic risks?

New Zealand's built environment

New Zealand's population is around 4 million people [6]. We live in about 1.3 million dwellings and work in around 67 000 commercial buildings [7, 8]. Eighty-five per cent of New Zealanders live in urban or suburban environments – one of the highest rates of urbanism in the world [9]. Because the tools under discussion in this paper were developed with New Zealand's homes and offices in mind, an understanding of what an average home or office comprises, and how it may or may not cope with future climate risks, is required.

The average house

With 1.3 million dwellings in a land area of 27 million hectares, New Zealand's urban pattern is one of low density with a relatively large geographical spread. Despite this spread and regional variations in climate across New Zealand, the housing stock is relatively uniform in style and design [10]. The average age of the housing stock (using 1999 figures) is 30–40 years, with an average floor area of 140 m^2 [11]. Most houses are detached and constructed with timber framing and weatherboard cladding.

The majority of New Zealand houses are not designed to deal with extremes of cold or heat [10], are rarely air-conditioned, and are only marginally heated [12]. For instance, a typical New Zealand house is heated only in the mornings and evenings often in a poorly controlled manner [13]. Furthermore, New Zealand homeowners do not appear to be strongly aware of ways to protect their houses from environmental risk, or how to reduce their environmental impacts in general [14].

The average office

Offices are commonly built of concrete and steel with a trend towards low-rise buildings (97% being less than or equal to three storeys), and have floor areas of more than 300 m^2. In terms of office space requirements, the average area has decreased from about 34 m^2/person in 1988 to 18 m^2/person in 1998 [15]. For new offices that are less than 300 m^2, mandatory thermal envelope requirements (the same as for housing) are in place. Recent New Zealand Building Code (NZBC) revisions now require similar energy performance requirements for larger commercial buildings [16].

Only 12% of office buildings have air-conditioning but most are heated [12]. In general terms, it is not known how New Zealand's offices will cope with climate change. Research exists that summarises potential changes in comfort levels in buildings that are air-conditioned and compares them with those that are not [12]. However, it is clear that if a particular building does cope well, it will be by accident rather than by intent [14]. It would appear therefore, that both homes and offices in New Zealand are some way from being climatically resilient.

New Zealand's political context

In global terms, New Zealand's greenhouse gas emissions are small. The economy is largely reliant on primary production and hence has much to lose (and gain) with

Figure 11.1   Cover of the mitigation tool.

climate change [17]. The New Zealand Government signed the instrument of ratification to the Kyoto Protocol in December 2002. As part of the preferred policy package to meet our obligations under the Protocol, it is clear that there is a strong focus on mitigatory measures, particularly energy efficiency, to meet New Zealand's targets.

However, as alluded to above, typical homes and offices in New Zealand are largely unprepared for climate change in general, and the public is unaware as to how to mitigate the likely impacts. The remainder of this paper describes the tool that BRANZ Ltd has prepared to provide support and advice to the public and the New Zealand building industry in this regard.

The mitigation tool

The mitigation tool, called the *Easy Guide to Being a Climate-Friendly Kiwi* (see Figure 11.1), is a paper-based publication focusing on greenhouse gas reduction measures for homes and small- to medium-sized offices. In addition to providing a wide range of practical tips to reduce greenhouse emissions, it includes a New Zealand-specific 'carbon calculator' so readers can work out their emissions on a monthly or yearly basis (see Figure 11.2).

The calculator was included as a 'measure to manage' instrument, based on the premise that you can't manage (or reduce) your emissions if you don't know what they are. Because the calculator is paper-based and thus relatively simplistic, it focuses only on CO_2 emissions from energy, heating and transportation, and paper waste.

The main objective of the guide is to support opportunities for the New Zealand public actually to do something about climate change in the areas where they spend

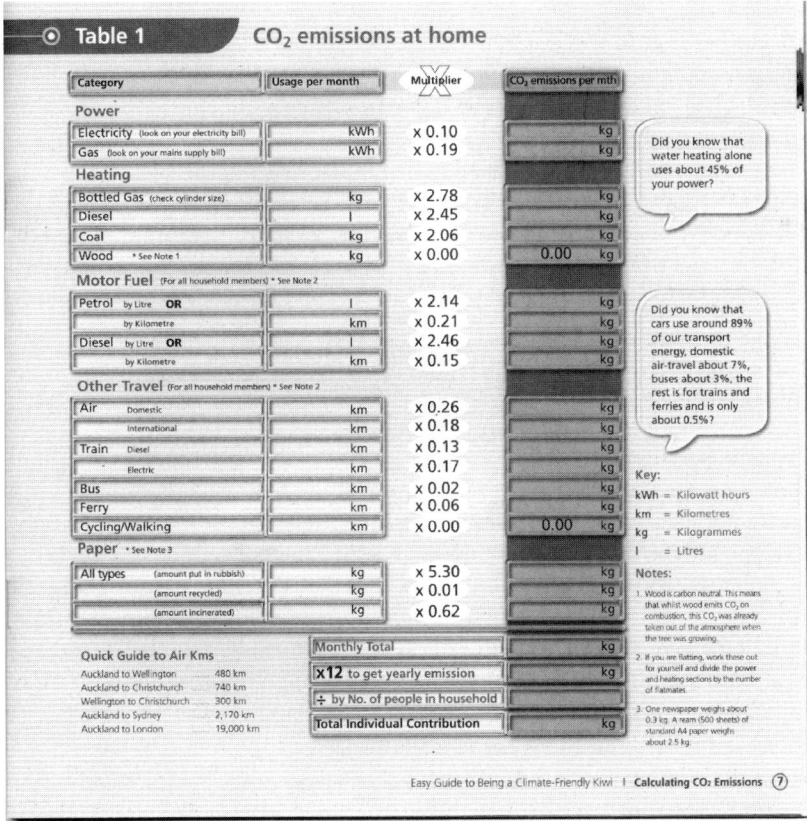

Figure 11.2   The carbon calculator sheet for residential dwellings.

most of their time – at home and at work. The main output is a user-friendly, fully illustrated colour booklet. The reader is taken through what climate change is all about, and how it relates to them. It describes greenhouse gases and where they come from. From here the focus shifts to the emissions for which the individual is directly responsible, namely energy use.

Following this, the remainder of the booklet is dedicated to taking action to reduce emissions. For electricity use, tips are provided for lighting, appliances, water heating, insulation and windows. Tips are also provided for transport, reducing paper use and other recycling opportunities. A 'Quick Tips' section is also provided along with lists of relevant web page links to further sources of information and web-based carbon calculators. The 'ten quick tips to reduce your CO_2 emissions' are:

- Insulate your hot water cylinder.
- Choose energy efficient appliances.
- Increase insulation in the ceiling, walls, floors and windows.
- Recycle paper waste.
- Support 'green' energy.

- Travel smart.
- Wash clothes in cold water.
- Switch off appliances when not in use.
- Rate the sustainability of your new home or office design.
- Plant trees and shrubs.

The booklet is part of BRANZ Ltd's 'Easy Guide' series of publications and these are available for downloading from the 'BRANZ bookshop' (www.branz.co.nz). In summary, this tool is a timely publication that provides easy-to-use, practical information for individuals on how to reduce their greenhouse gas emissions on a day-to-day basis.

Summary and future directions

New Zealand is a small island nation in the South Pacific with much to lose (and gain) from climate change. It would appear that, in general, New Zealand's built environment is currently not well-placed to minimise the identified risks or maximise benefits. The New Zealand Government has ratified the Kyoto Protocol sending a clear signal that climate change is an important issue for all sectors of the economy, not least the building industry.

As buildings have a relatively long life with slow turnover, tools that give specific advice on how to mitigate the predicted climate change impacts are both timely and relevant. One such tool has been described in this paper. It aims to assist the public to reduce their greenhouse emissions at home and at the office. While the tool is a relatively simple paper-based publication at present, possible future directions include the development of the carbon calculator to a web-based version for enhanced ease of use and increased sophistication, e.g. under 'kilometres travelled by car', the type of car, number of passengers, type of fuel, etc. could be included. There is also potential for a 'schools' version to be produced. The predictions and practical tips in the tool will need to be periodically revised as certainty in climate change scenarios and government policies evolve, and as new technologies emerge to help the adaptatory process.

Whatever the final direction, the aim of the mitigation tool has been to demonstrate practical solutions for the public and the building industry so that New Zealand's built environment is climatically responsive in the years to come.

Acknowledgements

This work was funded by the Building Research Levy and the Foundation for Research, Science and Technology from the Research for Industry Fund.

References

[1] Ministry for the Environment (1997) *The State of New Zealand's Environment*. Ministry for the Environment, Wellington, New Zealand.

[2] Intergovernmental Panel on Climate Change (2001) Third Assessment Report, www.ipcc.ch

[3] Wratt, D. (2002). Climate change and New Zealand – science, projections, impacts, uncertainties. NIWA. In *Adapting to the Impacts of Climate Change. What Role for Local government?* Ministry for the Environment, Wellington, New Zealand.

[4] Camilleri, M. (2000) *Implications of Climate Change for the Construction Sector: Houses.* BRANZ, SR 94. Judgeford, New Zealand.

[5] Kenny, G. (2002). *Climate Change and Land Management in Hawke's Bay.* Ministry for the Environment, Wellington, New Zealand.

[6] Statistics NZ. (2002) www.stats.govt.nz, accessed March 2002.

[7] Page, I. (2002) Personal communication.

[8] Quotable Value, (2000) Personal communication.

[9] Parliamentary Commissioner for the Environment (1998) *The Cities and Their People. New Zealand's Urban Environment*, Parliamentary Commissioner for the Environment, Wellington, New Zealand.

[10] Saville-Smith, K, (1998) *Implications of Climate Change for the Construction Sector: The Social Dynamics of Climate Change Responsive Housing*, Centre for Research, Evaluation and Social Assessment, Wellington, New Zealand.

[11] Clark, S. J., Page, I., Bennett, A. F. and Bishop, S. (2000) New Zealand House Condition Survey, BRANZ, SR 91, Judgeford, New Zealand.

[12] Camilleri, M. and Jaques, R. (2001) *Implications of Climate Change for the Construction Sector: Office Buildings*, BRANZ, SR 96, Judgeford, New Zealand.

[13] Stoecklein, A., Pollard, A., Camilleri, M., Tries, J., Isaacs, N. and Bishop, R. (2001) *The Household Energy End-Use Project: Measurement, Approach and Sample Application of the New Zealand Household Energy Model*, CIB World Building Congress, April 2001, Wellington, New Zealand.

[14] Saville-Smith, K. (2000) *Implications of Climate Change for the Construction Sector: Local Government Response to Climate Change*, Centre for Research, Evaluation and Social Assessment, Wellington, New Zealand.

[15] Jaques, R., Camilleri, M. and Isaacs, N. (2000) New Zealand's Initiatives in 'Greening' Commercial Buildings, APEC 'Green Buildings – Investing in our Future' Conference, Taipei, 16–18 October 2000.

[16] Stoecklein, A. (2002) Personal communication.

[17] Ministry for the Environment (2001) *National Communication 2001*, Ministry for the Environment, Wellington, New Zealand.

Part 3
Design, construction and operation issues

12 The potential for prefabrication in UK housing to improve sustainability

M. T. Gorgolewski[1]

Summary

Off-site factory production of dwellings has the potential to lead to significant improvements in both the quality and speed of construction compared with traditional site-based construction. The technical benefits include increased speed of production, reduced levels of defects and waste, greater efficiency in the production process, and improved environmental performance. These are all generated as a direct result of moving more of the construction process from the building site, where efficiency is poor and management difficult, into the relative safety of a factory, where working conditions, efficiency and control can be better managed. This can also lead to social benefits including improvements in health and safety, more stable employment, investment in machinery and the development of skills. Greater stability in the manufacturing process also generates potential economic benefits.

All of these contribute to the environmental, economic and social benefits of sustainability. The following is a review of the sustainability benefits of prefabrication, and the various alternative ways that prefabricated building systems are beginning to have an impact on the way that residential buildings are constructed in the UK.

Introduction

In the UK, government predictions suggest that 3.8 million additional new dwellings will be required over the next 20 years. Current replacement rates are minimal, with most new housing adding to the stock rather than replacing old, outdated housing. Within the social housing sector it is estimated that from the year 2000 to 2016 one million new dwellings will be required. These are expected to be mainly high-density, terraced houses and purpose built apartments. This presents a significant challenge to the UK construction supply chain with its diminishing labour force and increased business performance demands. Furthermore, client requirements for higher building standards and the industry's increasing regulatory improvements, particularly in

---

[1] School of Architectural Science, Ryerson University, Toronto, Canada.

thermal and acoustic performance, and health and safety issues are pushing the industry to reconsider on-site methods of construction and to investigate other ways of building homes.

Increasingly, concern about the impact of the construction process, and of the additional 3.8 million new homes, on the environment and local communities is beginning to push clients to take more account of the sustainability impacts of how we build, operate and maintain our buildings. For example, the Housing Corporation which funds much social housing now requires an Ecohomes [1] environmental rating for all housing schemes that they finance and 50% should achieve a rating of 'Good'.

A further factor leading to change in the industry is the UK government's construction policy, which is now dominated by the report of the Construction Task Force, *Rethinking Construction* [2], published in 1998 and the subsequent report, *Accelerating Change* [3], published in 2002. These encourage the industry to address market demands for improved efficiency, better quality, faster construction and better cost control. This has led to a greater interest in off-site manufacturing technologies and many house builders are currently investigating a variety of innovative ways of building dwellings. Recent reports [4] have identified considerable areas of overlap between the agenda of improved industry efficiency (through prefabrication and partnering) and the sustainability agenda.

Off-site manufacturing systems

To date prefabrication in housing in the UK has not been a commercial success. It has often been associated with a reduction in flexibility and choice for the designer, client and end user, and with higher costs. Until recently, perceived market resistance has prevented significant uptake of such technology in the UK despite examples illustrating its technical feasibility. However, off-site fabrication, both volumetric construction and panellised construction, has received a great deal of attention in recent years. The trend began with the hotel sector, where quality and repeatability of units lend themselves to volumetric buildings. This technology is now being increasingly applied to apartments, houses and sheltered accommodation.

In the UK, there are three principal approaches to off-site production of buildings using lightweight construction currently being used. These are discussed below.

Volumetric systems

Three-dimensional units are manufactured in the factory with a high degree of services, internal finishes and fit-out installed in controlled, factory conditions prior to transportation to site (Figure 12.1).

This has many benefits including improved quality, reducing defects and snagging on-site, increased speed of construction on site, better working conditions, increased predictability and efficiency in the production process. This approach is particularly suited to highly serviced areas such as kitchens and bathrooms, which have a high added value, and cause disruption and delays on site, but may be less appropriate for other rooms which have less internal fit-out.

Figure 12.1   Volumetric modules constructed in the factory.

Volumetric systems have the disadvantage that each unit has to be transported separately, and the maximum size of the unit is determined by the practical problems associated with transportation by road. Factories operate most efficiently when a large number of similar units are made to the same dimensions. Both of these factors work to reduce flexibility in layout and design. For these reasons most volumetric construction in the UK to date has been in the hotel, hospital and fast food chain sectors, where repetition of units is possible. Increasingly, they are also being used for student accommodation.

Since the units have to be self-supporting and are craned into place on site, the strength and rigidity must be sufficient to allow this process to occur without damaging the units. The strength requirements for the lifting operations may exceed those for in-use service, thus the structure may be over-designed for its end-use, leading to wastage of materials.

Panellised systems

Flat panel units are manufactured in a factory, and fixed together on site to produce the three dimensional structure (Figure 12.2). Services, windows and doors, internal wall finishes, and external claddings can potentially be installed in the factory but in most current systems in the UK they occur on site.

Panellised systems are more flexible and can more easily accommodate variations in unit plan and detail design than volumetric systems. Spaces such as bedrooms and living spaces lend themselves to panel construction systems, providing greater choice

Figure 12.2   Panels assembled on site.

to the client and designer, with few restrictions on room size and layout. Furthermore, the advantage of panellised systems is that they can be stacked flat, so more of the structure can be transported in one journey, reducing transport impacts. However, the level of finish and degree of services integration that it is practical to install into panels prior to shipping to site is less than for volumetric construction. This leads to more work on site and requires further deliveries of other materials, components and labour to the site. This may not be much of a problem for plain walling but it would be a disadvantage for highly serviced areas such as kitchens and bathrooms. Also, there is a greater likelihood of damage to the finishes applied to the panel during transportation or on site.

Hybrid (semi-volumetric) systems

A third option, which maximises the benefits of prefabrication, is to use volumetric units for the highly serviced areas such as kitchens and bathrooms and construct the remainder of the building using panels or by another means. This offers the opportunity of removing the highly serviced areas from the critical path of the project, and potentially brings together the benefits of different construction systems. It can also address the issues of providing flexibility and consumer choice.

Thus, some schemes have used volumetric modules for bathrooms and kitchens in hot rolled steel frame or concrete frame buildings. Alternatively, volumetric units are used in combination with panels for the less serviced areas. Also, volumetric units have been used to extend buildings and provide additional accommodation with minimal disruption.

Such an approach may combine the benefits of economies of scale and the economies of scope, using mass production, factory production and standardisation to provide flexibility of options offering customisation. A kit of parts can be used to provide flexibility yet maintain the benefits of standardisation.

Sustainability benefits

Site benefits

The key feature of prefabrication is that much of the process is removed from the site to controlled factory conditions. This reduces the amount of time spent on site, which leads to less detrimental impacts on the locality. Experience in the UK shows that prefabricated hotel buildings can be constructed on site in half the time (or less) of a traditionally built hotel of a similar size, and this could be reduced further with the use of factory applied claddings. In the catering industry clients have claimed a factor of ten improvement in installation and commissioning timescales for a typical fast food restaurant when using volumetric construction. This means that the locality around the site is disrupted for a shorter period so reducing noise, pollution emissions and local traffic disruption. In addition, the lightweight nature of the buildings can often result in smaller foundations and therefore less groundworks, also reducing local disruption as well as reducing the volume of materials used in the groundworks and having less spoil to be removed.

Furthermore, prefabrication generally leads to fewer deliveries to site compared with traditional construction methods. Some monitoring of a site in London suggested that deliveries to a volumetric site were reduced by up to 60% compared with a similar, but traditional, building site nearby. This will also reduce local disruption, although volumetric sites will generally have larger lorries delivering bulky, awkward volumetric units that require cranes to off-load. Thus, careful management is required to ensure minimum disruption when deliveries are made.

The wider transport implications of prefabrication are difficult to measure. Volumetric deliveries often travel considerable distances from the factory. However, there are generally fewer deliveries than with traditional construction. In addition, the shorter period on site and the nature of the work means that less labour is required on site and for a shorter period. Panellised construction can be more efficient in delivery to site, but more subsequent work on site to finish the building off can lead to additional transport movements. In general it is likely that a well managed site using prefabricated components can significantly reduce the impact of transport.

The additional transport movements related to the factory should be considered. However, the workforce in a factory is more likely to be local than at a building site, and thus will travel shorter distances, and is more likely to use public transport, where possible. Secondly, material deliveries to a factory can be planned so that full loads are always delivered, and local suppliers can be used.

Process benefits

A building site does not provide an ideal environment for achieving quality construction or safety. Construction work on site can be a dangerous activity and leads to significant numbers of casualties and fatalities. Allowing much of the process to be carried out in more controlled and comfortable factory conditions enables safety requirements to be more easily met and policed, and healthy and comfortable working conditions are more readily maintained. More demanding health and safety requirements are pushing many builders to consider off-site manufacturing techniques.

Conversely, the use of heavy lifting equipment to locate the prefabricated components on site requires careful management. The use of scaffolding is a particular safety concern. Some schemes in the UK have tried to eliminate the need for scaffolding completely by integrating claddings in the factory. The perceived market preference for brickwork is seen by many as a potential obstacle for further off-site manufacturing. Nevertheless, several companies are developing options for alternative claddings to traditional brickwork, with improved detailing that avoids the need for scaffolding, and some of these can be installed in the factory. Many UK schemes using prefabrication use alternative cladding systems such as cedar boarding, polymer based renders or rain-screen cladding.

Building sites are also notoriously inefficient in the use of labour and materials. Studies in the UK have shown that site based activities have a considerably lower efficiency in the use of labour compared with factory based activities. In some cases the useful activity on site can be below 50% of full potential. Furthermore, it is estimated the between 13% and 18% of materials delivered to UK construction sites are wasted and never used properly. Manufacture in a factory allows far better management of the waste stream with more efficient use of materials and ordering of exact amounts, more careful storage, and the possibility of design to suit standard sizes. In addition, any waste that occurs can be more easily collected and reused or recycled. Many off-site manufacturing plants have recycling facilities installed, as this reduces the costs of disposal of waste. There is further potential for reducing waste when using prefabrication if the designer is prepared to coordinate sizes so that materials such as timber and gypsum sheets are used in their standard sizes without generating many off-cuts.

Improved final product

Moving work off-site can also lead to various quality benefits resulting from better working conditions and an opportunity for better quality management. Factory based activities allow testing and checking procedures to be more easily implemented. For example, volumetric units can have electrical and water installations fully tested prior to leaving the factory. General experience in the UK suggests that far less call-backs are necessary to make good defects after completion for buildings using a lot of pre-fabrication. This is a significant cost and efficiency benefit to the builder and leads to satisfied customers. It also reduces wastage of resources.

The correct and careful installation of the elements of the fabric, in particular insulation materials and air barriers are important to the thermal and acoustic performance of the building in use. Thermal and acoustic performance is very dependent on the quality of workmanship and supervision. Factory manufacture allows operatives to be better trained and supervised in these tasks, and allows regular checking and testing of performance. Problems such as omitted insulation and badly fitted air barriers are less likely to occur. Comparisons from North America suggest that better energy efficiency in use is achieved in homes that use off-site manufacturing techniques with a similar specification to site built homes.

In addition, volumetric construction allows a building to be potentially dismantled and the modules reused at a different location. Modular hotels in the UK have been dismantled and removed to a different location when found to be uneconomic at their original site. Traditionally, many volumetric buildings were used as temporary buildings and removed for reuse when no longer necessary. Thus, the technology for reuse is well established. Many of the materials used in this type of construction such as the steel framing can also be extracted for recycling at the end of the life of the module. This is made easier by the lightweight, dry construction methods that are generally used. This is likely to become more significant in the future when EEC legislation about producer responsibility encompasses the construction industry.

Benefits to the workforce and community

Building sites are temporary employment locations, so they generally offer few long-term amenities for the local community. Manufacturers in factories are often closely linked with the local community, with much of the workforce coming from the locality. They provide a long term economic and often social service for the community. Many manufacturers of prefabricated modular or panel units in the UK are well-established in particular locations and have developed a highly trained, local workforce, and strong links with the local community.

Employment at a factory manufacturing prefabricated building components is generally more stable and long term than site based employment, which is intrinsically transient. As a result, factory based employers are generally more willing to invest in training for their workforce. Furthermore, to function efficiently prefabrication requires high levels of skill and flexibility in the workforce. This necessitates greater training by employers.

From a financial point of view, the shorter construction period allows a quicker return on investment by the client, and reduced overhead costs.

Conclusions

There is an increasing interest in off-site manufacture in the UK house building industry. House builders are beginning to realise that there is a need to improve standards and that new regulatory requirements such as the changes to the Building Regulations dealing with energy efficiency and acoustics will be more easy to satisfy by increasing the amount of off-site manufacture. Health and safety issues are driving many changes in approach.

There is an established sector of the construction industry manufacturing volumetric units for hotel construction, key worker accommodation, the education sector and budget restaurants. The economics of this type of construction relies on repeatable units so that large production runs can be set up. A figure of 40 repeatable units (of similar size and layout) is sometimes quoted as a minimum number required for such systems to be cost competitive. The volumetric manufacturers have developed technical systems and construction processes to maximise the benefits of this type of construction.

Panellised systems are more commonly in use in UK housing, but with relatively little finishing-off of the panels in the factory. Several builders are looking to move towards significantly more integration of finishes in the factory.

The management of the construction process has been identified as a particularly important area, and the interaction of the prefabricated components with site processes such as foundations, service connections and cladding are critical. It is essential to consider the process of manufacture and assembly when considering the building design. Technical decisions can lead to process problems, and so both must be considered together.

The objectives of sustainability and the 'Rethinking Construction' agenda of improving efficiency in construction overlap in several areas, notably waste minimisation, process integration, a commitment to people and a quality driven agenda. Prefabrication offers an opportunity to address both these agendas, and improve both efficiency and sustainability. However, the industry has much to learn to fulfil the potential of this technology.

References

[1] BRE (2000) Ecohomes – the environmental rating for homes, BRE report BR 389.
[2] Egan, J. (1998) *Rethinking Construction: The Report of the Construction Task Force on the Scope of Improving the Quality and Efficiency of UK Construction*, The Stationery Office, London.
[3] Egan, J. (2002) *Accelerating Change – A report by the Strategic Forum for Construction*, Rethinking Construction, www.rethinkingconstruction.org.
[4] CRISP, (1999) *Integrating Sustainability and Rethinking Construction*, Construction Research and Innovation Strategy Panel – Sustainable Construction Theme Group, London.

13 Estimating the increasing cost of commercial buildings in Australia because of greenhouse emissions trading

D. H. Clark[1], G. Treloar[2] and R. Blair[3]

Summary

The ratification of the Kyoto Protocol by most industrial nations will result in an international greenhouse emissions trading market by or before 2008. Calculating the quantity of embodied energy in commercial buildings has therefore taken on added significance because it is in the creation of energy that most greenhouse gas that causes global warming is released. For energy efficient commercial buildings in Australia, the embodied energy can typically represent between 10 and 20 years of operational energy.

When greenhouse emissions trading is introduced in Australia, the cost of energy will rise significantly, particularly electricity which relies primarily on burning fossil fuels for generation. This will affect not only the operating energy costs of buildings (light, power and heating/cooling) but also the cost of building materials and construction.

Early estimates of the potential cost of future greenhouse emission permits in Australia vary between AUS$10/tonne to AUS$180/tonne. This cost would be imposed primarily on the producers of energy and would be passed on by them to consumers via higher energy costs. For a typical commercial building this could lead to an increase in the total procurement cost of buildings of up to 20% due to the energy embodied during the construction or refurbishment of the building.

To assist in evaluating these potential cost increases McKean & Park, Sinclair Knight Merz and Deakin University have developed a web-based CarbonCost Calculator for commercial buildings.

[1] Sinclair Knight Merz, Melbourne, Australia.
[2] BERG, School of Architecture and Building, Deakin University, Geelong, Australia.
[3] Special Council, Future Law Team, McKean & Park, Melbourne, Australia.

Introduction

Australia is the world's highest per capita emitter of greenhouse gases at 27.6 tonnes of CO_2-e per person per year, compared with the USA at 21.1 tonnes and the European Union average of 10.3 tonnes [1]. This is primarily due to Australia's reliance on fossil fuels for power generation.

The Kyoto Protocol Australia signed in 1998 commits it to limiting its emissions in 2010 to an 8% increase above 1990 levels. Australia has agreed to comply with this but has not as yet agreed to ratify the Protocol.

Nations that have ratified will participate in an international emissions trading scheme from no later than 2008, an approach designed to allow greenhouse gas reductions to be made in the most economically efficient way. Localised schemes are already in operation in various countries including the UK.

Embodied energy is the energy required directly and indirectly to produce a product (which may be a physical entity or a service). Calculating the embodied energy in commercial buildings has taken on added significance because, if the cost of energy rises due to the introduction of emissions trading in Australia, then the procurement cost of all commercial buildings will also rise due to the increased cost of the embodied energy.

When designers consider energy in their building designs, they usually consider only the operating energy used to provide services such as ventilation, heating, cooling and lighting. However, given that the energy embodied in the materials used in construction and refurbishment can also be significant, both will need to be considered in the future, particularly since embodied energy primarily affects the up-front capital cost rather than the on-going operating cost of buildings.

Embodied energy in commercial buildings

Current knowledge

For commercial buildings in various climates, several studies have shown that embodied energy can be significant, typically equivalent to 10 to 20 years of operational energy depending on the operating energy efficiency of the building [2, 3].

It has also been found that, while the initial energy embodied in furniture was small (around 10% of the initial embodied energy of the building), due to high churn rates the energy embodied in furniture used over the building's life cycle represented about the same amount as the initial embodied energy [4].

Figure 13.1 shows the typical life cycle greenhouse emissions for a typical medium rise commercial building in Australia. It assumes a 'whole of building' Australian Building Greenhouse Rating of four stars, a fit-out churn rate of 7 years, and a major refurbishment approximately every 20 years. The embodied energy in the building (excluding furniture) accounts for approximately 25% of the life cycle greenhouse emissions of the building over a 50-year period.

The emissions are higher for the same building in Victoria than for NSW or Queensland due to the additional greenhouse emissions associated with brown coal electricity generation.

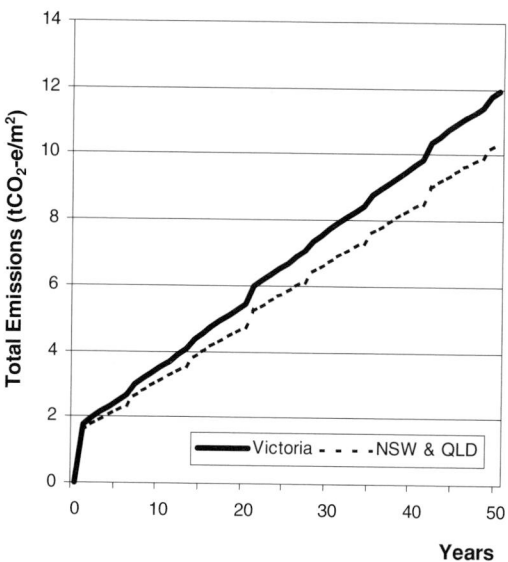

Figure 13.1 Life cycle greenhouse emissions for a typical commercial building in Australia with ABGR of 4 stars.

The embodied energy data used in Figure 13.1 is based on data used in the CarbonCost Calculator developed by the authors (refer below for details) and previous extensive embodied energy research by Treloar.

It is important to note that this data is being constantly evaluated and improved. The embodied energy values are likely to rise over time with additional information becoming available, including the energy to make buildings and equipment being amortised over the recent history of the manufacturing industry in Australia and incorporation of non-energy greenhouse emissions, for example from cement production, forestry practices and aluminium manufacture.

Development of embodied energy simulation models

The energy embodied in residential and commercial construction was recently identified as a significant area for the Australian Greenhouse Office to commission, develop and introduce a variety of greenhouse gas emissions abatement strategies.

Currently, however, there is no reliable method available for estimating the embodied energy of the construction of commercial buildings. Deakin University, with their industry partner Sinclair Knight Merz are currently engaged on a three year ARC Linkage funded research project, Embodied Energy in Commercial Buildings, which aims to contribute to the development of a more reliable method for estimating embodied energy in Australian buildings.

The key objective to be addressed in the research project is the application of currently available embodied energy values for building materials and products derived at the national and/or industry level to individual commercial buildings during the early to mid stages of the design process. As early as possible in the design process, embodied energy modelling allows changes of greater impact to be made more cost-effectively.

Separate to the above research project, McKean & Park, Sinclair Knight Merz and Deakin University have jointly developed the web-based CarbonCost Calculator for commercial buildings. This will be further developed in the future to cover other building types including residential (houses and apartments). The CarbonCost Calculator will enable preliminary assessments of the embodied energy, and the associated carbon costs, to be estimated from basic building construction information. The tool is discussed in more detail below.

Impact of greenhouse emissions trading on the cost of Australian buildings

Kyoto emissions trading

The Kyoto Protocol seeks to force the industrialized nations and subsequently other nations, to reduce the use of fossil fuels. In order to achieve this, all industrialized nations in 1997 agreed to targets (based on their respective 1990 greenhouse emissions) to which they would reduce those emissions by the first commitment period extending from 2008 to 2012. It is anticipated that further commitment periods will follow progressively and immediately, each commencing on the conclusion of its predecessor and each setting new and lower national targets for the nations to comply with.

To meet obligations under the Kyoto Protocol, an emitter of greenhouse gas would be required to lodge regular periodical emission returns and to acquit with each return sufficient emission permits or emission credits to equal the disclosed emissions. Failing this, it is anticipated that the government will purchase either the emission permits or emission credits necessary to meet the obligation and charge the emitter up to three times the cost of the purchase.

Emission permits, it is anticipated, will be issued by a national regulating authority and will equal the national target figure for that year. In other words, if Australia is entitled to emit 400 million tonnes of CO_2-e then it can issue up to 400 million $\times$ 1 tonne electronic permits. Emission credits will be issued to equal the CO_2 recovered (and retained) by 'so called' carbon sinks (currently – Kyoto-compliant forests) [5].

It is anticipated that the price paid will be passed on to consumers. Unlike the GST, price changes occurring as a result of the introduction of emissions trading will be anything but uniform. For example, where energy is produced from natural gas, less greenhouse gas is emitted and therefore less permits/credits are required and the price increase is substantially less. On the other hand, energy produced from brown coal will be considerably dearer because far more greenhouse gas is emitted in the process.

The Australian government is delaying taking the decision to embark on emissions trading as long as possible. It does this because of the pre-eminence of industries such as coal and aluminium that are likely to be worst affected by implementation. Nevertheless, the production of electricity in Australia is anticipated to increase by up to 30% by the year 2010. This will add considerably to the growth in energy consumption which is and has for some time been racing along at about 2% annually [6]. This is a

Table 13.1   Potential increase in building costs due to emissions trading.

	Greenhouse emissions (tCO$_2$-e)	Approximate building cost increase for different carbon permit prices		
		$10/t	$40/t	$180/t
Commercial building (10 000 m^2)	10 000–20 000	$100 k–$200 k	$400 k–$800 k	$1.8 m–$3.6 m
3-bedroom house (150 m^2)	70–160	$700–$1600	$2900–$6400	$12 600–$28 800

very substantial growth rate and is the key reason why Australia faces great difficulty in reducing its greenhouse emissions to Kyoto and post-Kyoto requirements.

However, since many of Australia's trading customers have ratified the Kyoto Protocol, it is likely that Australia must eventually have to face up to its responsibilities if it wants to continue in the trading markets it has built up over the years.

Cost implications for Australian buildings

The introduction of emissions trading in Australia would have a flow-on effect to all goods, services, investments and property. Because that effect must be sufficient to cause consumers to use other forms of energy, or to use less energy, or to use different products, the overall financial effect will be much greater than most people imagine.

Already the Australian Bureau of Agricultural and Resource Economics (ABARE) has estimated that up to approximately AUS$180 per tonne is the likely price for emission permits and emission credits [7]. Even at the more modest figure of AUS$40 a tonne suggested in the Allens' Report [8], this would increase the cost of buildings significantly.

The future cost of buildings and their major refurbishment will be affected by these changes in the price of energy and this must feature strongly in all future economic considerations. If the total of embodied energy in a commercial building is typically 10 to 20 years of the energy expended in operating the same building, and if the life expectancy of many commercial buildings (either to rebuild or for major refurbishment) is as low as 20 years, it is easy to understand why we should be looking at these issues now.

Typical embodied greenhouse emissions and the corresponding increase in building procurement costs due to the price of emission permits are shown in Table 13.1. Greenhouse emissions vary by State, due in part to electricity generation. The increase in the capital cost of buildings due to emissions trading could be up to 20% depending on the cost of emissions permits.

As a matter of practical necessity, if emissions trading forces energy prices to anything like the levels being predicted, then construction issues such as building design, longevity, planning, materials, construction and layout take on a totally new significance. The embodied energy cost is also an up-front capital cost, rather than an on-going cost, and this may influence the financial feasibility assessment of future building projects.

Who needs to predict future carbon costs?

Many people have a need to know now what the cost of embodied energy in commercial buildings in the future is likely to be.

Valuers, for example, are not only asked to give their opinions on present value but on future value as well. They are asked their opinions as to the likely re-sale value of property at some specified future date. If that date falls after 1 January 2008 then the additional energy costs associated with re-construction or major refurbishment of that building will need to be taken into consideration. That sort of question is already being asked.

There is a range of additional factors that will need to be taken into consideration in respect of all buildings. *Architects*, *engineers* and *quantity surveyors* will need to lead the way in advising as to the most cost-effective commercial buildings for the future. They will need to be capable of producing buildings using materials that require less energy to produce and which require less energy to refurbish.

Buildings will need to have a far greater life expectancy because it will be much less attractive to put up short-term buildings and to demolish them after (say) 20 years. Long-life buildings that are intended to have regular major refurbishments should be designed accordingly. It is possible that the cost effective building of the future will be considerably lower because the amount of energy embodied in a building radically increases the higher it rises and the greater depth it has to be bedded down.

Banks and *lenders* on the security of property will certainly need to reassess their mortgage books. There will be a sizeable proportion of buildings already constructed or constructed in the future without due consideration for their embodied energy content, that may become known as 'greenhouse lemons'. *Lenders* and *investors* must be wary of this type of building. *Unit purchasers* will also need to be concerned because the kind of elements discussed above will reflect on the value of the property in which they are investing, but without any ability to rectify the building as a whole.

Real estate agents and *superannuation funds* are among other groups that will need to be concerned with what is occurring. Superannuation funds currently invest in property on advice that has yet to come to grips with changing property values resulting from their embodied energy content.

CarbonCost Calculator

Introduction

The web-based CarbonCost Calculator is intended for use by building valuers, property developers, architects and many others involved in the planning of new constructions or in the purchase, sale or refurbishment of existing buildings. The initial planning for new constructions often starts 5 to 10 years before completion and the financial viability of the project is evaluated by calculating the likely rental returns and the site and construction costs.

The CarbonCost Calculator is designed to estimate the additional costs likely for such a construction as a result of the energy costs resulting from a greenhouse emissions market. These are not currently considered in future project evaluations, which try to estimate the financial viability of a project by considering the financing costs and rental returns against the assumed total cost.

Input

The user is required to input specific basic parameters about the building from which the embodied energy and associated greenhouse emissions can be calculated. The inputs are either numbers (e.g. floor areas, number of storeys, building perimeter) or selections from drop down menus (e.g. floor construction, standard of finishes, facade type). Separate inputs are required for office, retail and car park areas.

Embodied energy calculation

The embodied energy data used in the CarbonCost Calculator, was derived from a number of sources, including information on energy use and material production by industry (public domain, less than 10 years old), energy and greenhouse data from the Australian Bureau of Agricultural Resource Economics (analysed by the University of Sydney, Department of Physics), economic data relating to inter-sectoral transactions from the Australian Bureau of Statistics, National Accounts: Input–Output Tables. The data were combined using an innovative hybrid method developed by Treloar [9].

In order to limit the amount of information required to be entered by the website user, a simplified embodied energy calculation engine was developed by the authors. The calculation engine is based on assumptions regarding the construction and materials of typical commercial office buildings in Australia and the energy sources used during these processes. For example, it makes allowance for increased structure as building height increases and also converts the embodied energy to greenhouse emissions taking into account the fuel sources for electricity generation in each State.

The calculator currently does not include greenhouse emissions due to non-energy processes (e.g. liberation of carbon from limestone during cement production).

The calculation engine estimates the embodied energy content for the following items: structure (floors, columns, core, footings, basement, stairs and roof); facade; internal base building (core finishes and services), fit-out (partitions, ceilings and floor covering) and other (direct energy and other items).

Output

The output is in the form of a printed certificate that contains:

- a summary of input data
- an estimate of total embodied greenhouse emissions (in tonnes of CO_2-e)

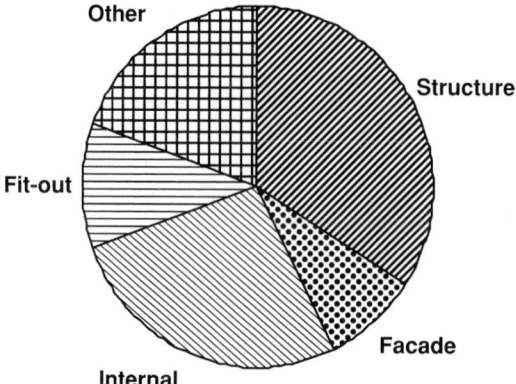

Figure 13.2   Example of embodied energy breakdown from CarbonCost Calculator.

- the suggested range and best estimate of likely additional carbon cost
- a breakup of embodied energy by building items (see Figure 13.2).

However, with such generic tools, the output can only be considered as an estimate or rough guide to the greenhouse emissions associated with the energy embodied in the building. More detailed assessments of individual buildings can be undertaken by Sinclair Knight Merz and Deakin University as required.

Conclusion

Anyone involved in the ownership, development or procurement of buildings in Australia will need to assess the risk on future costs and the consequent value of buildings due to the introduction of emissions trading. The CarbonCost Calculator is a web-based tool that can assist in evaluating the potential increase in procurement or refurbishment costs in commercial buildings due to the embodied greenhouse content.

References

[1] Turton, H. and Hamilton, C. (2001) *Comprehensive Emissions per Capita for Industrialised Countries*, The Australia Institute.

[2] Treloar, G. J. (1996) *The Environmental Impact of Construction – A Case Study*, Australia and New Zealand Architectural Science Association, Monograph 001, Sydney, 89 p.

[3] Foster, R., Harrington, L. and Treloar, G. J. (2000) Lifecycle energy consumption and greenhouse gas emissions inventory for Queensland building and construction industries sector, for the Built Environment Research Unit, Public Works Department Queensland, January, 94.

[4] Treloar, G. J., McCoubrie, A., Love, P. E. D. and Iyer-Raniga, U. (1999) Embodied energy of fixtures, fittings and furniture in office buildings, *Facilities*, **17**(11), 403–409.

[5] Australian Greenhouse Office (1999) *National Emissions Trading: Designing the Market, Discussion Paper No. 4*, Commonwealth of Australia, Canberra.

[6] Wilson, N. (2003) Energy shortfalls on horizon – MacFarlane, *The Australian*, February 24.

[7] Jakeman, G., Heyhoe, E., Pant, H., Woffenden, K. and Fisher, B. (2001) The Kyoto Protocol: economic impacts under the terms of the Bonn agreement, *International Petroleum Conservation Association Long Term Carbon and Energy Management – Issues and Approaches*, Cambridge, Massachusetts.

[8] The Allens Consulting Group (2000) Greenhouse Emissions Trading, prepared for the Department of Premier and Cabinet, Victoria.

[9] Treloar, G. J. (1997) Extracting embodied energy paths from input–output tables: towards an input–output-based hybrid energy analysis method, *Economic Systems Research*, **9**(4), 375–392.

14 Raised floor systems for the sustainable fit-out of office buildings

J. Yang[1] *and G. Zhang*[1]

Summary

The traditional fit-out of office buildings typically involves suspended ceiling, ducted supply of heating, ventilation and air conditioning (HVAC), conduits and cabling for power and data, and dry wall partitions. These 'fixed' methods and products cannot cope with the new and emerging business trends that are symbolized by changing business operating patterns, shortened intervals for technology upgrades, widely ranged tenancy requirements, frequent office reorganization, etc. Raised floors were brought into use as a result of the growth of computer applications in office buildings. In spite of their unique advantages, they have not received sufficient recognition as a valid fit-out option for general office space, as demonstrated by their lack of implementation in new built facilities. Modern raised floor systems consist of improved raised floor structures along with an array of innovative ancillary components. Can they meet the requirement of new office buildings? Will they be recognized as the next generation of office fit-out options?

This chapter introduces a research project focusing on the application of the raised floor system to office buildings. The project integrates products specification and design, constructability study, and life cycle costing analysis, with surveys of the industry, on-site observations and laboratory experiments, in an aim to identify the true merits of raised floor systems, their associated problems and possible solutions to these. The literature study, product specification, survey design and research methodology will be presented in the chapter, along with the initial findings of this on-going research.

Introduction

Raised floors were introduced to Australia for computer room applications in the 1980s. Over the years, raised floors have experienced a gradual change of role from specific use (e.g. in dealer rooms for a stock exchange) to more general office space applications. The product development has also gone through a great deal of innovation. Modern

---

[1] School of Construction Management and Property, Queensland University of Technology, Australia.

raised floor systems (RFS) can exhibit a wide range of floor panels, understructures and ancillary components. Incorporating under-floor services, i.e. underfloor HVAC system, power, video/voice and data (PVD) distribution system, RFS is perceived to be able to promote workplace flexibility, good ergonomics and productivity, and to accommodate organizational, technological and environmental changes, all of which are key characteristics representing sustainable use and operation of office space [1–3]. Despite positive feedback from researchers around the world, the implementation of RFS in office buildings has not been widespread in real world practice, particularly in Australia [4]. What are the real reasons for this? Does RFS have the potential to become the next generation of office fit-out methods?

This chapter introduces an independent research project focusing on RFS applications in Australian office buildings. It discusses the conflicts between traditional fit-out and changing office space requirements before introducing some of the sustainable fit-out strategies. RFS products and their potentials are then described followed by an attempt to identify possible barriers to the application of RFS. The RFS research model, preliminary findings and on-going work of this research project are also outlined.

The conflict between expectations and current practice

The needs of building occupants are one of the major drivers in reshaping today's work environments [5]. New and emerging IT and telecommunication technologies enable individuals to have control over where, when and how they work. In addition, business operations are becoming so customer-focused and adaptive to organizational changes that workplaces are facing frequent refurbishments and high churns (workplace relocation). More importantly, rising expectations of thermal comfort and indoor air quality, reduction of energy use, and lifecycle building costs, are now growing concerns for building owners and occupants. As a result, there is a need for building professionals to apply sustainability principles, e.g. the triple-bottom line approach, to office building fit-out.

In contrast, the current fit-out practice adopted by most new build facilities or refurbished projects in Australia is still traditional in style, symbolized by suspended ceiling, sheet metal overhead ducts for HVAC supply, wall-mounted power and data cable conduit, and dry wall partitions. These 'fixed' methods cannot cope with new and changing business operating patterns, shortened intervals between technology upgrades, a wide range of tenancy requirements, and frequent office reorganization, in a cost-effective and time-saving manner [6, 7]. Typical restrictions associated with the operations of two traditional building service systems are indicated in Table 14.1.

With the growing number of conflicts between traditional fit-out methods and ever-changing office space requirements, innovations for sustainable fit-outs of office buildings are imperative.

Sustainable office fit-out strategies

Over the last two decades, sustainable development in the building practice has been increasingly highlighted because of shrinking natural resources and the negative impact

Table 14.1   Typical limitations in building service systems under traditional fit-out.

Service systems	Limitations
Ceiling-based HVAC system	Uncomfortable temperature
	Poor indoor air quality
	Inability to control individual temperature for workstations
	Restriction on future re-fit, thus resulting in additional costs
	Excessive ductwork installation, thus high material and labour costs
	Unsafe overhead work environment for HVAC unit/ducts installation
	High running cost due to low temperature supply air and idle time.
PVD distribution system	Interior artistic design restricted by outlets, switches and wires in dry walls
	Fixed features resulting in high relocation cost
	Limitations for new technology upgrades
	High labour costs due to awkward working conditions

of various construction processes on the environment. It is one of the essential goals for designers to create a healthy, economic and productive built environment [8]. When it comes to office space, the building sustainability concept must be applied to the whole life-cycle of built facilities. According to an AGO report [9], the first cost is less than one-fifth of the operating cost of a building over its lifespan. More importantly, the economy of building operation is dependent on the fit-out approach. Accordingly, in order to deliver a better performing workplace, sustainability principles on the fit-out process deserve in-depth study.

At a minimum level, sustainable office fit-out strategies should respond to three issues of economy of resources: life-cycle design, and human and ergonomic factors:

- *Economy of resources* can be defined as the ability to achieve the desired goals by effectively managing limited resources. To enhance the economy of resources in office fit-out projects, considerations should be given to the whole-of-life building costs to ensure that fit-out methods, materials and approaches contribute to the minimization of on-going building and operational costs, such as development, operation, service, maintenance, upgrading and refit, etc, as well as cost savings from reduced system/staff downtime and improved working efficiency.
- *Building life-cycle design* describes a new philosophy that does not solely focus on the current needs of office fit-out, but also actively plans for evolving requirements for future workspace operation, business growth, technology adoption and upgrade and asset management with a holistic and systematic approach.
- *Human and ergonomic factors* consider issues such as occupants' wellbeing, safety, access, efficiency, productivity, esteem, equity, culture and aesthetics in the workplace environment. A sustainable fit-out needs to respond to these issues through project conception, design, construction, maintenance and future refurbishment of the office space.

Table 14.2 Composition of a typical raised floor system.

Structural units	Accessories	Service systems
Panels	Floor coverings	Underfloor HVAC system
Understructures	Cable carriers	Air terminals
Pedestals	Cable grommets	Fan coil units
Stringers	Floor service outlet/Viewers	Underfloor PVD system
	Perforated panels/Dampers	
	Ramps, steps and fascias	
	Skirting boards	
	Panel lifters, etc.	

Owing to their inherent limitations, traditional office fit-out methods cannot cope with the demand rising from the above issues on a deliberate and long-term basis. Deployment of appropriate construction technologies, innovative building design and flexible facility management patterns needs to be done. As an alternative fit-out method and product, RFS needs to be comprehensively studied for its potential to meet the sustainable fit-out strategies for office buildings, as well as for its limitations.

Opportunities presented by RFS fit-out

A typical RFS consists of structural units, accessories and service units, as shown in Table 14.2. The structural units, including panels and understructures, provide the basic platform of RFS, supporting loads above. The panels will form the raised floor surface on which office activities are carried out. Understructures supporting these panels create a plenum between the sub-floor and the panels that allows the distribution of building service systems and provides easy access to maintenance. The accessories, such as floor coverings, cable carriers and ramps, are indispensable to many flexible RFS functions, structural integrity, and improvement of workplace environment. The service systems include underfloor HVAC and underfloor PVD systems, both of which can enhance the versatility of RFS products for office buildings.

The literature study suggested a range of potential benefits of RFS using an underfloor HVAC system and a PVD distribution system for office building fit-out [1, 2, 5, 10], as indicated in Figure 14.1.

The RFS technology has been explored by many industry practitioners and applied to a number of green and smart building projects [1, 11, 12]. While many merits and potentials have been identified, as with the introduction of many new technologies, some of the earlier applications have experienced problems and encountered barriers preventing more effective implementation.

General barriers to the application of RFS in Australia

Although the introduction of RFS applications to the Australian building industry was as early as the 1980s, the lack of widespread adoption in new build facilities in Australia

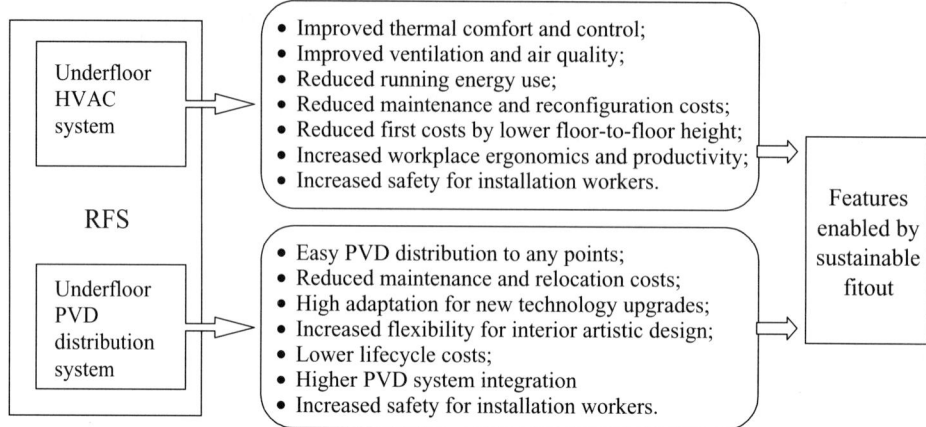

Figure 14.1   Potential benefit of RFS for office building fit-out.

indicates a degree of problems in their implementation. According to literature studies [2, 5] and the initial analysis of a survey to developers, clients, design consultants and contractors of office buildings, conducted by the Queensland University of Technology (QUT), some of the barriers to RFS application (existing and perceived) are identified as in Figure 14.2.

The existing barriers indicate problems experienced by industry practitioners when dealing with RFS-based office fit-out projects. The perceived barriers represent subjective opinions of building professionals. A comparison between the RFS merits and potentials discussed above, and the barriers in particular the perceived ones, indicates a degree of lack of understanding in the industry. Much work is needed to promote RFS products and to assess inherent problems. Ideally this should be carried out free from the influence of associated product vendors or professional prejudice and unwillingness for innovation. It will also be feasible to establish showcase projects, or role models of RFS based fit-out of office buildings.

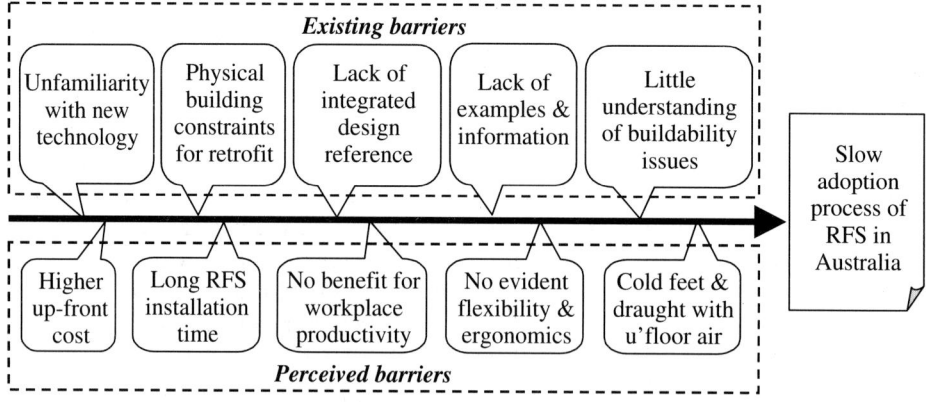

Figure 14.2   Existing and perceived barriers in the way of RFS application.

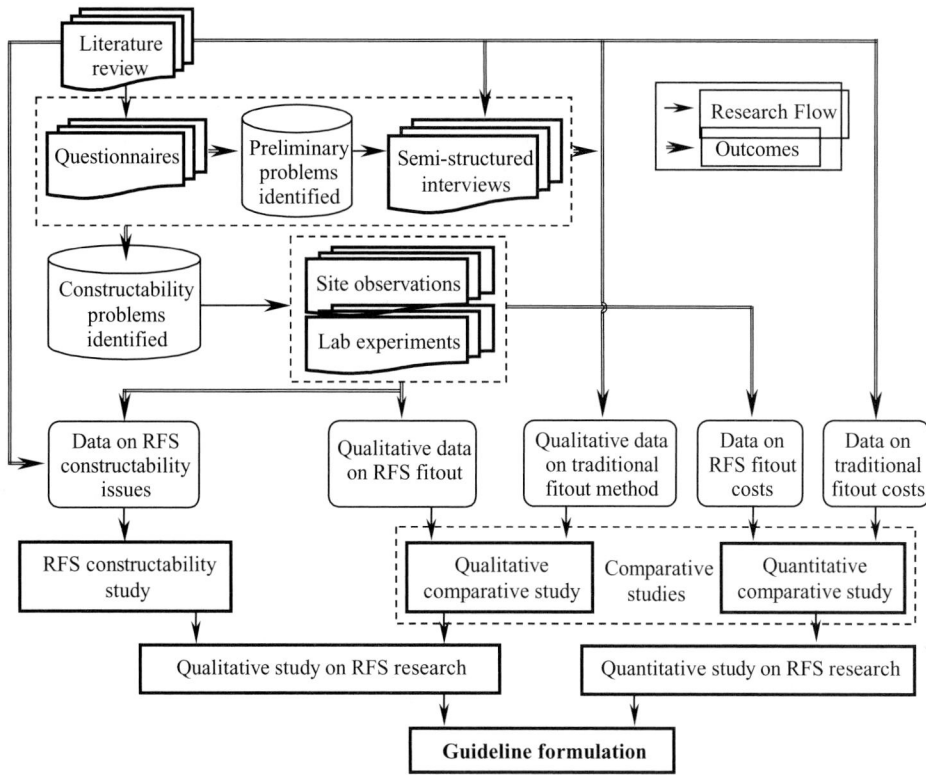

Figure 14.3   Research model for RFS study.

Optimizing sustainable outcomes through RFS enabled fit-out

As part of the efforts to justify RFS suitability and identify and remove current and potential barriers of RFS based office fit-out, QUT is conducting a research project with Australian characteristics. It also aims to optimize sustainable outcomes through RFS enabled fit-out for office buildings. The research will specify appropriate products, identify implementation problems, study RFS constructability, justify cost benefits of RFS on a comparative and lifecycle basis, and seek potential strategies for the effective implementation of RFS. The RFS study combines qualitative and quantitative aspects of research. The research and development processes and data flow is shown in the research model in Figure 14.3.

Data used in this research are being collected through three main stages. The first stage is literature reviews, which help provide the focus and scope of study, define research objectives, obtain essential information on traditional fit-out methods, and prepare for survey questions. The second stage is the survey process, which begins with mailed questionnaires to identify generic RFS problems, then follows on to semi-structured interviews to determine RFS constructability issues. The third stage consists of site observation and laboratory experiment (coinciding with an office retro-fitting project in QUT), in which specific constructability problems are investigated and solutions tested, and realistic data for RFS fit-out are collected. Data collections will enable qualitative and quantitative studies on the RFS. The former includes RFS constructability study and a qualitative comparison between processes of the RFS based and traditional fit-out methods, while the latter deals with fit-out costs and lifecycle cost comparisons. As one of the final outcomes, the research will also formulate a RFS implementation guideline for the fit-out of Australian office buildings.

Preliminary findings and outlook of on-going research

The RFS research has been ongoing since November 2001. To date, preliminary results can be summarized as below:

- A comprehensive literature review has been completed. It included findings on 'both sides of the story', i.e. studies commissioned by raised floor manufacturers and reports from occupants and facility managers who had to live with the products. Contexts for this research, including scopes, problems and objectives were defined and the limitations of traditional fit-out methods were initially assessed and summarized. Moreover, the basis of survey questions has been identified.
- A classification of RFS products suited for office buildings was made. Key characteristics of RFS structural units, accessories and service systems were summarized. Specific RFS products for different office spaces were identified.
- General RFS enabled fit-out specifications have been synthesized for common office environments – general offices, computer/equipment rooms and lift lobbies. For each environment, a typical RFS fit-out specification considering design and performance requirements, products, examination and installation was developed.
- A processed-based conceptual model on the selection of RFS products was developed. The model starts from the identification of a particular fit-out environment, and then goes on to a pre-conformity test in different environments (e.g. general offices and computer rooms) in an office building. A two-stage selection process incorporating technical, economic and strategic factors will follow. The most appropriate products for all RFS components can be decided upon after comprehensive conformity examination in all specified environments.
- Questionnaires were mailed out to Australian office building owners, developers, contractors, and commercial real estate agents to ascertain their level of knowledge of RFS products and application potential as well as their experiences of RFS problems. The survey consists of 5 'A' sections, namely, Awareness, Advantages, Accessibility, Adaptation and Adoption. 132 questionnaires were sent out by mail and 46 useful feedbacks were returned, representing a total of 35% response rate. While detailed analysis is still to be completed, some initial results have been summarized in Table 14.3 and Figure 14.4 to illustrate the focused areas.

The results of Table 14.3 obviously show that industry sectors have relatively high awareness of RFS structure and components, but inadequate knowledge of RFS

Table 14.3   Industry's perception on key issues of RFS implementation [rating on perception (1: least, 7: most)].

Key RFS issues	Respondent categories			
	Occupants	Owners/ developers	Contractors	Real estate agents
Awareness:				
RFS structure and components	5.67	4.36	5.30	1.86
RFS application scenarios	3.83	3.24	3.67	2.35
Accessibility:				
Procurement issues of RFS	3.22	3.45	4.17	1.22
Project team skills and resources	2.89	3.21	3.90	1.28
Adaptation:				
RFS construction	3.64	3.45	5.30	1.78
RFS subsystem adaptation	3.74	3.51	3.71	2.13
Adoption:				
RFS reconfiguration and maintenance	4.78	3.91	3.90	1.31
RFS limitations	4.53	3.20	4.78	2.15

applications. This is symbolized by most respondents' consideration of RFS as something for computer/equipment rooms rather than general office/commercial buildings. At the same time, most respondents are not familiar with RFS accessibility and adaptation issues, with contractors being the only exception possible due to their professional interest. Furthermore, the occupants seem to be most concerned with RFS adoption because of the reconfiguration and maintenance work. As an influential but often-neglected industry sector, the commercial real estate agents demonstrated the lowest familiarity with all issues. This is a very alarming problem as these agents' perception can often impact on the decisions of owners and occupants.

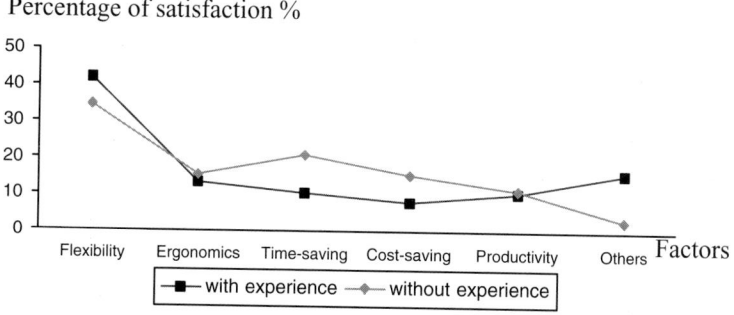

Figure 14.4   RFS advantages identified by respondents with or without RFS experience.

Figure 14.4 shows that all respondents agree that the most important advantage of RFS is its flexibility. However, respondents who have actually used RFS rated flexibility at an even higher satisfaction level than those without RFS experience. Comparatively, users of RFS are more realistic about time and cost savings than those without actual RFS experiences. Some advantages of RFS, e.g. improved system integration and a lower travel of sound, could only be identified through users' own experiences. However this category represents a total 20% that warrant further investigation.

The above analysis indicates that most Australian professionals are not informed of the latest development on RFS. The scarcity of RFS applications in realistic building projects offers no evidence and support for issues in relation to RFS planning, procurement, construction, operation and maintenance.

With the feedback from survey analysis, the remaining development work will first engage semi-structured interviews to identify specific constructability problems associated with RFS implementation, and seek advice on possible solutions from industry practitioners. Site observations and lab experiments will be performed on experiment solutions and alternatives of the identified problems as well as to collect data to support the qualitative study on RFS constructability and the comparative studies between RFS and traditional fit-out approaches. Potential outcomes may come from four aspects as follows:

- the highlighting of problems associated with RFS constructability by systematic literature reviews, questionnaires and semi-structured interviews
- the identification of potential solutions to these problems and the exploration of innovations on RFS constructability, aided by site observations and laboratory experiments
- the justification of the true merits or shortcomings of RFS, including lifecycle cost analysis, flexibility, ergonomics and reliability through comparative studies between RFS enabled and conventional fit-out methods
- the formulation of an implementation guideline to facilitate RFS application for office building fit-out in Australia.

Conclusion

The ever-changing requirements of business operation and increasing public emphasis on sustainability issues call for many features that are additional to those that a conventional office workspace can provide. Flexibility, adaptability, low churn costs, high levels of occupant comfort, an individually controllable work environment for better health and productivity, are just a few of the new demands placed upon sustainable office building partitions. Fit-out options incorporating raised floor systems possess many unique characteristics and the ability to integrate with innovative services to meet these new challenges, when traditional fit-out approaches tend to fall short. Before all of the office building industry seriously considers RFS options, much work needs to be done to improve raised floor products, educate construction professionals, overcome implementation barriers, and promote 'role model' installations. Very importantly, these efforts need to be made on an independent and unbiased basis, must be sufficiently comprehensive to cover all fit-out development aspects, and must be able to respond to sustainability concerns. Doing this will ultimately help reveal areas for improvement and justify the worthiness of raised floor systems to be considered as a viable fit-out alternative for office space of today and tomorrow.

References

[1] York, T. R. (1992) Access floor – don't plan your building without it, *FM Journal*, Nov/Dec.

[2] Ellison, J. and Ramsey, B. (1989) Access flooring: comfort and convenience can be cost-justified, *Building Design & Construction*, April.

[3] Tate. (2001) Project spotlights: www.tateaccessfloors.com/prod_app_office.html, Tate Access Floors, Inc. MD.

[4] Zhang, G. and Yang, J. (2002) Raised floor system: a paradigm of future office building fit-out? *Proceedings of the ABT Conference*, Hong Kong, 1177–1184.

[5] Bauman, F. S. (1999) Giving occupants what they want: guidelines for implementing personal environmental control in your building. *Proceedings of World Workplace 99*, October 3–5, 1999, Los Angeles, CA.

[6] BOMA and ULI (1999) *What Office Tenants Want: 1999 BOMA/ULI Office Tenant Survey Report*, Washington, DC: BOMA International and ULI – the Urban Land Institute.

[7] Terranova, J. P. E. (2001) Underfloor Ventilation, *HPAC Engineering*, March.

[8] CIB (1999) *CIB Agenda 21 on Sustainable Construction*, CIB Report Publication 237, July.

[9] AGO (2000) *Building for Energy Efficiency – Reducing Greenhouse Gas Emissions from Australian Buildings*, Australian Greenhouse Office, 2000.

[10] Chiang, C. M., Lin, F. M., Chung, S. C. and Chung, C. M. (2001) Prediction of the reduction of impact vibration in raised access floors, *Building Acoustics*, **8**(3), 199–211.

[11] Becket, E. (1992) *GSA Access Floor Study*, The U.S. General Services Administration, Washington, D.C.

[12] Guttmann, S. (2000) Raising the bar, with raised floors, *Consulting-Specifying Engineer*, October, 50–54.

15 Sustainable deconstruction of buildings

F. Schultmann[1]

Summary

In this chapter, an approach for sustainable deconstruction of buildings is presented. To anticipate a sustainable end-of-life management, numerous different objectives in environmental, social, technical and economic means have to be taken into account. Thus, alternative scenarios for the deconstruction of buildings are considered on a strategic and on an operational level. Based on a material flow management approach, sophisticated planning models are presented for strategic as well as for detailed planning. Different objectives are modelled using extended time-based or cost-based objective functions, whereas different alternatives to meet certain targets in the field of sustainability are modelled by using multiple modes. The proposed models also consider limited financial or technical resources and therefore allow the calculation of enhanced solutions for different conditions in deconstruction management. The implementation of the methodology within a planning system is addressed. Finally, a case study illustrates the application of the approach.

Introduction

In future, decision-making in construction management of buildings will not only focus on the genuine aim of profitability but will also have to meet the criteria of sustainability, e.g. limiting the discharge of pollutants into the environment over the whole life-cycle of buildings. The latter can be supported by applying material flow management, which has been proven as a suitable approach to meet prerequisites for a sustainable development in the construction industry. Material flow management covers the entire value chain of quarrying, production and transport of building materials as well as the construction process itself, followed by the use of buildings and finally their deconstruction and recycling. However, the rapid development of ideas for the end-of-life treatment of complete buildings or components has so far resulted in only a few planning systems for the final phase of the building life cycle. Nevertheless, in recent years there have been various attempts to set up advanced recycling technologies for construction and demolition waste. However, as further improvements in processing

---

[1] University of Karlsruhe, French–German Institute for Environmental Research, Karlsruhe, Germany.

are technically limited, future efforts will have to concentrate on improving the methods of deconstruction. The conventional demolition of buildings, carried out by pulling down a building with a backhoe for instance, often results in a mix of various materials and the contamination of non-hazardous components. Hence, advanced approaches aim at deconstruction or selective dismantling of buildings, disassembling a building into its various parts. Deconstruction, rather than demolition, helps to separate different building materials and to reuse recycled materials in superior utilization options.

The purpose of this chapter is to investigate how sophisticated planning approaches can be applied to plan and optimize the environmental-friendly deconstruction and recycling of buildings. The concept is based on material-flow management, which ensures that certain environmental requirements are met. Based on these requirements, mixed-integer programming formulation is proposed for the planning of optimal deconstruction schemes on a strategic planning level. Given these results, project scheduling models are developed in order to optimize deconstrucion processes on the site. The approach is applied to the deconstruction of residential buildings and the results show that tremendous improvements in the management of deconstruction projects can be realized.

The chapter is structured as follows: the next section is devoted to a material-flow analysis, which serves as a framework for further planning. The following section covers a strategic planning approach consisting of a mixed-integer programming formulation. An approach for detailed site planning and scheduling is then introduced. In the next section, some results are outlined. Finally, conclusions are drawn.

Material-flow management for deconstruction projects

Problems with material-flow management in the construction industry mainly arise because of the long time lag between construction and end-of-life of buildings. Because the composition of buildings at the end of their life cycles are generally not known, the first step for dismantling and recycling planning is a sound pre-deconstruction survey, also called a building audit. The building audit mainly aims at identifying and quantifying materials in order to give decision support on the dismantling procedure. Based on the documents of the building (e.g. construction plans, descriptions, history), detailed data on the composition of the building have to be collected and analyzed. During this audit, indications of substances in the building that may influence the quality of the materials must be determined. The audit results in a bill of materials, also called an inventory, that contains details of the construction elements and the corresponding building materials [1, 2]. In order to model the relevant material-flows on deconstruction sites, material-flow graphs can be developed. The construction materials are the sources of this graph. By applying a set of dismantling activities the corresponding building is dismantled into various parts. Depending on the stage of dismantling, the dismantled components consist of either a single construction element or a mix of various building materials. To avoid a mix of toxic and non-toxic materials, the environmental compatibility of various components has to be determined [3].

One should be aware that there is a risk that both plain and mixed grades of building waste might contain pollutants that can harm the environment, especially by leaching,

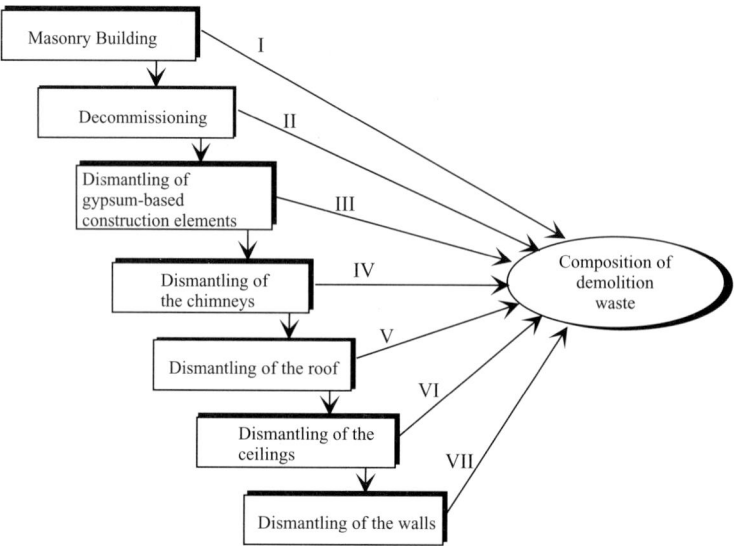

Figure 15.1   Dismantling cascade for a domestic building (simplified).

during storage or re-use. Since generally only a very small part of the building materials contains pollutants, it is essential to identify these before dismantling starts, in order to avoid mixing a small amount of toxic material with a large amount of non-toxic material. Relevant shares of pollutants are contained both in building materials and surfaces. Pollutants are included in construction materials as part of their natural material composition, or may have been artificially added during manufacture, for example in the form of additives. Furthermore, a large proportion of pollutants is caused by surface area treatment. Based on the bill of materials, the content of the pollutants can be represented by a so-called pollutant vector for each material and for each surface. The content of pollutants in building materials can then be described by a pollutant matrix (c.f. [4]).

  Based on the bill of materials and the information about pollutants, material and pollutant balances for different dismantling steps can be established. For details on the allocation procedure see Reference 5. Figure 15.1 illustrates a simplified hierarchical structure of how the composition of demolition waste can be influenced by performing seven alternative ways of dismantling a domestic building (Alternative I represents a demolition of the building without any dismantling, Alternative II reflects decommissioning followed by demolition, etc.).

  It can be shown (c.f. [6]) that the demolition of a building without any dismantling (Alternative I) leads to cause significant amounts of heavy metals and organic compounds in the building waste, which will result in recycling problems. The reduction of the amount of heavy metals (e.g. cadmium or zinc) in the remaining building waste requires at least decommissioning (Alternative II), whereas the content of polycyclic aromatic hydrocarbons (mainly found in the chimneys) can only be significantly reduced after dismantling the chimneys (Alternative IV).

Material and pollutant balances may serve as a framework for the deconstruction work to be carried out in order to guarantee a certain quality level of recycled materials. Using these results, detailed planning of deconstruction can start. However, detailed planning has to include not only material- or pollutant-flow aspects but also the necessary technology and resources for the deconstruction work. The next two sections will concentrate on how this can be done.

Strategic planning of deconstruction activities

When approaches for the deconstruction planning of buildings are developed, the typology of construction and deconstruction systems has to be considered first. According to the common manufacturing typology, the deconstruction and recycling of buildings (as well as the construction of buildings) represents a make-to-order production [7, 6]. Considering the internal structure of the manufacturing system, the deconstruction and recycling of buildings is regarded as on-site manufacturing because all resources needed for dismantling have to be transferred to the construction site, instead of vice versa. Usually, on-site manufacturing requires more planning than other types, such as job shop or flow shop manufacturing [8]. While the erection of buildings can often be characterized as more or less hybrid, using certain prefabricated components (produced in flow shop or job shop structures) that are finally assembled on the construction site (on-site manufacture), the demolition or dismantling of buildings at the end of their life can take place only on the deconstruction site.

In contrast to assemble-to-order-systems, prevailing for instance in the automotive industry, in make-to-order systems even subassemblies and components are manufactured only if they are required for the production of a customer-ordered final product. One disadvantage of the several planning approaches developed for make-to-order production is that resource capacities are often not considered explicitly. However, in reality resources are normally scarce. As a consequence of a limited resource availability, revisions in the scheduled production plans are usually necessary, often resulting in large delays in the scheduled activities. Recently, new capacity-oriented concepts have been suggested to overcome this problem.

Production at the beginning of a product's lifetime, as well as disassembling at its end is determined by the characteristics of the product. Like consumer products, such as cars, buildings have finite lifetimes. However, the lifetime of a building usually ranges between 50 and 150 years. Furthermore, buildings can be characterized as meta-products, since they represent an assembly of multiple products all with their own characteristics, combined in unique and complex ways [9], which has led to concepts for modeling building structures by several layers. Both the attribute meta-product with unique characteristics and the long lifetime impose severe problems for deconstruction planning. The unique combination of products or components, also called construction elements of a building, requires an approach that considers the individual building's properties at each time when dismantling and recycling are planned. This uniqueness is one of the motivations for the approaches considered in the next section. On a strategic planning level, interdependancies of the dismantling on site and the recycling processes have to be observed. Consequently, an integrated approach is needed. Such an approach is realized by an integrated dismantling and recycling planning system [10]. The system has been implemented and validated for domestic buildings in various regions. For each of the predominate buildings in the considered region, a representative building has been chosen and a detailed bill of materials has been prepared. On the basis of these bills of materials, the different deconstruction techniques have been analyzed and aggregated into so-called dismantling activities. For these dismantling activities the associated dismantling cost and the precedence relations have been determined, taking consideration of technical aspects, thus a dismantling-precedence-graph can be developed for each class of buildings identified [11].

Following the lead of the engineering industry, where the concept of precedence-graphs is used in Materials Requirements Planning (MRP), the precedence-graph is the starting point of the dismantling planning process. By applying of a set of dismantling activities $v_1, \ldots, v_J$ the corresponding building is disassembled into construction elements and parts. The decision as to whether a certain dismantling activity will be applied or not depends on the cost-comparison of the following alternatives: demolition and recycling of the partially dismantled building on the one hand, or dismantling and recycling of the dismantled construction elements and parts on the other.

Integrated dismantling and recycling planning

The dismantling planning process can be carried out only if the recycling cost for all construction elements and building materials are already determined. Hence, the recycling planning process has to precede the dismantling planning process. On the other hand, the recycling planning process requires information about the amount and composition of the dismantled construction elements and materials. Because the recycling capacities of the different reuse options are limited in the geographical region considered, an optimal allocation of the dismantled construction elements and building materials to the available reycling or reuse options is necessary in order to determine the minimal recycling cost. Following this idea, the dismantling planning process has to precede the recycling planning process. Because of these interdependancies, the dismantling and recycling planning processes cannot be carried out independently, but an integrated approach is desirable [12]: The dismantling-precedence-graph of the geographical region considered is then composed of the dismantling-precedence-graphs that are derived from the predominant types of buildings. These buildings can be dismantled by the application of J different dismantling activities v_j. Consequently, the quantities of the dismantled components and materials can be represented by the vector $y^T = (y_1, \ldots, y_m)$ (cf. Figure 15.2). Taking into consideration all feasible recycling and reuse options as well as the capacity constraints, an optimal allocation has to be determined. Because of the great complexity of the planning problem described, the system is based on a sophisticated operations research model. The model includes all deconstruction sites in the geographical region considered during a certain planning period. If new sites arise in the following planning period, the model is evaluated again, taking into account the construction elements

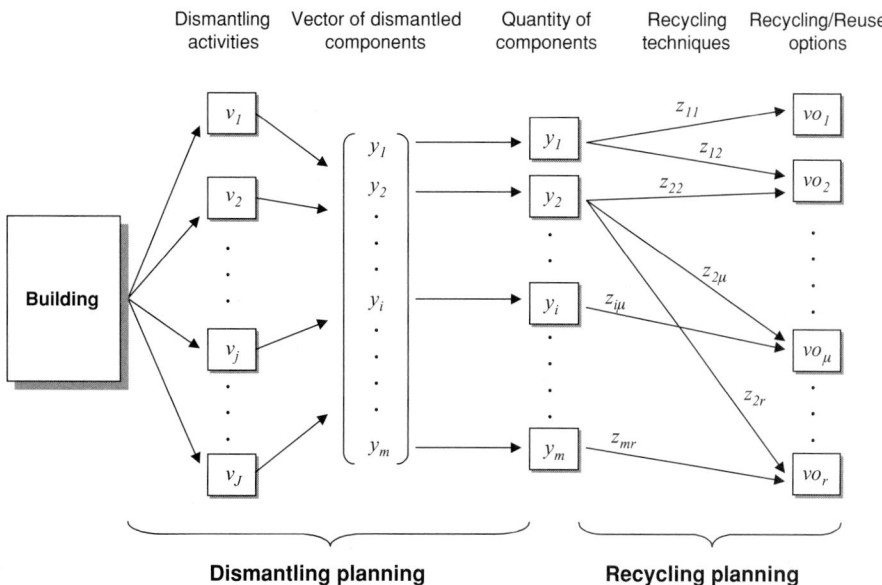

Figure 15.2   Structure of the integrated dismantling and recycling planning problem.

that are already deconstructed and thus scheduled for recycling in the actual planning period (a model formulation for this detailed scheduling will be given in the next section).

Model formulation and solution procedure

The integrated dismantling and recycling planning problem can be formulated as a mixed-integer linear programming model (MILP) [12]. The subject of the optimization problem is the maximization of the total achievable marginal income (i.e., reuse proceeds reduced by the dismantling cost) subject to certain constraints. This problem can be solved by the application of various kinds of solution procedures for combinatorial optimization problems, such as cutting-plane-algorithms, branch-and-bound-algorithms, dynamic-programming-algorithms, or decomposition algorithms. Solution procedures for the problem sketched are proposed in Reference 12. General mixed-integer problems are extremely complex (NP-complete) so that the required solution time will grow exponentially with the complexity of the considered planning problem. Standard optimization software packages (e.g. LINDO™, CPLEX™) generally use sophisticated branch-and-bound-techniques. Upper bounds are generated by an LP-relaxation of the mixed-integer problem. The upper bounds computed in this way are often not strong enough to reduce the required solution time drastically. In this case it can be advantageous to introduce so called upper-bound-functions such as the Lagrangean function. However, branch-and-bound-algorithms do not take any advantage of the special structure of the considered dismantling and recycling planning problem, which can be partitioned into an all-integer linear dismantling problem and a linear recycling problem. A decomposition algorithm developed by Benders [13] uses this special feature in order to find an optimal solution by the iterative solution of the all-integer and the linear problem.

With the help of this model, the cost and interactions of dismantling and recycling can be evaluated and cost-efficient combinations of dismantling and recycling techniques for various types of buildings can be determined subject to technical, environmental and capacity constraints [10].

Detailed planning and scheduling of deconstruction activities

The results of the strategic dismantling and recycling planning as sketched in the last section, i.e., cost-efficient combinations of dismantling and recycling techniques subject to the type of building, the geographical region considered and environmental constraints, serve as a framework for detailed (or short-term) planning. Now, the time horizon is limited to the duration of deconstruction sites, generally ranging from some days to several weeks. Consequently, in addition to strategic planning issues, time-aspects have to be integrated, when detailed planning of deconstruction takes place. When dismantling activities on the deconstruction site are planned, the degree of dismantling, i.e., the number of dismantling activities j to be carried out, is already given by the outcome of the strategic planning. A promising methodology to improve the management of deconstruction sites is the use of resource-constrained project scheduling, which originally gained particular attraction in make-to-order production. Resource-constrained project scheduling problems take into account the fact that resources are limited and therefore they help to overcome these shortages. Jobs, also known as activities, can be carried out in different ways, that is, using different techniques and different resources.

During the last few years, several researchers have concentrated on modelling and algorithms for resource-constrained project scheduling. For a survey on modelling concepts as well as scheduling algorithms for resource-constrained project scheduling used in deconstruction see Reference 4. In this section, concepts to apply resource-constrained project scheduling for deconstruction operations are outlined.

Keeping in mind the results of the material-flow analysis as well as the strategic planning as introduced in the sections above, detailed deconstruction planning aims first at setting up a detailed technological and environmentally oriented order for the dismantling activities to be carried out. Similar to the strategic planning, where dismantling-precedence-graphs were introduced, the technological precedence-relations, and in certain cases also certain environmental constraints of the deconstruction process, have to be respected. For this purpose, a topologically ordered activity-on-node network is used, where the nodes represent the dismantling activities and the arcs the precedence relations between these activities. Regarding the scheduling model to be formulated, the network contains one unique source and one unique sink. This can always be guaranteed by introducing a dummy source and a dummy sink, respectively. Different networks may have to be defined accordingly, including the type of building under consideration, the deconstruction techniques applicable, or the objective of

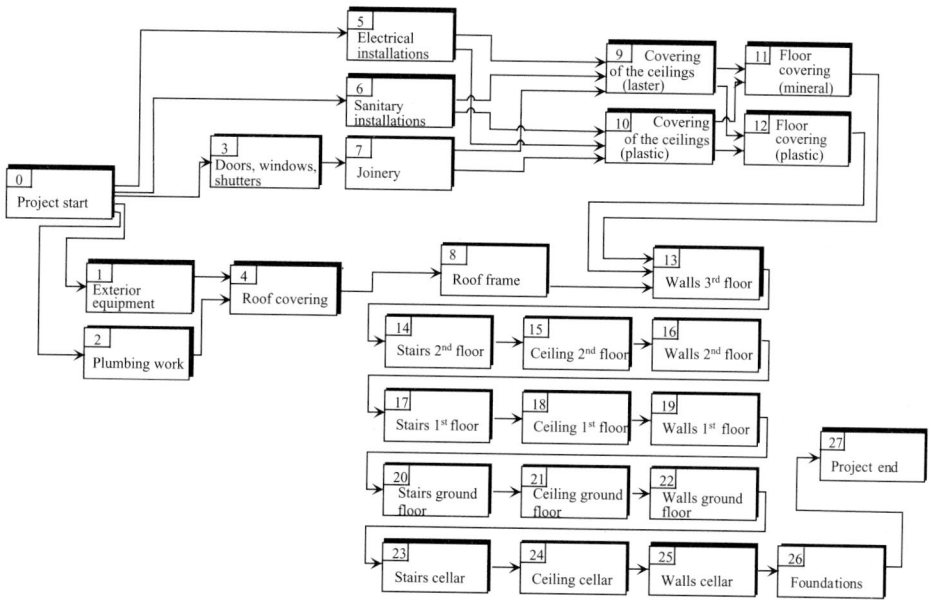

Figure 15.3 Deconstruction network for a residential building.

the deconstruction effort. Also, different environmental constraints, such as obligatory levels of separation (i.e. materials containing asbestos) can lead to different networks. Figure 15.3 gives an example of a topologically ordered deconstruction network for a three-storey residential building with 28 dismantling activities. This network served as a basis for a pilot project on selective dismantling carried out in practice [14].

Modelling of different techniques and resources

After determining the precedence relations, the dismantling activities have to be specified in detail. This comprises the determination of the resources needed as well as the duration of the activities. Usually each activity can be processed in different ways, e.g. using different dismantling techniques that can be expressed by different resources. Moreover, each different technique may result in different processing times. For instance the deconstruction of outer walls can be carried out by dismantling, using pneumatic hammers, by a grabbing bucket, or by demolishing the wall with a hydraulic excavator, each resulting in a different processing time [6]. Several alternatives of ways in which a job can be carried out can be modelled by introducing different modes. Changing the number of activities in a project or altering the precedence relations leads to different networks that cannot be incorporated into a single scheduling model. However, altering modes, such as the consideration of the activity duration as a (discrete) function of the resources and/or amounts of the resources allocated, while keeping precedence relations and the number of activities constant, can be modelled in an integrated model.

In order to formulate the resource-constrained scheduling model, the resources needed have to be classified. Performing an activity in a certain mode is associated with the usage of renewable resources and the consumption of non-renewable resources. While renewable resources (e.g. machines, workers) are constrained on a periodical basis only (possibly varying from period to period), i.e., after a dismantling activity is accomplished, the renewable resources used by that activity are available to process another activity, non-renewable resources (e.g. the financial budget for the deconstruction work) are limited with regard to the entire duration of the project. Consequently, the consumption of a non-renewable resource by an activity reduces its availability for the rest of the project.

Model formulation and solution procedure

In order to reduce the number of variables in the programming formulation, time windows with earliest and latest finishing times (or earliest and latest starting times) for each dismantling activity can be derived (neglecting resource constraints). This can be done by using the well known critical path method (CPM) [15]. Critical path analysis requires an upper bound for the makespan of the project. Scheduling on deconstruction sites can now be formulated as a binary linear program (cf. [6, 5]). Environmental constraints, such as different levels of sorting materials or certain recycling paths, are reflected by the project network, in the precedence relations as well as the configuration of activities.

The objective function of the model minimizes the completion time for the deconstruction work. Several further criteria, such as minimizing the tardiness of the activities, as well as minimizing the net present value or leveling of resources, might also be considered [5]. But, since minimizing the makespan is the most important objective of deconstruction project scheduling by far, the discussion of other objective functions is omitted. Model-constraints ensure that each activity is processed exactly in one mode and that one completion time is assigned. Moreover, dismantling precedence relations have to be respected and the capacity restrictions for renewable and non-renewable resources have to be met.

The scheduling model sketched is one of the so-called Multi-Mode Resource-Constrained Project Scheduling Problems (MRCPSP). It belongs to the class of combinatorial optimization problems. The incorporation of different modes for each activity enlarges the solution space, which means that practical production situations can be modelled and ways to better solutions are given. However, this makes the MRCPSP very complex. In fact, the problem is NP-hard which results in intensive computation if the project scale increases. In the example given, the model is solved by a branch-and-bound algorithm. The branch-and-bound algorithm generates exact solutions as far as regular measures of performance (cf. [15]) are considered. A slightly modified version of the algorithm proposed by Sprecher is used here and, due to the NP-hardness, the computation time for certain problems is limited (for details see Reference 5).

	Costs						Time			
	Deconstruction		Recycling		Total		Duration			
							Deconstruction site		Deconstruction steps	
	[EUR]	Δ [%]	[EUR]	Δ [%]	[EUR]	Δ [%]	[h]	Δ [%]	[man h]	Δ [%]
Status-quo	44378	0	11593	0	55971	0	408	0	1390	0
Complete deconstruction / different deconstruction scenarios										
Scenario										
No. Characteristics										
1 Parallel work	36535	-17.7	16906	45.8	53441	-4.5	218	-46.6	1382	-0.6
2 Manual deconstruction	30110	-32.2	16913	45.9	47023	-16.0	221	-45.8	1398	0.6
3 Partially automated deconstruction	23576	-46.9	16910	45.9	40486	-27.7	168	-58.8	1091	-21.5
4 Material oriented deconstruction	29942	-32.5	17094	47.5	47036	-16.0	192	-52.9	1234	-11.2
4a Material oriented deconstruction + scaled costs for disposal/recycling	29942	-32.5	6734	-41.9	36676	-34.5	192	-52.9	1234	-11.2
Complete deconstruction / detailed planning under different site-conditions										
space limitations on site										
1-3 low	18188	-59.0	16910	45.9	35098	-37.3	163	-60.0		
1-3 high	18233	-58.9	16910	45.9	35143	-37.2	167	-59.1		
4a low	27415	-38.2	6734	-41.9	34149	-39.0	182	-55.4		
4a high	27702	-37.6	6734	-41.9	34436	-38.5	186	-54.4		
Combinations of deconstruction procedures under different conditions										
Grade of de-	Costs						Recycling rate			
construction Recycling Manpower										
3 low low low	19091	-57.0	23767	105.0	42858	-23.4	43%			
3 low high low	19104	-57.0	30225	160.7	49329	-11.9	43%			
3 low low high	24665	-44.4	23744	104.8	48409	-13.5	43%			
3 low high high	24651	-44.5	30234	160.8	54885	-1.9	43%			
3 high low low	23325	-47.4	18324	58.1	41649	-25.6	97%			
3 high high low	23339	-47.4	23356	101.5	46695	-16.6	97%			
3 high low high	41329	-6.9	18321	58.0	59650	6.6	97%			
3 high high high	41316	-6.9	23360	101.5	64676	15.6	97%			

Figure 15.4   Deconstruction scenarios and planning results for a residential building.

Consequently, in some cases only heuristic solutions were generated. Although is was accepted that in 25% of the cases considered only suboptimal solutions could be found, feasible solutions were generated in all the cases considered.

Application of the operative planning model

The quality of project scheduling by using the detailed planning model presented is evaluated in terms of improvements that could be achieved, compared with conventional deconstruction. In Figure 15.4 computational results for a domestic building are compared with the results obtained in practice (depicted as 'status-quo').

As illustrated, computational results indicate that the site duration, i.e. the project's makespan, as well as the costs can be drastically reduced. When performing scenario 1, which considers nearly the same resource utilization and constraints as in practice, the total dismantling time can be reduced from 408 to 218 hours if work is carried out simultaneously wherever possible, i.e., according to the precedence-relations given in the dismantling-network while respecting resource constraints. In coherence with the reduction of the duration, the cost of technical equipment decreases because the cost-intensive devices remain on the construction site for only very short periods. Detailed planning under different site-conditions allows one to optimize different scenarios (1–3) simultaneously, which leads to a further reduction in the duration and costs due to the possibilities of accelerating certain activities by using sophisticated machinery. Depending on the economic framework and the grade of deconstruction, a recycling rate of 97% can be achieved. Associated with that rate is not only the portion of the materials recycled, but also a high quality of recycling. In fact, taking the material-flow approach, it can be shown that certain scenarios enable a high proportion of materials to be reused for high-grade applications [3].

One major advantage of scheduling on a detailed planning level is that the solution allows detailed planning, such as the determination of start and finish times, for all deconstruction activities as well as logistics management on deconstruction sites [7, 16].

Conclusions

In this chapter, models for dismantling and recycling planning have been presented using material-flow management as a basis. For several examples of deconstruction it can be shown that efficient planning helps to support environment-friendly deconstruction strategies, which are not necessarily disadvantageous from an economic point of view. Compared with results in practice, impressive reductions in dismantling times and cost can be achieved by optimization techniques. In addition to environmental aspects, the results presented here can also help to develop and assess future concepts for cost-efficient project management, not only for deconstruction, but also for construction purposes. Nevertheless, the models presented here cover only a selection of some of the limitations relevant in practice. For instance, the results presented here are based only on the resources that are needed for the deconstruction on the site, i.e. workers and technical equipment. Moreover, non-renewable resources such as the financial budget have to be considered in the computation. Future work will concentrate on methods covering uncertainties and weak data with methods such as stochastic or fuzzy scheduling [17].

References

[1] Schultmann, F. and Rentz, O. (1999a) Material flow based deconstruction and recycling management for buildings, *Proceedings, R'99 International Congress Recovery, Recycling, Re-integration*, Geneva, Switzerland, February 2–5, 1999, Vol. 1, 253–258.

[2] Schultmann, F. and Rentz, O. (2000) The state of deconstruction in Germany, in *Overview of Deconstruction in Selected Countries*, Kibert, C. J. and Chini, A. R. (eds), CIB Report No. 252, International Council for Research and Innovation in Building and Construction (CIB), Rotterdam.

[3] Schultmann, F., Sindt, V., Ruch, M. and Rentz, O. (1997) Strategies for the quality improvement of recycling materials, *Proceedings, Second International Conference Buildings and the Environment*, Paris, France, June 9–12, 1997, Vol. 1, 611–618.

[4] Schultmann, F. and Rentz, O. (2001b) Environment-oriented project scheduling for the dismantling of buildings, in *OR Spektrum*, **23**(1), 51–78.

[5] Schultmann, F. (1998) *Kreislaufführung von Baustoffen – Stoffflussbasiertes Projektmanagement für die operative Demontage- und Recyclingplanung von Gebäuden*, Erich Schmidt, Berlin.

[6] Schultmann, F. and Rentz, O. (2002a) Scheduling of deconstruction projects under resource constraints, in *Construction Management and Economics*, **20**(5), 391–401.

[7] Schultmann, F. and Rentz, O. (2001a) Sophisticated planning models for construction site management, *Proceedings (CD), CIB World Buildings Congress*, Wellington, New Zealand, April 2–6, 2001, Paper High Performance Teams (HPT) 19.

[8] Gaither, N. and Frazier, G. (2002) *Operations Management, 9th* edn, South-Western/ Thomson Learning, Cincinatti, Ohio, 583 ff.

[9] Fletcher, S. L., Popovic, O. and Plank, R. (2000) Designing for future reuse and recycling, *Proceedings, Deconstruction – Closing the Loop*, Building Research Establishment (BRE), Watford, United Kingdom, May 18, 2000.

[10] Spengler, T., Püchert, H., Penkuhn, T. and Rentz, O. (1997) Environmental integrated production and recycling management, in *European Journal of Operational Research (EJOR)*, **97**, 308–326.

[11] Schultmann, F., Ruch, M., Spengler, T. and Rentz, O. (1995) Recycling and reuse of demolition waste – a case study of an integrated assessment for residential buildings, *Proceedings, R'95 International Congress Recovery, Recycling, Re-integration*, Geneva, Switzerland, February 1–3, 1995, Vol. 1, 358–363.

[12] Spengler, T., Nicolai, M. and Rentz, O. (1994) Planning models for the optimal dismantling and recycling of domestic buildings with regard to environmental aspects, in *Operations Research Proceedings*, Heidelberg, Germany, 1993, 306–313.

[13] Benders, J. F. (1962) Partitioning procedures for solving mixed-variables programming problems, in *Numerische Mathematik*, **4**, 238–252.

[14] Rentz, O., Ruch, M., Schultmann, F., Sindt, V. and Zundel, T. (1998) *Déconstruction sélective – Etude scientifique de la déconstruction sélective d'un immeuble à Mulhouse*, Société Alpine de Publications, Grenoble.

[15] Meredith, J. R. and Mantel S. J. (2003) *Project Management: A Managerial Approach, 5th* edn, John Wiley & Sons, New York, 443 ff.

[16] Schultmann, F. and Rentz, O. (1999b) Logistikmanagement auf Basis kapazitätsbeschränkter Projektplanungsmodelle – dargestellt am Beispiel der Baustellenfertigung, in *Logistik Management: Intelligente I+K Technologien*, Kopfer, H., Bierwirth, C. (eds), Springer, Berlin, Heidelberg, New York, 51–60.

[17] Schultmann, F. (2003) Dealing with uncertainties in (de-)construction management – the contribution of fuzzy scheduling, in *Deconstruction and Materials Reuse*, Chini, A. (ed.) Proceedings of the 11th Rinker International Conference on Deconstruction and Materials Reuse, University of Florida, Gainesville, USA, May 7–10, 2003, published as CIB Report 287, Rotterdam.

Part 4
Evaluating past experiences and strategies

16 Time and the sustainable development agenda

P. S. Brandon[1]

Summary

This chapter examines the issue of 'time' in sustainable development and in particular how time horizons impact on the evaluation of sustainability. Despite the massive interest in issues related to sustainable development the questions of evaluation are still under development and within this context the issue of over how long a period we should view the future in terms of sustainable development is poorly served. A quick review of textbooks dealing with this topic within the built environment showed that although the topic was implicit in all the discussions within the texts the question of 'time' was addressed infrequently and seldom explicitly.

This chapter seeks to provide a questioning framework within which this issue can be addressed and it will call on some specific case studies to explore the topic. It will look at some of the work undertaken on lifecycle analysis and in scenario planning to see whether there are lessons to be learned for sustainable development. It will pose some questions with regard to how time should be considered in evaluation, the constraints to implementation and the differentiation of 'time' related to the various aspects of sustainability.

Introduction

At the heart of sustainable development are some assumptions about how long a development is expected to be sustainable. Over what period are we considering the issue? One answer might be 'for ever', another might be 'over a human lifetime' and another might be 'until something comes along that is better or changes the reason for trying to sustain the development'. Underlying all the assessments and evaluations of sustainable development must be some consideration of the time period over which we are making the assessment. Some might argue that as sustainable development is thought to be a *process* then it is not necessary to pay too much attention to this matter. It is part of getting all the stakeholders to think in a certain way about the future to avoid leaving future generations in a worse position than we are today. It is therefore as

---

[1] Research Institute of the Built and Human Environment, University of Salford, UK.

much about culture and the creation of a learning environment as it is about calculation and prediction.

However true this might be, at some stage, decisions have to be made about what to build, how to build and how to use the built environment. Finance houses, clients, local authorities and all the other participants in the process who have some power, or require accountability in the process, will want to know over what time period these assessments have been made. Every decision is made within the context of an assumed time period. It influences the choice of material, the speed at which development occurs, the response to market forces, the design and layout and a whole host of other factors that make up the complexity of the built environment. Whilst our horizon might be the long term future we have to make decisions in the here and now.

Strangely, it appears not to be something that is a major issue in the literature on the subject. It is hidden from view but is an implicit assumption in many of the techniques that are employed. A quick review of some text books on sustainable development in the built environment has revealed that few have a reference to 'time' in their index. This may be a reflection of the nature and youth of the subject. It may reflect the imprecision in the definitions of the term sustainable development or it may be that the lack of structure underpinning the subject prevents us from getting to this level of detail in general discussion. After all, the time period over which the stakeholders will view a decision will vary from one to another. For example:

- Political support for a development in an area might be limited to the time of a term of office of an elected politician or party.
- Finance houses may view the development over the time required to get a pay back on their investment.
- Retail clients may view the development over the number of years they believe they have left before the market moves on elsewhere or the market has grown to the point where they need a new store or a major extension.
- A group of citizens may be interested in the development over their lifetime or the lifetime of their children.
- Planners may see the development within the lifetime of their 'master plan' or other such strategic document.
- Developers may view the development from the financial point of view but also what is happening in adjacent sites, regions, and even other countries and therefore a response to market conditions (in the markets in which they work) over the time it takes to create the development.
- Experts in demography will be interested in the changing age patterns around the development over a specified period related to, perhaps, government horizons.
- Lawyers might at one level might be interested in the development for the time it takes to sign off a contract, and/or at another level the length of time new legal business will exist and at another level the implications of changes in the law over a much longer time period.
- Valuation surveyors may be interested in the time taken to create an increase in property and land values.
- Architects will be interested over the lifetime of their commission but also the long term impact of their design as expressed in the building.

It can be seen that there is a variety of views of the time dimension from even this short list of potential stakeholders. If the aim is to create a harmony of view among all the participants then these different levels of interest over different time periods should be recognised as an essential aspect of the sustainable development process. It raises many questions of course, such as:

- Whose view should take priority in the case of a dispute? Is it the person or organisation which has the longest time interest in the development?
- Should it be the financiers, who take the major financial risk, that are considered to be pre-eminent in the decision-making process? If they are not, then will the finance become available to undertake any development?
- Should market forces be challenged as, in time, the markets will adjust to the new situation that faces them. However the time-lag may be too great to avoid irreparable destruction to the environment and is this acceptable?
- Is it the aim of sustainable development to avoid negative influences on the environment or is it to provide positive influences on what is believed to be a better way of living?
- Are our techniques for evaluation sufficiently sensitive to the way society views sustainable development?
- Is it easier to identify potential critical failure points in the quest for sustainable development rather than or as well as critical success factors?

Each of these questions contains the essence of a research question that at this stage of the topic is yet to be answered.

Time and innovation

Brand, in his book *The Clock of the Long Now* [1] provides an insight into the innovation cycle and the ability to respond to change. He describes a series of turning concentric circles, going from 'fashion', commerce, infrastructure, governance, culture and nature (see Figure 16.1). The outer layers show the fast cycles exemplified by fashion and technology where change is almost demanded by those engaged in the process. The lower layers show those areas that provide some stability and cannot quickly respond to change. The time scales for these lower layers can be centuries or indeed millions of years. For almost all of the life of mankind on earth, these matters have been in harmony and have balanced each other. It is only recently that these have become out of harmony and a potentially unsustainable environment has arisen. This does not mean that things did not previously change over time but only that they emerged and evolved without rapid deterioration of the environment.

The speed at which innovations occur in the outer layers provides much energy to those below, which can then speed up the whole. The fast layers innovate, the slower systems stabilize and the whole combines learning with continuity. In days gone by this may have created a harmony or balance that today is lacking. Changes in each system have varying timescales and it is time which becomes the moderator. It is essential to our understanding of what makes a development sustainable.

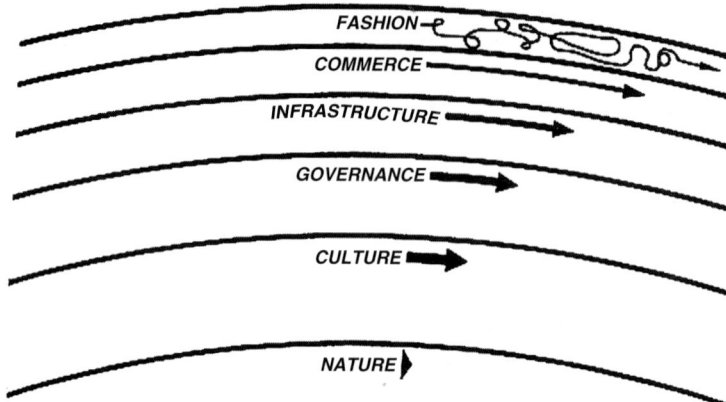

Figure 16.1   The Order of Civilisation. The fast layers innovate: the slow layers stabilize. The whole combines learning with continuity. Adapted from Brand, S., *The Clock of the Long Now*.

At the present time we may be experiencing the kind of malaise which Alvin Toffler called 'future shock' [2] caused by the future intruding too fast. The built environment can be found in all these layers because we impact or assist in the creation of them all, but our major contribution would be to 'infrastructure'.

Perceptions of sustainable development

There is within the human psyche a latent model of the world and the future that believes within a closed system, such as the Universe, that as time progresses less energy becomes available to be used and the system falls into decay. Entropy seems to be the fate of all closed systems. This model pervades our thinking and we think in terms of something being created, existing for a finite time and during that period of existence it will probably increase in energy before reaching a peak and then moving into decline. Figure 16.2 shows this in graphical form.

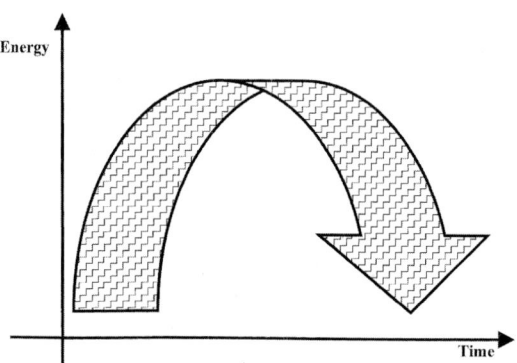

Figure 16.2   Entropy in closed systems.

Conventional wisdom within sustainable development seems to have this model behind the thinking. A development is undertaken, there is growth in that development, both in physical and social terms, and then it reaches a peak. For a further period of time it remains at this level and then it begins to decline for a wide variety of reasons and eventually it disappears as a recognisable development. This process may take thousands of years or it may be measured in tens of years or even less if a major catastrophe should fall upon the development. The purpose of sustainable development is to halt the downward decline and, if possible, to increase the availability of energy represented by social cohesion, physical well-being, bio-diversity, appreciation of the habitat and so forth that go to make up a sustainable community that, in turn, creates the sustainable physical environment in which the community lives.

Evidence for this pattern of events can be seen in a large number of the cities we see around us. They start as small settlements, they grow into larger conurbations with a strong social activity and then they decline, often as the result of the economic well-being of the country or context in which they find themselves. Often this pattern is repeated at the sub-city level with certain suburbs going into decline as crime and poverty begin to establish themselves. Others continue to rise as they become fashionable and they sometimes create barriers to entry from the unwelcome influences that exist in the poorer suburbs beyond even the financial cost of entry. In time, two societies exist side by side with tension between them and in some cases this tension is so great that it creates a social breakdown that can lead to the demise of both. These events are almost unpredictable until they are well into the decline phase of the graph. The potential for breakdown can be articulated but it is much more difficult to know exactly *when* this might occur in time.

If we are to address the Bruntland definition [3] of sustainable development then we have an obligation to leave the environment in at least the same position, and if possible, a better position, for future generations. We should not compromise their ability to make decisions about their future even if it means some short term sacrifice in the way we behave now. The problem is that it is difficult to get people to accept the concept of self-sacrifice when they are not the beneficiaries. Even in the short term we know that this is true because governments that tax to provide something better a few years ahead, or to aid the redistribution of wealth, often find themselves unpopular and voted out of office. This is where education and public participation have a major part to play. Education is required to develop a different culture with a set of values that reflect sustainable development, and public participation to enable all stakeholders to be informed and engaged in the planning process adopting these values.

In many situations there is considerable inertia to change. Plans are made, budgets are set, political mandates are established and, together with the desire of many for certainty and routine, there is a reluctance to alter the status quo. It is not until a real breakdown of social cohesion or security or quality of environment occurs that we see a willingness to alter direction or to make substantial investments. The danger here is that the breakdown may be irreversible and significant damage may have been done that may destroy the community and the stakeholders may well not be interested in doing anything about it. In the wider dimension of the earth's natural resources, for those resources that cannot be replaced i.e. they are non-renewable, it will be impossible within any time frame to do anything about it. In other instances, such as the destruction

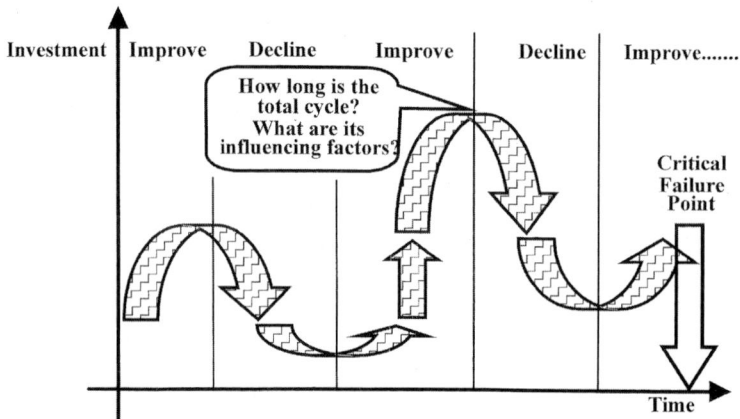

Figure 16.3   Hypothetical pattern of investment in the built environment over time.

of rainforests and other habitat, it may be possible to reverse the trend but there must be the political will. In the built environment it is difficult to envisage a total loss of the urban infrastructure as it can rise again as it has done for centuries – often one on top of the other! Its nature may be forced to change because of the scarcity of non-renewable resources that make up its physical presence but its ability to emerge again is always possible. What will be lost is some of the less physical aspects of the built environment such as its historical and cultural value, its use as a social integrator, its ability to be a focal point of religious significance, etc. The pattern of sustainable development in the urban context is often cyclical like the graph in Figure 16.3.

Of course the investment is not necessarily so abrupt or the decline so rapid as envisaged above but nevertheless the pattern is recognisable and can even be seen in our personal investment in our homes. We do not usually invest in a new washing machine when the original is still serviceable. Then, later, we might decide to buy a better one if our finances allow it or the decision is forced upon us because the original has completely broken down and is not worth repairing.

In cities the decisions are similar but of course they are much more complex and bigger in their impact. The decline of docklands in many parts of the world because of a change to container traffic and other forms of transport (a change in the technology) has resulted in considerable expanses of blighted urban landscape. More recently these landholdings adjacent to the docks have been seen as an opportunity by developers and have been rapidly developed as new conurbations, revitalising a derelict area. The London Docklands, Salford Quays in Manchester and the Albert Dock in Liverpool are all prime examples in the UK. In Bilbao it has been the Guggenheim museum that has revitalised the whole city and in Sydney, Australia, the Opera House has become an icon for the whole country. These transformations are almost always the result of a political will to get things done. In the early days of Salford Quays, which was originally the old Manchester docks, the local authority tried to sell the site without success. No developer would invest; in fact the developers approached were asking for money to take it off the authorities' hands. It was not until a new vision was created by a small number of like-minded individuals with influence, and the government changed its planning policies and began to invest in urban infrastructure that a new and successful life was given to the area. Now it is a prime development site and has wonderful new cultural buildings with each new development reinforcing the others' success.

The question for all these developments is how long will they last before they move into decline? In fact, of course, nobody knows. A serious downturn in the economy creating a lack of tenants followed by a lack of maintenance and security could quickly see the beginning of a demise. If war should break out on a large scale then again it is difficult to predict what might happen. Sustainable development can only survive while all the external factors that bear upon the development are in harmony. A failure in any of the major factors could well bring the whole of the development into crisis. The aim of sustainable development seems to be to ensure that the overall pattern of investment into an area continues in an upward direction even if we have to accept that there will be fluctuation in the upward graph caused by normal investment cycles. It is worth noting here that investment in this context is being used in its widest sense to include any input of resources, whether it be labour, finance, infrastructure, arts, social welfare or whatever is required to sustain or improve the built environment.

Critical failure points

In the majority of decision-making strategies relating to the built environment the people making the decision are driven by 'critical success factors' (CSFs). They look for the returns and the key ingredients that will make or break the development. This is the basis for some of the sustainability indicators that are used. In sustainable development these positive attributes still hold good but at the same time it may be important to give equal attention to the critical failure points. These are the factors that if they fail or do not exist could lead to a rapid decline in the sustainability of development in general and possibly the demise of the whole scheme or area. The type of issue which may be of this nature includes:

- The loss of a key resource such as water. In India the city of Fatapur Sikri near the Taj Mahal lasted fifteen years because the water supply dried up and the technology was not available to do anything about it.
- The loss of the major employer in the region can destroy the local economy and the ability of the community and its infrastructure to survive. Examples of this include some of the towns built around coal mines that closed, or steel works that became part of a concentration of production elsewhere.
- Pollution of air, land or water, if on a long time scale, can mean that an area becomes uninhabitable. Examples are toxic chemicals in the land, pollution of sea water depriving the fishermen in an area of their livelihoods, or acid rain destroying forests.
- A breakdown in law and order, which can mean that property values fall and the residents become trapped in a cycle of decline or, if they are financially able, they move to other places and no one wishes to take the place they have vacated.

- A breakdown in the commitment of a community due to a challenge to the faith that has been practised there. Towns in the former Yugoslavia identified as Muslim or Christian or towns built around a religious order find themselves vulnerable if the basis of that community is challenged.

It can be seen from the above that most of these issues are related to well-being and the quality of life. Some of them such as the pollution or loss of a key resource are secondary to the need to have a quality of life, however that is defined, although ultimately these issues impinge on the enjoyment of life anyway. It should be possible to overcome these matters but it requires substantial resources or a level of technical competence that the community might not have. It is, therefore, better for the community to move elsewhere and thus avoid the problem. By moving, their quality of life is expected to improve. In the context of this discussion on 'time' in sustainable development they complicate the issue. We do not know when these movements might occur and in many cases we will not know with certainty the underlying causes. It is, therefore, difficult in developing a model of sustainable development to prejudge when we can expect a critical failure point to manifest itself. All that we can do is ensure that, as far as our knowledge exists today, the circumstances which might lead to such an eventuality are avoided or mitigated through the process of development.

The majority of 'failures' are not critical in this respect. They do not result in sudden failure. In general there appears to be a spiral of decline, a vicious cycle, where a lack of investment, a period of disinterest by the current and potential stakeholders or a lack of economic well-being, in particular, can result in gradual decline. Eventually the possibility is that it is no longer feasible to create a virtuous spiral that will build the community again and result in a sustainable solution. Again, the time scale for this is unknown. We cannot predict with certainty how long this will take and we often do not know when the stage of non-renewal has been reached. So here we have a strong psychological urge to ensure that we create something sustainable and yet we cannot predict the events that will create the environment where this demise will take place nor can we predict the timescale over which it is likely to happen. The two are, of course, related.

Kohler [4] in his work on the life cycle analysis of cites has suggested that there might be a corridor of solutions that need to be examined and evaluated over time within which any decision-making process should be aware and keep within (see Figure 16.4). It is possible to overshoot as much as undershoot and the job of the decision maker is to keep in balance all the contributing factors. Critical success and critical failure are therefore built within this framework. Of course, even with dereliction there is usually the opportunity to build again but the economic, social and other costs are that much larger.

Another view on this problem is the changing time scales for renewal within a process of an emerging and evolving city, which can complicate the time when decisions *can* be made. In a complex organism such as a city they make the task of addressing the sustainability problem that much more difficult. (See Figure 16.5.)

This suggests that we have to address 'time' in some other way, not as a measure but as a continuum within which we learn and improve. This is not unlike the arguments

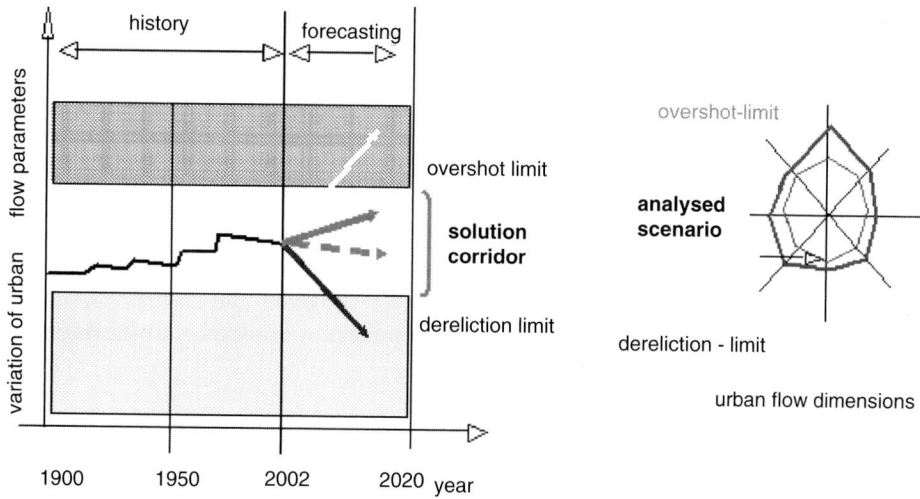

Figure 16.4   Solution Corridor for the Development of a Modern City (Kohler, 2003).

being put forward by those advocating the concept of the learning organisation within business. Senge in *The Fifth Discipline* [5] describes 'learning organisations' as

> *'organisations where people continually expand their capacity to create the results they truly desire, where new and expansive patterns of thinking are nurtured, where collective aspiration is set free, and where people are continually learning how to learn together.'*

It appears that the act of learning and sharing the results of learning can lead to a corporate view of the problem and the solution that allows for more creative ideas and

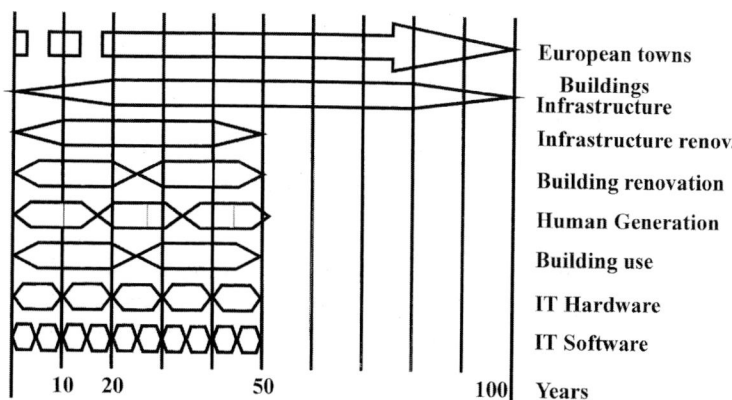

Figure 16.5   Cycles of Transformation (Kohler N., IntelCity Workshop Sienna, 2003).

a positive attitude to the aims of the organisation. He also goes on to say that

> *'learning disabilities are tragic in children, but they are often fatal in organisations. Because of them, few corporations live even half as long as a person – most die before they reach the age of forty.'*

This approach requires a move to systems thinking that we shall address later. For now, it is worth noting that a focus on working and learning together is thought to be beneficial to organisations and as the built environment is an organisation of a sort, there may well be lessons that can be learnt for achieving a sustainable development. If we do not do this then the cycles described are likely to continue and we can expect failure on a regular basis.

Time in evaluation

Even with a learning organisation approach and the focus on the process it will not be possible to ignore the effect of time in our evaluation and assessments. As we have said earlier, most of those authorities with financial or political power will want to have proposals justified in order to persuade committees or shareholders or boards or whatever group to which they are accountable. This inevitably means that some form of risk assessment has to be made and this is a recognition that we cannot predict or control all future events.

In economic evaluation the concept of discounting is used to take account of the effect of time on the view of the investor at the present day. In simple terms the view is held that the value of a payment or receipt in the future is worth less now because, in the case of a payment, a smaller sum of money could be set aside now, which could grow over time to meet the needs of that payment at the time specified. In the case of a receipt, then the value to the recipient now is worth less because of the interest lost on that amount of money while they are waiting to receive it.

What this means is that if you are set to receive $100 in 25 years' time, its value to you now is $29.50 if you expect your investments to produce a return of 5% and $9.20 if you expect a return of 10%. (See formula later.) In other words, if you invested $9.20 now in a bank or other financial concern and a return of 10% was guaranteed then it would accrue, with interest added, to $100 in 25 years' time. Notice that the higher the interest rate, the lower the value to you now or the smaller the amount that needs to be invested now to accrue to the same figure. The time remains the same but the effect on the supposed value is quite different depending on the interest rate. It follows that if we were to use a higher rate of interest in our calculations we would be discounting the effect of future transactions more than if we used a lower rate. If in sustainability we want to take a long term view and encourage this within our calculations then we would use a small rate of interest because this would make future activities appear more important in financial terms.

The choice of interest rate is therefore critical and is more complex than it first appears. No account has been taken of inflation and this might have a substantial impact on the calculation. It might, for example, eat away at the real benefit from the investment over time and some would argue that the 'real' rate of interest that should be used is the assumed rate less the inflation rate. Others would argue that inflation can be ignored as it affects both income and expenditure equally. This may or may not be true as differential inflation is quite common. It is also quite clear that long term periods of stable interest rates and inflation are almost non-existent within the time scales of the built environment.

The time element is also a major consideration. The formula for computing the present value is based on the compound interest formula and is presented as

$$\frac{1}{(1+i)^n} \tag{16.1}$$

where i is the interest rate divided by 100 and n is the number of years.

It follows that there is an exponential curve that rapidly discounts future values as time increases. The result is a model of the world upon which decisions are made that is based on a view that suggests that the future is something to be discounted and is of considerably less value than the present. It is mechanistic and uses few variables in its operation. Nevertheless, much of financial investment is based upon it. It has replaced some of the other models such as 'pay back' where it is the length of time required to pay off the original investment that is the criterion, because it is thought to reflect the logic of the financial markets more accurately. However, even in this assumption it may not reflect the real values that investors adopt within their decision-making processes.

There is a desperate need for a new economic order that will value the human and environmental capital alongside the economic one, be found on company balance sheets and be valued by society as a necessity.

Future aversion

It could be argued that when time enters a calculation then most people are likely to prefer present over future gains (this could be termed 'future aversion') and future losses over present losses (this could be termed 'future seeking'). However, there is unlikely to be symmetry between the two and it is plausible on intuitive grounds that a postponed loss is less aversive then a postponed gain of a similar amount is attractive. This can be shown diagrammatically as in Figure 16.6. In sustainable development

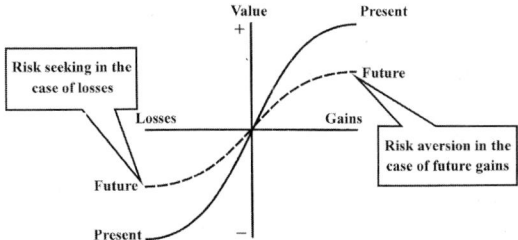

Figure 16.6   Risk seeking in the case of losses and risk aversion in the domain of gains [6].

where the emphasis is on reducing future losses this asymmetry could be important in reflecting the psychology of the decision maker within the technique. (See Kahnemann and Tuersky [6].)

If there is such a view at work in the mind of the decision-makers, then it does have relevance to sustainable development and the way in which those who are encouraging sustainable development are prepared to argue for different models that allow a longer term perspective to be addressed. It would suggest a move away from the conventional economic models to the adoption of a moral imperative that will demand that future values are given significant weight. This could be done in some cases by legislation and regulation that require minimum standards to be kept – say the reduction in major pollutants, or it could be that business advantage is achieved by taking the long term view.

There are already instances in banking where banks that take an ethical stance in their investments have managed to increase their performance substantially. However, this may be the absorption of a niche market of those sensitive to these issues. Nevertheless, it is a start and with further education in these issues it may be that the minority niche market becomes the main stream. Directives such as those contained in the Agenda 21 documents and adopted by many authorities throughout the world will hasten the take-up of a longer term assessment. There is little doubt that it will require a variety of approaches to ensure that the concepts of sustainable development are included as the norm in addressing decisions in the built environment.

Clever or wise?

Patricia Fortini Brown in *Venice and Antiquity* [7] draws to our attention to the fact that the ancient Greeks distinguished two kinds of time, 'kairos' meaning opportunity or the propitious moment and 'chronos' meaning eternal or ongoing time. 'While the first ... offers hope, the second extends a warning'. *Kairos* is the time of cleverness, *chronos* the time of wisdom. Our dead and our unborn reside in the realm of *chronos*, murmuring warnings to us presumably, if we would ever look up from our opportunistic, *kairotic* seizures of the day. Today we live in the golden age of *Kairos*, where opportunity is all and where the cult of the individual is paramount and where the corporate sense that will allow us to engage with time is hard to come by.

Practical assessment of 'time'

This discussion has revealed some of the issues relating to 'time' within decision-making. It has not, however, put forward a proposal that can be used on a day-to-day basis to address the matter. This is because no one method exists. In fact when the situation is analysed fully, it is realised that it is difficult to obtain a universal view of time scales for something as varied and complex as the built environment. Boulding [8] diagnosed the problem of our times as 'temporal exhaustion' – 'If one is mentally out of breath all the time from dealing with the present, there is no energy left for dealing with the future.' She proposed a simple solution: expand our idea of the present to two hundred years, a hundred years forward, a hundred years back. A personally experienceable, generations-based period of time, it reaches from grandparents to grandchildren – people for whom we feel responsible, thus allowing human nature to support the longer term perspective.

Whatever scale we choose it would seem that a philosophy is more appropriate than a range of techniques. It is more about a way of behaving within a framework that is conducive to the objective to be achieved. One of the attempts to outline a philosophy such as this in a simple way that all could understand, was that of Alex Gordon's 3 L's concept, 'Long life, loose fit, low energy' [9] when designing buildings. It had no quantitative measures but it provided a frame of reference within which it was possible to begin to collect quantitative evidence and then seek improvement. It then enters the realm of the learning organisation where the call for improvement becomes the watchword of the organisation – in this case society. The questions asked are then: 'Will this building development last a long time or longer than previous developments?' 'Will it be easily adaptable to change in the future to avoid using up non-renewable resources either in extraction or use?' and 'Will it use less energy in manufacture and operation than similar types of building?' Once we get into this frame of thinking we begin to devise the techniques and measures that are appropriate to this view of the world. It provides a belief system to which those who adhere to this belief can respond and justify their behaviour. If it becomes the mantra of the many, then it becomes politically unacceptable to follow a different path and it is adopted within the culture. In the 3 Ls concept, two of the drivers have time as a key feature and so it begins to permeate the thinking of the many. What seems appropriate for a building soon becomes the view of the planners and the local authorities and begins to have significance for the district and then the city. The formation of a virtuous circle has begun.

References

[1] Brand, S. (1999) *The Clock of the Long Now: Time and Responsibility: The Ideas Behind the World's Slowest Computer*, Basic Books, New York.
[2] Toffler, A. (1985) *Future Shock*, Pan.
[3] World Commission on Environment and Development (1987), *Our Common Future*, Oxford University Press, Oxford.
[4] Kohler N. (2003) Presentation: IntelCity Workshop Sienna (under the auspices of University of Salford, UK).
[5] Senge, P. M. (1990) *The Fifth Discipline*, Doubleday Publishers, 3.
[6] Kahnemann, D. and Tversky, A. (1984) Choices values and frames, *American Psychologist*, 39(4), 341–350.
[7] Fortini Brown, P. (1996) *Venice and Antiquity: The Venetian Sense of the Past*, Yale University Press, New Haven, CT, 6.
[8] Boulding, E. (1978) The dynamics of imaging futures, *World Futures Society Bulletin*, September 1978, 7.
[9] Gordon, A. (1974) The Economics of the 3 L's Concept, *Chartered Surveyor B & QS Quarterly*, RICS, Winter 1974.

17 Is 'smart' always 'sustainable' in building design and construction?

J. Bell[1]

Summary

The terms 'smart' and 'sustainable' are frequently used together in the context of buildings. However, smart buildings and smart construction are not necessarily synonymous with sustainable buildings and sustainable construction. Smart is often used to refer more to responsiveness of the building, in particular through the use of information technology and control systems than to using materials and design in 'smart' ways.

The key focus in this chapter is sustainability – how can we achieve sustainability in construction of buildings and in the operation of buildings. However sustainability goes further than environmental sustainability, and includes the economic sustainability and social sustainability of the building (the triple bottom line). Within this framework, ways in which smart technology, and smart processes, can be used to support sustainability in any of its manifestations will be explored. There are also aspects of smart technology that may not support sustainability, and examples where the two notions are in conflict will be discussed.

Introduction

Both 'sustainability' and 'smart' as terms and concepts today signify a range of meanings, attitudes and frequent misappropriations. It is useful therefore to start by briefly reviewing the origin of the terms in order to clarify some 'bottom line common denominators' on which this chapter is based.

Sustainability

As an international concept the notion of sustainable development first came to prominence with the publication of the report, *Our Common Future*, which in turn generated the UNCED Conference in Brazil in 1992, The Rio Declaration and the accompanying implementation plans of Agenda 21. *Our Common Future* clearly articulated that all was

---

[1] Queensland University of Technology, Australia.

not well with first-world development processes and that new concepts and forms of development were required to meet future human needs and to ensure the maintenance of the bio-physical systems of life on which we all depend. The core principles of sustainable development which were spelt out at that time – such as the need to preserve biological integrity, social equity and intergenerational equity – were encapsulated in the definition of sustainable development as defined by the Brundtland Commission, 'Development that meets the needs of the present without compromising the ability of future generations to meet their own needs' [1].

The accepted use of the term sustainability is evolving as we begin to understand the very broad impacts of the Bruntland definition of the term. The definition of sustainability does not discuss any specific technologies, or modes of development, but in order for us to implement sustainability we need to construct more specific understandings of the concept in order to, for example, construct a 'sustainable building'.

In the case of buildings, there is now a concept of environmental and ecological performance – or the footprint of a building – that is being used to define sustainability in a way that allows us to implement construction of sustainable buildings. In reality our ability to determine an ecological footprint for a development is really just an early stop on the road to sustainability, which requires a comprehensive cultural change to be fully realized.

To put this last point in perspective, we consider energy use in buildings. This is a major source of greenhouse gas emissions in Australia, and the majority of energy use across Australia is in residential buildings (12% of total energy use, compared with 6% for commercial buildings). How can we reduce the environmental impact of this energy use? The simplest way is to reduce the amount of electricity and gas used to heat hot water (30% of domestic energy use), and for lighting, heating and cooling. The technologies to achieve this exist and are commercially available – solar hot water systems, energy efficient lighting, improved building design and insulation to reduce the heating and cooling needs, etc. However these solutions are not yet widely adopted, and it is through widespread adoption of these technologies that we can make an impact. If every household in Brisbane (just one city of 1 million residents) replaced a 60 W incandescent globe that was commonly used (say 3 hours per day) by a compact fluorescent with equivalent light output (11 W), the total saving in energy would be 13 MWh per annum, with a value to the consumers of $1.3 million annually. Why is this not done? One reason that has been identified is that the initial cost of the compact fluorescents is significantly higher than the incandescent globe, and this is paid up-front, while the saving in energy bills for each consumer is small, and is seen only quarterly – once one has forgotten the initial cost of the compact fluorescent. One method of changing this cultural behaviour will be explored later in this chapter.

Smart

One of the first applications of the term 'smart' as applied to a technological product was to materials [2]. 'Smart materials' are seen as materials that can respond to the environment and provide some means of feedback or control on the environment (see, for example, Reference 3). A very good example of this is the Smart Window. The term was introduced in the mid 1980s by Claes Granqvist to describe optically switchable electrochromic glazings. These devices exemplify the fundamental characteristic of all 'smart materials': controllable variation of some property of the material (in this case the optical transmittance of the window). The control of the process is accomplished by the application of a small voltage (1–2 V) (or the passing of a small current), leading to the ability to control the transmittance of, for example, a building window [4].

The term smart building is now widely used (see, for example, Reference 5). The term is used to describe complex control systems integrated into building operations. The real benefit of a complex and integrated control system can only be realized if the appropriate network of sensors and actuators (for lighting, HVAC systems, fire and emergency systems, etc) is also in place, and this is the trend in 'smart buildings'. Alternative terms, such as 'electronically enhanced buildings' [6], have also been suggested. However the addition of individual smart technologies to a building is in itself not sufficient to make a building smart, and the degree of integration of these smart components is an important component of 'smartness' [7].

Paevere and Foliente [8] have identified four primary elements that should work in an integrated way in a smart building system: sensors; integrated information management system and performance models; actuators; and, importantly, the backbone or nervous system (i.e. the communications infrastructure that connects the sensors, actuators and control systems together).

'Smart houses' have also emerged in the last few years. These houses are often built as demonstrations of sustainable house design, emphasizing 'smart design' in the orientation, materials, and architectural design features of the house. Examples include the Tweed Shire Council Smart House in New South Wales [9] and the Research House in Rockhampton, Queensland [10]. However neither of these houses demonstrate one essential feature of smart buildings as discussed above – the use of complex control systems and sensors to enable the building to respond to changes in the external environment or to user needs. Indeed, one definition of smart housing [11] is quite explicit:

'Smart Housing is good practice in designing, planning and building homes to make them more socially, environmentally and economically sustainable.'

This use of the term 'smart' does not converge with either the ability of a building, component or material to change in response to an external stimulus. It rather refers to the ability of people (the owner, architect/designer and builder) to use human intelligence to create a house exhibiting best practice across a number of dimensions, including environmental, social and economic dimensions. Good building design is clearly critical to create true smart buildings, as it has been identified many years ago [6] that technology should add value to a building, and this cannot be achieved by adding even the best technology to a poorly designed building. However, a well-designed house that uses passive technology to achieve sustainable outcomes is not a smart house.

One component of smart housing which is becoming more common is 'smart metering', where individual energy-consuming features of a house are monitored separately to identify to the occupants their energy use [12]. Modern smart meters are able to display cumulative energy use for monitored items, and these definitely use the features of smart technology, and can contribute to the ongoing sustainability of a building [12].

More recently, the term smart has been applied to a much wider range of products and concepts. The term 'smart glass' has been used to describe a range of glazing products that do not show any variable or controllable characteristics, but the system devised by the manufacturer to select the glazing appears to contain the 'smartness'. Even more tenuously, many manufacturers are defining consumer products that incorporate computer technology as smart, even if the most identifiable feature is the ability to switch the product on or off remotely using a mobile phone! These technologies and products are not part of smart buildings unless they are linked into a wide range of technologies in the building and their operation is linked to the overall building performance.

Convergence of concepts

Can the terms smart (even using the restricted definition above) and sustainable really coexist comfortably in describing buildings? Or are the concepts so different that we should not use them together?

The remainder of this chapter clearly describes situations in which the two terms do fit together very well. Identifying these situations and expanding the use of the smart technology that enhances sustainability should be a focus of our efforts.

This section will identify three quite different aspects of building performance that can dramatically benefit from the addition of smart technology. All the examples also address sustainability.

Example 1: Greenhouse emissions

Greenhouse emissions from the Australian building sector are substantial and rapidly rising, particularly in the commercial building sector. This is clearly unsustainable in the context of global warming and the imperative to reduce greenhouse emissions to mitigate global warming. Leadership and innovation by the engineering and design community is urgently required to reduce further greenhouse emissions growth within the sector, as the following data highlight [13].

Under business-as-usual assumptions, building operating energy emissions are projected to increase to 148% of their 1990 levels by 2010. This is much higher than the 'Kyoto target', which allows an increase of only 108% on 1990 levels by 2010. Emissions from commercial building energy use are projected to increase by 95% between 1990 and 2010, compared with 17% for residential building energy use. The greatest increases in emissions are projected to come from commercial HVAC and lighting.

The challenge as presented above is formidable because it represents a national problem, and therefore a broadly-based response. The good news is that dramatic reductions in energy consumption and therefore greenhouse emissions can be achieved by the design and engineering community, with current technology. In the commercial building sector, industry experience clearly indicates that operating energy improvements of between 30 and 70% can be readily achieved on current standard industry practice. For example, in Australia the average lighting energy in commercial office buildings is about 16 W/m^2, but could be reduced to 8 W/m^2 with current technology, and 5 W/m^2 with some emerging technologies, with no reduction in lighting quality. This would amount to cost savings of millions of dollars to building owners, with substantial reductions in greenhouse emissions.

One approach to reducing lighting energy is the use of illuminance sensors, occupancy sensors, and advanced dimming or switching mechanisms in conjunction with high efficiency lights, and good design to maximize daylight penetration into buildings. An illustration of the benefits that can be derived from application of smart control technology, and the use of dimmable lights (activated by the interior illuminance levels) and switchable windows is shown in Figure 17.1. This shows results of simulated building energy usage for an office module 4 m wide × 6 m deep and with a 3 m ceiling height in Perth, Australia [14]. The graph indicates both the complex interplay between the different elements in the building – lighting and cooling energy and, as one of the control parameters, the irradiance on the switchable windows at which the windows switch between their clear state (low irradiance) and their dark (low transmittance) state. Approximately 50% of the lighting energy (full power 12 W/m^2) is saved by the linked dimming of the lights and switching of the windows under optimum conditions. However, neither the cooling energy nor the lighting energy is at its minimum level, indicating the complex interplay between the parameters that needs to be considered to achieve the optimum energy performance.

Much of this technology has been available for many years, but it has not often been used successfully because of the limited number of sensors, or large switching

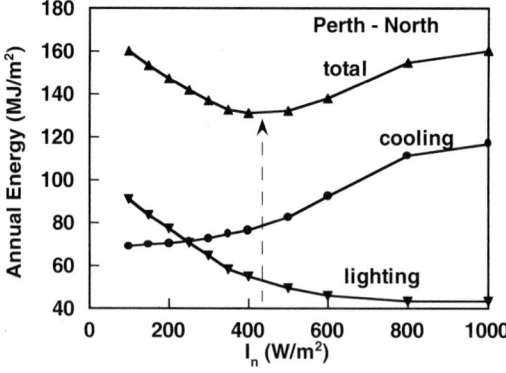

Figure 17.1   Results of simulation of office energy use with an electrochromic smart window, showing the effect of one possible control parameter (the normal irradiance on the window, I_n) on range of total energy use (see [14]).

regions in buildings. New developments in sensor technology, leading to larger numbers of networked sensors, controlled by much more sophisticated computers, and switching or dimming individual luminaires is leading to more significant reductions in lighting energy. To achieve maximum savings the use of embedded sensor and control technology and network protocols will be required, and many of these are now commercially available [15]. An example of these is the BACnet standard, recently adopted as Standard 16484-5 by the International Standards Organization (ISO, Geneva, Switzerland).

The implementation of smart lighting technology leading to a (conservative) 50% reduction in lighting energy in commercial buildings as demonstrated above would result in a cumulative saving of 40 MT CO_2 between 2003 and 2010. This is sustainability in action – using good (but not really smart) lighting technology, and a smart control system.

Similar savings can also be achieved through control of HVAC systems on a fine-grained scale, and using multiple control strategies. For example, appropriate use of outside air (when the external temperature is below the interior temperature, but cooling is still required) can minimize cooling energy requirements. However it does require appropriate sensors and damper controls, managed by an intelligent control system.

Example 2: The Pentagon

A more interesting example of the use of a smart building control system is the protection of the building infrastructure. The Pentagon Centre in Washington had a new Building Operations Command Centre commissioned in June 2001 [5]. This provided a control mechanism for thousands of sensors and actuators in the HVAC, lighting and safety systems of the complex. In the aftermath of the September 11 terrorist attack, when an aircraft was crashed into the building, the control system was able to prevent the spread of fire by closing dampers and turning off fans.

Is this an example of sustainability? The answer is clearly 'yes'. Damage to the fabric of the building and the fittings was reduced by the ability to contain the fire. This resulted in limiting the replacement of materials in the building. Material usage is a key factor in the environmental footprint of a development, so this use of smart technology definitely enhanced the sustainability of this facility. In addition the Pentagon BOCC is projected to reduce energy use by 35% by 2005, contributing to reductions in the operational component of the Pentagon Centre's environmental footprint.

Example 3: Building integrated photovoltaics

The use of photovoltaic (PV) cells on buildings is clearly sustainable (provided the lifetime of the cells is sufficient). The PV cells generate renewable energy on-site, eliminating greenhouse emissions produced in electricity generation and energy losses in distribution. However is this smart technology? The PV cells produce energy that can be used as part of the building's operational energy demand, and therefore also can

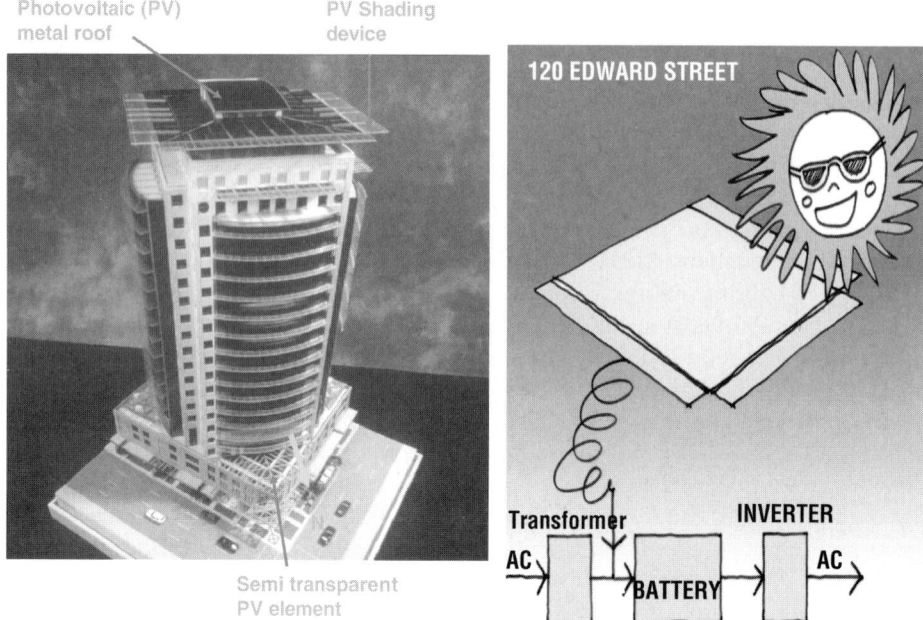

Photovoltaic (PV)
metal roof

PV Shading
device

120 EDWARD STREET

Transformer   INVERTER

AC   AC

BATTERY

Semi transparent
PV element

Figure 17.2   120 Edward Street building integrated photovoltaic system.

be (and arguably should be) linked into the building energy management system. If further the PV cells are integrated into the fabric of the building then they can offset the cost (both financial and environmental cost) of the building components that they replace. This adds additional dimensions to the sustainability of the PV cells.

An example of building integrated PV is the 120 Edward Street building in Brisbane, Australia, where photovoltaic cells are integrated into the roof and shading elements of the building [15], as illustrated in Figure 17.2. The energy provided by the PV cells is further integrated into the energy distribution system within the building by feeding the direct current output of the PV cells into the DC Bus of the buildings Uninterruptible Power Supply (UPS) [16]. While this installation is very well integrated with the building and its systems, the annual energy generation of approximately 80 MWh is only about 4% of the actual savings generated in this building through good design and other technological improvements.

An alternative approach to PV integration in buildings is illustrated in Figure 17.3, where an entire façade of the CSIRO Division of Energy Technology Building in Newcastle has been glazed using semi-transparent dye-sensitised photovoltaic cells [17].

Example 4: Smart technology and feedback

The examples above suggest that smart technology supporting the goals of sustainability is available and sufficiently well understood to enable large scale use of smart

Figure 17.3   The dye-sensitised solar cell facade at the CSIRO Division of Energy Technology in Newcastle, Australia.

technology in the home to lead to the sort of savings indicated above, through demand-side management of both energy and water consumption in houses. However this is not happening. Barriers to demand-side management are many and varied, ranging from policy barriers, such as inappropriate subsidies to retailers to encourage increased use of utilities, to lack of information about the impacts of certain behaviour [18].

A detailed study by Wood and Newborough [19] of household behaviour under the influence not only of smart metering, but specifically-designed feedback on energy use has shown that with appropriate feedback households were able to reduce their energy consumption. Feedback has several functions that make it much more effective than information. If it is sufficiently immediate and specific [20], then feedback can [19]:

- have a learning function
- be habit forming
- internalize behaviour.

By developing smart technology that provides feedback in a suitable way about consumption of electricity or water – and this really means in a format that can be understood by the average householder – then the smart technology can change individual behaviour and take on the biggest challenge to implementing the principles of sustainability – the challenge of changing our behaviour.

Conclusions

'Smart' and 'sustainable' can be a very appropriate combination of terms to describe buildings, as is demonstrated by the above examples. Sustainability is a very complex

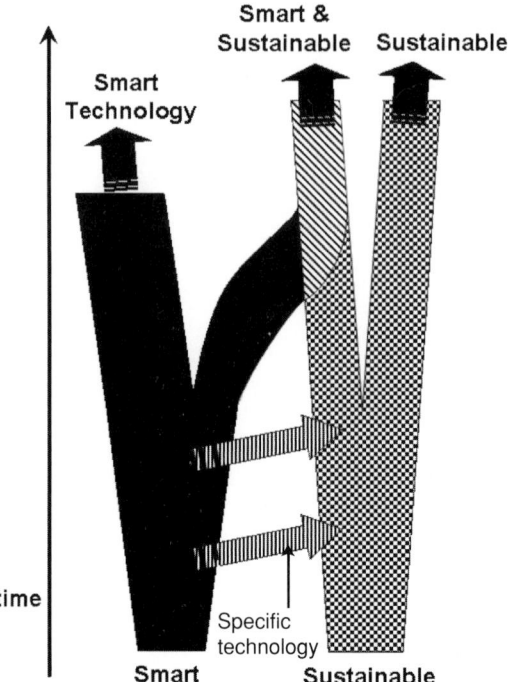

Figure 17.4   The convergence of parts of the smart and sustainable domains.

concept, with many facets, and many interrelated objectives. Sustainability does not require the use of smart technology, and buildings that demonstrate sustainability (or in more explicit language, that minimize the impact of the building on future generations) do not need to be smart buildings. All sustainable buildings require good design, based on sound architectural principles and incorporating basic features such as appropriate solar access, insulation and a good choice of materials.

The convergence of these terms is illustrated in Figure 17.4. This shows how some aspects of smart technology have delivered specific ideas and technology to deliver sustainable outcomes and how, more recently, there has been a much wider convergence of the terms in relation to some aspects of sustainability.

Significant value can be added to such a building, particularly a commercial building, through the use of smart technology to control and manage the complex interplay between the various subsystems in a building, including the occupants. Smart technology is not however sufficient, even when added to a well-designed building. Models of performance, and complex control systems are critical parts of smart buildings that enable them to improve all aspects of the sustainability of a building – the economic, social and environmental sustainability.

There is still much for us to understand in the ways that all different aspects of a building function, and improvements in building sustainability will continue to be found through these improved models coupled with smart technology. In particular, as discussed above, smart technology also has the potential to start influencing the behaviour of building occupants in the way that they use resources. Changing behaviour is the only way to deliver real improvements in sustainability, as users already have the choice of sustainable technology, but are not taking it up as rapidly as other forms of technology.

References

[1] Greene, D. (1997) *Towards Sustainable Engineering Practice – Engineering Frameworks for Sustainability*, prepared for the Institution Engineers Australia Task Force on Sustainable Development, Publication no. 18/97, IEAust, Canberra.

[2] Newnham, R. E. (ed.) (1994) *Smart Materials*, MRS Bulletin, Volume 18.

[3] Schwarz, E. (ed.) (2002) *Encyclopedia of Smart Materials*, Wiley, New York.

[4] Bell, J. M., Matthews, J. P. and Skryabin, I. L. (2001) Smart windows, in *Encyclopedia of Smart Materials*, Schwarz, M. (ed.), Wiley, New York, 1134–1146.

[5] Snoonian, D. (2003) Smart buildings, *IEEE Spectrum*, August, 18–23.

[6] Kroner, W. M. (1997) An intelligent and responsive architecture, *Automation in Construction*, **6**, 381–393.

[7] Arkin, H. and Paciuk, M. (1997) Evaluating intelligent buildings according to level of service systems integration, *Automation in Construction*, **6**, 471–479.

[8] Paevere, P. and Foliente, G. C. (2003) Smart buildings for healthy and sustainable workplaces: scoping study report, CRC for Construction Innovation, draft report.

[9] Tweed Shire Council Smart House, http://www.tweed.nsw.gov.au/agenda21/SmartHouse.htm

[10] The Research House, http://www.housing.qld.gov.au/builders/research_house/index.htm

[11] Queensland Department of Housing http://www.housing.qld.gov.au/builders/smart_housing/index.htm

[12] Engineer with a target in his sights, (1998) *Australian Energy News*, Issue 8, June (see also: http://www1.industry.gov.au/archive/pubs/aen/aen8/8barton.html)

[13] Australian Greenhouse Office (1999) *Australian Residential Building Sector Greenhouse Gas Emissions 1990–2010*, and *Australian Commercial Building Sector Greenhouse Gas Emissions 1990–2010*, Australian Greenhouse Office, Canberra.

[14] Bell, J. M., Moore, I. and Skryabin, I. L. (1999) Optimal control of electrochromic windows for minimum energy use in commercial buildings, *Proceedings of Solar 99 – Opportunities in a Competitive Marketplace*, ANZSES, CD.

[15] Encelium Technologies Incorporated, http://www.encelium.com (November 2003).

[16] Ridgeway, J. (2001) 120 Edward Street, paper presented at 'Facades' Workshop, July 2001, Queensland University of Technology.

[17] Tulloch, G. E. (2003) Light and energy dye solar cells for 21st century, Osaka Symposium on Dye Solar Cells, Osaka, Japan, 25 July 2003.

[18] Energy Futures Australia (2002) Mechanisms for promoting societal demand management, Research Report No 19, prepared for IPART (NSW), (see also http://www.ipart.nsw.gov.au).

[19] Wood, G. and Newborough, M. (2003) Dynamic energy-consumption indicators for domestic appliances: environment, behaviour and design, *Energy and Buildings*, **35**, 821–841.

[20] Seligman, C. and Darley, J. M. (1977) Feedback as a means of decreasing residential energy-consumption. *J. Appl. Psych.*, **62**, 363–368.

18 Balanced value – a review and critique of sustainability assessment methods

Tom Woolley[1]

Summary

This chapter grew out of an aspect of ongoing UK research, funded by the UK Department for Trade and Industry, which aims to produce a framework and toolbox to assist in the selection of materials, components and construction systems. The framework will address the issue of balancing sustainability and functional performance, considering environmental, social and economic issues, the 'triple bottom line'. It is not intended to produce yet another methodology, but to draw on existing tools and to signpost designers and specifiers to the wide range of advice and information that is available.

However it has become apparent, when reviewing over 100 tools and assessment methods worldwide, that there is both a great deal of commonality and a great deal of disparity between them. It will be argued in this paper that a halt should be called to the continued proliferation of yet more and more environmental assessment methods without some kind of international agreement on common baseline standards and methods.

Currently there is a great deal of variation in methodologies, assumptions and measurements while there is a dearth of readily accessible databases of basic information. Many tools cherry-pick criteria and standards are even presented as absolutes when they have been mediated by local political and economic pre-requisites.

The issue is explored further by examining a number of recent pieces of work in which we have compared different assessment methods and tools.

Introduction

As 'sustainable construction' achieves worldwide acceptance as a policy direction, defining what it means becomes increasingly important. Substantial credit for this globalization of green building has to go to Green Building Challenge in that GBC has encouraged individuals and governments to submit schemes for international

---

[1] School of Architecture, Queens University, Belfast, UK.

recognition. However, even for the most casual observer, it has been clear from the GBC exhibitions that there is a substantial difference between what might be regarded as sustainable in one country and what is sustainable in another. Some are comfortable with this variation, arguing that local conditions vary and different criteria will apply according to local circumstances. However we all live on just the one planet, even if we are trying to consume three or four! Excessive consumption of energy and resources in Paris will have just as much impact on pollution of the seas and depletion of the ozone layer as a building in Japan or Rio de Janeiro.

Given this rather obvious truism, it seems rather surprising that there has been little harmonization of international standards for sustainable building and it might appear that a new benchmark or tool for sustainable buildings appears every week somewhere in the world. This paper is concerned with what can be done about this proliferation and what it means in simple practical terms for designers, clients and governments who want to improve their environmental performance. I will then try to unpack some of the reasons for the plurality and discuss what might be done about it.

Variation of interpretation

A review of the literature on green building makes it clear that there are a wide variety of interpretations available. Some are very technocratic, whereas others are more concerned with *cosmic forces* and *energy fields* [1]. Few adopt a holistic approach, seeing environmental issues as largely about energy or building services [2, 3], while others focus largely on low impact materials such as cob or earth [4]. This plurality of approach can also be found when looking at environmental assessment tools and methods.

The literature on comparison of environmental standards is surprisingly thin. A review of Green Building Challenge [5] is an important attempt to compare a small number of tools, BREEAM, LEED, Eco profile, ESCALE, HK-BEAM and Ecoeffect. These were selected because they were similar in intent to Green Building Challenge, others such as Eco Quantum were not considered as they were regarded as too narrow, as being concerned only with building materials and products, for instance. This was probably fair enough for that paper, but there is a tendency to view assessment systems for buildings and for materials as two different things, whereas they are completely inter-related. The other difference is whether assessment tools are for designing buildings or to assess them once they are built. If assessment systems are not useful as design tools, then they are essentially measuring once the damage has been done. How useful are these tools and assessment methods for people when they embark on a building project and have to decide what to do?

Environmental toolbox

In a research project in which I am engaged with Taylor Woodrow Construction [6] and a number of other partners, funded by the UK government through its Partners in Innovation programme, we set out to compare a much wider range of tools and assessment methods. One aim is to produce a framework and toolbox to assist in the selection of materials, components and construction systems. The framework will address the issue of balancing sustainability and functional performance, considering environmental, social and economic issues, the 'triple bottom line'. It is not intended to produce yet another methodology, but to draw on existing tools and to signpost designers and specifiers to the wide range of advice and information that is available. We set up a simple Excel spreadsheet and did a very superficial assessment of a range of tools and published methodologies from around the world. The assessment involved looking at what the tools did or claimed to do and checking them off against a long list. We got to about 100 tools and didn't go any further, even though we were aware that there were many more. We only looked at those available in the English language; see Table 18.1.

Table 18.1   An abridged list of the criteria that each tool was reviewed against in the balance value project.

Tool form	e.g. book, software, web based, etc.
Specified use of tool	e.g. self assessment, government agency, external assessor
Training needed to use tool	By assessment organization, licensed assessors only
Tool applications	Planning, design, whole life, building materials, operation of buildings manufacturing, etc.
Country of origin, date and cost	
Source of data	Organization, research, experience
Tool's social criteria	Access, health, human capacity, cultural heritage
Tool's environmental criteria	Land use
	Natural resource
	Climate change
	Acidification
	Eutrophication
	Waste
	Ozone depletion
	Ecology
	Pollution
	Biodiversity
	Processes
	Energy
	Water
	Recyclability
	Other
	General
Tool's economic criteria	
Tool's performance/function criteria	
Tool status	Established, experimental, etc.
Discussions with tool developers	

It soon became clear that there was a wide variation between different systems as to what was and was not included and how it was handled. Our list of tools and how to access them should be available in the Balanced Value Toolbox, which will be an output of the project. We are not trying to make judgements in this project as to whether one tool is better than another or more complete, but simply to show potential users what they appear to contain. Of course, some tool developers may feel that we have not done justice to their particular tool. That is a risk we will have to take. The project also goes beyond signposting people to environmental assessment tools, but is trying to create a framework that will help in making rational decisions and choices. Anyone who has operated in the real world knows that all decisions are based on trade-offs and environmental choices will be balanced against cost, performance and wider social considerations. We have been experimenting with multi-criteria decision analysis software to see whether providing a clear and rational decision making process will facilitate taking on board environmental issues.

To publish a consumers' guide to environmental assessment tools would be a major task as there are so many tools and it might take some time to contact all the tool developers and get their agreement and approval. However it is a task that needs to be undertaken. As long as what is said about them remains within the bounds of 'fair comment', then there should be no restrictions on its publication.

One aim of the Balanced Value project is to provide signposts for a wide range of people within the construction sector to the wide range of environmental tools available. They may be able to use the toolbox to scan the systems available to see whether they do the job that they need doing. This may help them through the confusing array of tools. In countries such as the UK and USA market leaders such as BREEAM and LEED dominate the field. Many other countries have one well-known system that has been developed to suit particular local circumstances. However we have anecdotal evidence to suggest that many people would like to consider tools other than the primary ones in their country.

Sustainable housing

Using a spreadsheet to compare different environmental criteria had been a useful exercise that helps to reveal the differences between different tools. I carried out such an exercise for a working group set up by the UK Town and Country Planning Association (TCPA) [7]. This group has been looking at how Planning Law could be amended to enforce sustainable standards in building. See Table 18.2 which shows my earlier analysis.

The TCPA study has been closely linked with the World Wide Fund for Nature (WWF) million sustainable homes campaign that is beginning to attract attention from the UK government. Obviously if every scheme that requires planning permission had to meet, or at least make a declaration of environmental impact, this would be a way to enforce political decisions to reduce carbon emissions, pollution and so on. However in order to implement this in a robust way, that would stand up to scrutiny during planning appeals, requires a universally agreed methodology for defining sustainable building. The main achievement of the TCPA report will be to get this on to the political agenda, but it cannot bring about agreement on environmental assessment criteria.

Table 18.2 Extract from the author's paper to the TCPA working group.

	Eco-Homes	EGBF	WWF	Irish Energy Centre
Energy eff. building	✓	✓	✓	✓
Energy eff. services	✓	✓	?	✓
High insulation	✓	implied	implied	✓
Transport issues	✓	✓	✓	×
Polluting emiss. global	✓	✓	✓	?
Polluting emiss. local	✓	✓	✓	?
Materials	✓ (FSC)	✓	✓	✓
Waste/Recycling	×	×	✓	Optional
Water	✓	✓	×	Optional
Land use and ecology	✓	✓	(Planning)	✓
Health and IAQ	Implicit	✓	✓	✓
Comfort/Fuel poverty	×	✓	✓	✓
Building performance Life cycle	×	✓	×	×
Equity/Ethics	×	✓	✓ (FSC)	×
Community	×	✓	✓	×
Economic cost benefit	×	✓	?	✓
Sewage/Waste water	×	×	×	Optional
Passive solar	×	×	×	?

Finding the right tool for the right project

Another angle on tool evaluation is to embark on a real project and look around for a way to benchmark the proposals that are being made, drawing on the available tools. We found ourselves in this position recently when we were approached by clients in Northern Ireland that wanted to do a mixed-use development of 100 houses and some industrial and office units at Killaughey Road. They wanted to make sustainability claims that would form part of the planning application. This presented us with an interesting challenge because we wanted to find sustainability and environmental criteria, which the scheme could aim to match, that provided both a guide to design and a way of comparing the project with environmental best practice elsewhere. A further challenge in Northern Ireland is that there are virtually no sustainable developments and sustainability is not recognised as a criterion for planning approval! We found ourselves in the situation of 'shopping around' much as many others must have to do to find good tools and projects that would have credibility with the planning authority.

We looked at a development at Battery Park in New York City [8], which is a particularly well-documented example of applying LEED principles. We also considered the Austrian TQ system [9]. The main initial attractions of both of these were the apparently holistic and yet commercially realistic characteristics of both. We also used a sustainability checklist developed by the London Borough of Brent [10], which is intended as a method to assess planning applications. The Brent checklist, developed by Delle Odeleye is particularly thorough and comprehensive and gives a point scoring system

Table 18.3 Draft paper prepared by the BRE Centre for Sustainable Construction for the TCPA.

Criteria	Planning	Build. regs	Eco homes	Sustainability checklist	Best practice	Recommendations
Energy: CO₂ emissions	✓	✓✓	✓✓	✓✓	✓✓	• Increase Building Regulations requirements for CO₂.
Energy: Insulation	×	✓✓	✓✓	✓✓	✓✓	• Increase build quality (to meet above).
Materials: Low environmental impact	×	×	✓✓	✓✓	✓✓	• Use of A rated material. • Use of FSC timber. • Introduction of materials Approved Document.
Materials: Recycling, reuse & waste	✓	×	✓✓	✓✓	✓✓	• Waste recycling/sorting on site. • Provision of recycling and composting storage.
Pollution: Ozone depletion	×	×	✓✓	✓✓	✓✓	• None – *from 2004 all construction materials to have zero ODP.*
NOₓ emissions	✓	×	✓✓	✓✓	✓✓	• Maximum NOₓ emissions
Pollution: External air	✓	×	×	✓	×	• Planting to minimize pollution.
Pollution: Ground/Water	✓	✓	×	×	×	• Compliance with best practice pollution prevention.
Transport: CO₂ emissions	✓✓	×	✓✓	✓✓	✓	• Sustainable transport plan required.
Transport: Vehicle alternatives	✓✓	×	✓✓	✓✓	✓✓	• Recommendations of above to be adopted. • Provision of cycle storage.
Site issues: Ecology	✓✓	×	✓✓	✓✓	✓✓	• Appoint registered ecologist. • Enhancement of ecological value. • Focus on local native species.

					Recommendations
Site issues: Land use	√√	√	√√	√√	• Review land type classification. • Focus on ecological value.
Water: Low use	×	×	√√	√√	• Set maximum water consumption, greywater and rainwater recycling.
Water: Flood Risk	√√	√√	√√	×	• Sustainable drainage requirement on all sites.
Daylighting	×	√√	√√	√√	• Introduce into Building Regulations.
Health/Wellbeing IAQ	×	√√	×	√√	• Specify non-toxic and non-allergic materials.
Noise transmission	√√	√√	√√	√√	• Increase enforcement of Building Regulations Part E.
Private outdoor space	×	√√	√√	√√	• Planning requirement for private outdoor space.
Equity	×	√	×	×	• Introduce requirement for 'lifetime homes'.
Access	×	√	×	×	• All housing to be designed to DDA requirements.
Safety & security	×	√√	√√	√√	• Provide motion detector security lighting. • Design to 'Secured by Design'.
Community: General	√√	×	√√	√	• Community consultation. • Provision of community facilities.
Community: Employment	×	×	√√	×	• Provide local employment.
Flexibility	×	×	×	×	• Housing to be designed to allow future flexibility.
Procurement	×	×	×	√	• Procurement of locally produced materials.
Value for money	×	×	×	×	• Use of whole life costing to achieve best value.

Not covered ×   Partially covered √   Well covered √√

Figure 18.1   Housing project in South Africa.

for assessing proposals. Brent Council has now adopted this and many other planning authorities in England are developing similar systems. Needless to say they are all developing their own systems, so each one is different, some much more comprehensive than others. If the Killaughey Road scheme gets planning permission then we may go on to use a design tool such as 'Eco Tect' or 'ENVEST'; we will also carry out a BRE Eco Homes assessment for the scheme but none are ideal for the project. Finally, in order to influence local political opinion, we managed to find reference to an eco-housing project in South Africa which the Northern Ireland First Minister, David Trimble, had visited. While this may just have been a photo opportunity, he was quoted as supporting sustainable housing and this may eventually be the most useful benchmark we have found; see Figure 18.1.

Such an eclectic way of benchmarking a scheme is far from satisfactory. Inevitably, being confronted with a range of different standards and approaches bemused the client. It is intended to achieve an Eco Homes rating for the houses on the site, but the BRE Eco Homes standard does not deal with materials (except timber) and includes nothing about indoor air quality. It is also weak on construction methods and post construction testing for air-tightness etc. While Eco Homes has these weaknesses, in the UK context we have to support and promote it as its adoption would ensure greener standards than conventional construction. When the World Wide Fund for Nature (WWF) announced its million sustainable homes campaign for the UK at the Johannesburg summit, their concept of sustainable housing was already based on 'Eco Homes'.

In another project, the 'Rediscovery' Centre in Ballymun, Dublin [11], a centre for recycling, which will be built largely from recycled materials, we were again faced with the question of how to justify the strategy for the building in a feasibility study for Ballymun Regeneration Ltd. Essentially we have relied on the strength of argument for each specification decision, with a pragmatic model for assessing the impact of each choice. The credibility of the proposals depends as much on the experience and expertise of the design team as it does on measurements of carbon emissions. When developing an innovative and unusual building, most environmental design and assessment tools are found wanting, because they have been compromised to work with relatively conventional buildings and materials. Some seem more designed to greenwash projects than provide a critique of how bad they are for the environment. Furthermore most are based on designing new buildings, whereas the adaptation, re-use and conversion of existing structures is a much more complex art. Tools that deal with rehabilitation should surely be a priority from an environmental perspective.

Off-site construction

I was confronted with similar problems when approached by a leading UK manufacturer of portable and modular buildings. They felt that off-site construction could, by its very nature be regarded as sustainable, when compared with conventional on-site construction: less waste, higher standards of air-tightness with factory control and the ability to recycle and re-use the buildings. These arguments are supported by others [12], but only in general terms. When it came to establishing some form of rigorous benchmarking or evaluation, the marketing department soon became bamboozled by the apparent complications of considering the problem in a holistic way. Given the use of environmentally negative materials such as isocyanate based polyurethane insulation, PVC coated steel and some wasteful manufacturing processes it wasn't a simple black and white task to give the products a green stamp of approval [13].

The company had assiduously trawled the world of universities and government agencies looking for a system that would give them their green labelling but had largely been offered ISO 14001 based environmental management advice, which would have told them little about their manufacturing processes and the design quality of their products. They have explored whole life costing and life cycle analysis and eco-profiling, but even for a middle sized enterprise the cost of carrying out such work, and the time involved, is off-putting. Is it worth making such a huge research investment in detailed appraisals of all the materials, supply chains and processes involved to produce a fully detailed LCA analysis when at the end of the day it would be questioned by competitors and customers who would not be able to understand the rigour and detail of the process? The lack of universal agreement and understanding again becomes a problem.

Interestingly this company had an ethical commitment to sustainability that was not apparent from their marketing. They took CFCs and then HCFCs out of their insulation many years before legislation required them to do so. They were even buying FSC certified timber that I only came across by chance as part of my research. They were not willing to make green claims based on partial good practice, as this would leave them open to criticism for the other aspects of their production which were not so green. Essentially their choice of specification was based on performance and reliability of supply where no greener alternatives were obviously available. However the company has adopted an environmental policy and this may eventually be pushed out into their supply chain, making it easier to source greener products. We will have to see.

As part of the study I came across the work by J. G. Vogtlander and Ch. F. Hendricks [14] of the University of Delft, in which they have been able to draw on a wide range of databases to propose a cost based form of evaluating projects and activities. Their Eco-costs/Value ratio model can give an answer in terms of Euros that is a 'virtual' cost 'related to measures which have to be taken to make (and recycle) products in line with Earth's estimated carrying capacity'. This model permits greater fine-tuning when looking at a sector such as construction, whereas ecological footprinting [15] or mass balance accounting is much more broad brush. However it would require government and international agreement to encourage private companies to open their books and share environmental impact information before the model could be put fully into practice.

SMES and sustainable products

In a study that we carried out at Queens University, funded by the UK Engineering and Physical Sciences Research Council (EPSRC), we looked at the opportunities and obstacles facing a group of smaller companies who were trying to build up a market for green products [16]. Some of the partner companies were simply importing and selling on products manufactured in Germany or other European countries, where an eco-building market has already been established. Other partners had invested in developing green products (mortgaging their houses to do so) and were taking on a huge challenge to get their green products accepted in the market.

The difficulties facing such companies were not simply those of anyone starting a new business, but the companies were also faced with the problem of challenging the bogus claims of other products and the lack of any publicly available system for validating or benchmarking green products. They were also able to observe at first hand how their products were specified in high profile demonstration green building schemes only to find the contractors substituted them with conventional materials. Such building projects were still able to achieve good BREEAM or other environmental accreditation, however.

Some of our partners also identified wider difficulties when considering assessment methods. For some companies that would describe themselves as 'ethical', assessment methods would need to include physical, social and health considerations, controversial as these may be. They further deplored the lack of transparency and accountability of research institutions, consultants and other accreditation bodies who charge high fees and who use so-called 'black-box' methods, so that data and information are not readily accessible. Of course a great deal of environmental information is commercially sensitive and expensive to accumulate. Charges have to be levied to fund further research in a commercial climate where there is little state aid in this sector.

Environmental Building News refers to these problems as the search for the 'Holy Grail':

'Problems arise concerning the quality, consistency and availability of data on products and processes; the methods used to compile inventories and especially the assumptions and systems used to translate inputs and outputs into measures of environmental impact.' *Environmental Building News* [17]

European progress?

It is possible that there are people striving away in darkened rooms to come up with a perfect international standard for green buildings, products and materials. I looked at the ISO TG59 web site but there was nothing publicly available. Work is going on in Europe at the CEN and the British Standards Institute but it is painfully slow. The European Construction Products Directive was published in 1988 and contains environmental criteria, but it has still not been agreed how these will be defined and enforced. In Brussels most of the main manufacturing federations, cement and concrete, mineral wool and a range of chemical manufacturers, have offices and invest millions in lobbying the European commission to postpone the day on which environmental standards will be clearly defined and enforced, effectively putting them out of business. The latest skirmish is over the European commission approval of a German Government decision to subsidize renewable insulation materials. The mineral wool manufacturers are appealing against this. Conventional building materials manufacturers are able to use restraint of trade laws to frighten public sector bodies into maintaining a free market and to outlaw environmental restrictions. However experts in supply chain management can demonstrate how writing good specifications for materials or even buildings can ensure that high environmental standards can be achieved without running into obstacles from the free trade lobby [18].

Ethical issues

At a global level, pressures for free trade are even more powerful and lengthy supply chains can obscure the origins and thus the environmental provenance of materials and products. The UK and Western Europe import vast amounts of illegally logged timber from both the tropics and the former soviet block. Often respectable businesses are involved in rebranding materials. Architects constantly tell me that the timber they have specified is from sustainable sources without them having a clue where it has come from. Interestingly, ethical issues are rarely included in environmental assessment criteria, even though they underpin so many aspects of how we assess the environmental impact of materials in terms of resource depletion, health and so on [19].

In our work in the UK to try and develop alternative building systems and methods which have low impact in terms of resource depletion, pollution, energy and carbon emissions, we have found it difficult to get R&D funding and Government support. The oil and petrochemical companies have a disproportionate influence in UK universities and the bodies that fund research. This was recently exposed in an important report [20]. The vested interests involved in keeping alive solutions to carbon emissions using fossil fuel based products undermine attempts to develop genuinely renewable alternatives, particularly those that cannot easily be exploited by big business. In our EPSRC study, the SME partners had received little support from Government until recently when 'Natural Building Technologies' [21] has received substantial funding from the UK

Carbon Trust. Anyone active in this area of work is constantly caught between the two stools of trying to widen interest in sustainable good practice in the mainstream, while not compromising on ethical and technical principles.

Conclusions

These issues require political analysis to understand what may happen. Yet one thing clearly emerges from a familiarity with the sustainable building literature: just how non-political is it as a movement? Driven by a need to achieve political acceptance and respectability and a need to attract funding from government and big business, most research in the field is painfully non-political and uncritical of the mechanisms, which prevent real progress in this field. It will take political will to drive moves to establish international harmonization of environmental standards for buildings and building products and construction systems. It should therefore be a key priority for us to become more politically engaged and involved in getting sustainable construction higher on the international political agenda. This doesn't just mean convincing political parties, but even influential environmental NGOs do not have a clear understanding of the importance of sustainable construction.

Professionals, who are responsible for design and specification decisions, rarely prioritize environmental and energy questions, which remain in the background [22]. Certainly within the UK, after a decade of work by environmentalists, we still have to work hard to persuade people on the deck of Titanic Earth that it is really sinking. Fossil fuel energy is running out and we will have to become more autonomous within a couple of decades [23] A key factor, however, in the failure to address this holistically, is the confused diversity of environmental criteria, tools and assessment systems that make it easy for big business and government to escape from their responsibilities. As strongly argued by Williamson *et al.* [24], the reductionist approach adopted in many environmental standards ignores many of the contextual issues that surround sustainability. Thus a key priority for us all will be to work towards an international harmonization of standards. In doing so it is critical that we do not water things down for the sake of comfortable compromise and easy agreement, but instead push for the most radical and holistic and comprehensive standards that we can achieve.

References

[1] Saunders, T. (2002) *The Boiled Frog Syndrome: Your Health and the Built Environment*, Wiley-Academic, Chichester.
[2] Roaf, S., Fuentes, M. and Thomas S. (2001) *Ecohouse. A Design Guide*, Architectural Press, Oxford.
[3] Smith P. F. (2001) *Architecture in a Climate of Change. A Guide to Sustainable Design*, Architectural Press, Oxford.
[4] Kennedy J. F., Smith, M. G. and Wanek, C. (2002) *The Art of Natural Building*, New Society Publishers.

[5] Todd, J. A., Crawley, D., Geissler, S. and Lindsey, G. (2001) *Comparative Assessment of Environmental Performance Tools and the Role of the Green Building Challenge*, Building Research and Information, **29**(5), 324–335.

[6] Balanced Value (2003) Taylor Woodrow Construction, London, http://www.taylorwoodrow.com/3/balanced_value.html

[7] Town and Country Planning Association (2003) *Building Sustainably – How to Plan and Construct New Housing for the 21st Century*, report of Sustainable Housing Working Group, October 2003, http://www.tcpa.org.uk

[8] McCourt, P. and Gelb, S. (2002) Hugh L. Carey Battery Park City Authority High Rise Residential Environmental Guidelines, *Sustainable Building International Conference Proceedings*, Oslo.

[9] Geissler, S. and Bruck, M. (2002) Total quality (TQ) Assessment as the basis for building certification in Austria, *Sustainable Building International Conference Proceedings*, Oslo.

[10] London Borough of Brent (2003) http://www.brent.gov.uk/planning/sustainableguidelines

[11] Howley Harrington Architects and T. Woolley (2003) Ballymun Rediscovery Feasibility Study, September 2003, unpublished.

[12] Gorgolewski, M. (2002) The potential for prefabrication in UK housing to improve sustainability, *Sustainable Building International Conference Proceedings*, Oslo.

[13] Woolley, T. (2003) Sustainability review for a UK modular construction company, Queens University Belfast, unpublished.

[14] Vogtlander, J. G. and Hendriks, Ch. F. (2002) The eco-costs/value ratio, *Materials and Ecological Engineering*, Netherlands, Aeneas.

[15] Wackernagel, M. and Rees, W. (1996) *Our Ecological Footprint Reducing Human Impact on the Earth*, Gabriola Island New Society Publishers.

[16] Woolley, T. and Caleyron, N. (2002) Overcoming the barriers to the greater development and use of environmentally friendly construction materials, *Sustainable Building International Conference Proceedings*, Oslo.

[17] Environmental Building News (2002) March 2002, http://www.BuildingGreen.com

[18] IEMA (2002) Environmental Purchasing in Practice: Guidance for Organisations, The Institute of Environmental Management and Assessment. Best practice Series Volume 2 September 2002.

[19] Fox, W. (ed.) (2002) *Ethics and the Built Environment*, Routledge, London.

[20] Simms A. (2003) Degrees of capture, universities, the oil industry and climate change, New Economics Foundation, Joseph Rowntree Trust and Greenpeace, London.

[21] Natural Building Technologies (2003) http://www.naturalbuilding.co.uk

[22] Guy, S. and Shove, E. (2000) *A Sociology of Energy, Buildings and the Environment Constructing Knowledge*, Routledge, London.

[23] Douthwaite, R. (ed.) (2003) *Before the Wells Run Dry; Ireland's Transition to Renewable Energy*, FEASTA, Dublin.

[24] Williamson, T., Radford, A. and Bennetts H. (2003) *Understanding Sustainable Architecture*, Spon Press, London.

19 Sustainability criteria for housing – a whole-of-life approach

F. Martin[1] and A. Pears[2]

Summary

While numerous examples of 'sustainable houses' have been designed and constructed, typically as one-off projects by architects, the volume housing market has been slower to respond to an emerging interest in green housing. However, large-scale transformation of housing stock will not be possible without the engagement of this industry segment.

This chapter outlines a recently commenced research project to support the volume housing market in moving towards more sustainable practice. The project aims to determine the life cycle environmental performance of the 'EcoHome', a relatively standard home designed for the volume housing market but with a range of easily-adopted modifications incorporated to improve its environmental performance.

A comparison of a range of environmental rating tools for houses was conducted to identify the most commonly used indicators of sustainability. These will inform the development of a methodology for rating the EcoHome's performance. Monitoring of a range of parameters will be conducted post-construction and the results will also contribute to an overall evaluation of performance. Life cycle assessment and embodied energy analysis will also be used to quantify impacts.

Introduction

There is little argument over claims that the construction industry needs to ensure that new buildings are more sustainable than current stock. More contested is what attributes define 'sustainable' buildings and which trade-offs are acceptable. As with the term 'sustainable development', there are numerous interpretations of sustainable building, with similar potential for rhetoric and 'greenwash' in the absence of a clear and united vision and direction.

There are numerous recent case studies of typically architect-designed houses that incorporate various sustainability features in response to client requirement [1]. However, while these houses provide useful examples and lessons, the techniques they have

---

[1] RMIT University Centre for Design, Melbourne, Australia.
[2] Sustainable Solutions, Melbourne, Australia and RMIT University.

espoused have not yet been widely adopted by the volume market – although there is a recent increase in interest. Any genuine transformation in the construction industry will need to focus on adoption of sustainability principles by the volume building sector.

The EcoHome project

A recently commenced, multi-disciplinary research program at the RMIT Centre for Design aims to support the construction industry, particularly the volume housing sector, in their shift to more sustainable practice. There are three distinct yet inter-related strands of research focusing on design, social and technological aspects related to sustainability and the volume housing market. The technological research is the focus of this paper. It should be noted that this research is still in the preliminary stages and the purpose of this paper is to outline early observations and future directions.

A demonstration house, known as the 'EcoHome', will be the focus of much of this research, providing empirical data to support model development. This house is under construction at the time of writing and is anticipated to be completed in January 2004. (It was, in fact, completed in 2004.) The home will then serve as a display home for twelve months prior to sale and occupation by the residents. A condition of sale will be that the residents agree to participate in the research for up to two years.

The EcoHome is a relatively standard, four bedroom, two bathroom, off-the plan home, but has had a number of modifications made to ensure that it is more sustainable. Several people with expertise in sustainable housing worked with the building company, Metricon, and the former Urban and Regional Land Corporation (now VicUrban) during the design stage to develop strategies for modification that would allow the house to perform more sustainably.

The overarching aims of this activity were to construct a marketable sustainable display house that could be reproduced with minor adaptations and to encourage visitors to the home to incorporate more environmentally responsible features when buying or retrofitting homes.

The house will also act as a marketing device to showcase some of the readily available sustainable building products, appliances and landscape elements. At a more technical level, the objectives were defined as follows:

- obtaining an energy FirstRate rating of 6 stars
- minimizing water use
- minimizing indoor air pollution
- maximizing energy efficiency through orientation.

In brief, some of the more significant sustainability features of the home include:

- installation of a greywater system to collect and treat water from the washing machine for subsurface garden irrigation
- zoning daytime areas of the house to the north
- provision of a gas-boosted solar hot water service
- installation of a rainwater tank, which will be used for toilet flushing
- use of low-water-consumption appliances and fixtures

- installation of a ventilation heat recovery system to improve indoor air quality while maintaining energy efficiency
- heating and cooking by gas
- no air-conditioning but installation of ceiling fans for both cooling and recirculation of warmed air in winter
- use of water-based paints and other materials chosen to reduce emissions and off-gasing
- construction materials such as insulation batts selected for a high recycled content
- landscaping with a mix of native plants, fruit trees and a vegetable garden
- provision of recycling and composting facilities
- provision of home office facilities to facilitate telecommuting
- provision of a bedroom with en-suite that is detached from the other bedrooms to better accommodate an extended family living within the one residence.

A major task of the technical research programme is to determine the sustainability performance of the home over its life cycle, both against the original objectives set, and also in more relative terms. The home will be monitored prior to occupation to provide baseline data, and then for up to two years after occupation. Monitoring will include parameters related to thermal comfort, indoor air quality, and greywater and rainwater quality. Obviously, behavioural aspects will play a major role in the overall performance of the house. For example, there would not necessarily be reductions in energy consumption regardless of the house design if lighting and appliances are used more intensively than average. To this end, it is intended that technical monitoring will be complemented by periodic interviews and surveying of the EcoHome occupants, to attempt to correlate monitoring trends with occupant behaviour.

Rating sustainability of housing

Critical in allowing determination of the EcoHome's sustainability performance over its life cycle is the ability to define its performance at a point in time. In order to do this, a selection of existing environmental ratings tools for houses were examined and compared to identify which sustainability criteria are most commonly identified.

There are a growing number of environmental rating tools for buildings, both nationally and internationally. At the time of writing, several of the tools developed specifically for rating houses are still in draft form (and in some cases with revised versions anticipated to be significantly different from earlier drafts). Tools currently in draft form which have been analysed to date include:

- The draft first version of NABERS, the National Australian Building Environmental Rating System (developed by the Commonwealth Government) [2]. This was a project aspiring to be the 'world's best' and was designed primarily to rate the environmental impact of existing buildings, both residential and commercial. It was not intended to be a design tool. The second version of NABERS [3], whose content has changed significantly from first draft, was publicly released for comment in late October 2003, but had not been analyzed in depth at the time of writing.

- Draft LEED Residential (Leadership in Energy and Environmental Design), a ratings tool being developed in the United States by the Green Building Council [4]. Unlike NABERS, LEED Residential is a tool aimed primarily at the design phase of construction.
- The draft version 6 of the Sustainable Housing Code, developed by the South East Queensland Region of Councils [5]. This is proposed to be adopted as a model code under the Queensland Development Code, part of the State Building Regulations, for new buildings.

EcoHomes (which is unrelated to the EcoHome in Melbourne) is the British environmental rating scheme for homes developed by the Building Research Establishment [6], and also intended primarily as a design tool. This rating tool was also analysed and compared with those described above.

BASIX (Building and Sustainability Index), currently under development by the NSW Department of Infrastructure, Planning and Natural Resources [7], was not available for analysis at the time of writing but will also be critically examined when the final version is released as part of this research project.

The process of analysing and comparing these various ratings tools has proven challenging, given their complexity and the different approaches and philosophies adopted by their developers. Because some of the tools are aimed at building designers while others focus more on existing buildings, the criteria are not always directly comparable.

However, it was felt that in consideration of the EcoHome's life cycle, it was important to give consideration to both design aspects of the home and longer-term operational aspects.

Of the tools analysed, LEED Residential was overall the most comprehensive, but arguably also the most prescriptive. The first draft version of NABERS stated that it wanted to avoid prescriptive measures, and instead where possible focus on outcomes, allowing the home occupant to determine the most appropriate means of optimizing outcomes. It did not always live up to this claim.

In the ratings tools compared, sustainability criteria were grouped into overarching themes. There was considerable overlap between these themes, which in part reflects their interconnectivity (for example, criteria related to public transport accessibility were contained under 'transport' in some tools and under 'sustainable site' in another).

Broadly speaking, the themes in the various tools can be aggregated into the following categories:

- land/site (including site ecology in some instances)
- water
- energy
- materials
- transport
- interior/indoor environmental quality (with related criteria also covered in one tool under the theme of health and well-being)
- waste
- social.

Of the four tools analyzed, only the Sustainable Housing Code had a set of criteria that were explicitly called 'social sustainability', although it is understood that BASIX also has 'social' as one of its nine themes. In other tools, criteria related to social issues were embedded in other themes to varying degrees.

Broad criteria that were common to more than one of the tools were identified. In many instances, the exact intention was slightly different, as were the point scores awarded.

These criteria included:

Land/Site

- Selection of sites that are less environmentally sensitive or have lower ecological value (for example, selecting urban demolition or brownfield sites over agricultural land or purpose-cleared land) and/or sites that do not worsen urban sprawl (i.e. favouring denser urban environments).
- Maintaining and/or planting as high a ratio as possible of either native/localized and/or productive plants (fruits/vegetables) on unbuilt areas of the site.
- Reducing the percentage of impermeable surfaces as much as possible to minimize impacts to hydrological cycles.
- Reducing floor plan of building as much as possible.

Water

- Higher scoring for reduced water consumption per person/per bed space.
- Use of water conserving fixtures (for example AAA-rated shower heads).
- Collection of rainwater on-site.

Energy

- Greenhouse gas emissions of the building (higher scores for lower emissions).
- Encouraging the highest possible use of renewable energy.
- Ensuring building envelope performance is optimal.
- Provision of a clothesline/drying space.
- Provision of energy efficient appliances (such as white goods) or, if these are not provided, providing education material on energy efficient appliances to occupants.
- Reducing energy use for hot water production.
- Provision of energy efficient lighting.

Materials

- Choosing building materials from 'better' sources; that is, in accordance with relevant publications in that country (for example, the first draft of NABERS refers to the *Your Home Technical Manual* [1]).

- Using timber from reliable sustainable sources (preferably with Forestry Stewardship Council certification).

Transport

- Building within close proximity to convenience store and other facilities (such as a post office, bank, medical facilities, parks/play areas and so on).
- Proximity to adequate public transport for travel to work/school.

Interior/Indoor environmental quality

- Use of low volatile organic compound (VOC) products such as paints.
- Management of combustion products (such as appropriate use of flues).
- Consideration of daylighting in the home.

Waste

- Provision of recycling facilities.
- Provision of composting facilities.
- Reducing sewage production and potable water use (such as waterless sewage treatment systems).
- Wastewater treated on-site and used for landscaping.
- Waste management strategies in place during construction.

Other

- Provision of adequate outdoor space (such as patios).

From this exercise, some observations can be drawn that provide pointers for development of a strategy to determine the sustainability performance of the EcoHome. First, an assumption that the most commonly appearing criteria are more 'important' is not necessarily appropriate. The criteria that did not appear in more than one of the tools considered may be equally telling. For example, social sustainability criteria in the SEQ draft code such as street surveillance and design for disabled access were introduced as a result of consultation with stakeholder groups, and do seem to be quite relevant to the principles of sustainability, yet these have not been addressed in other rating systems. This situation suggests that the coverage of rating tools is still evolving.

The large discrepancies between criteria used in the various tools suggest that the same home evaluated by different tools may produce very different final scores. This poses problems for industry practitioners who in good faith attempt to operate in accordance with the content of a particular rating tool, only to perform poorly when rated by another. This may possibly result in a lack of confidence in these tools, and a reluctance to use them while they are only used voluntarily until better mechanisms exist to provide similar ratings across a range of tools. Once final versions of several rating tools are available, the EcoHome will be rated with a range of these tools to determine the degree of difference between various scores obtained.

This dilemma may be explained to some extent by the observation that many of the tools do not adequately return to the fundamentals of sustainability. Requirements are stated or encouraged through provision of higher scores, but a clear explanation of the exact reason for their selection does not always accompany these criteria. For example, provision of composting facilities is a reasonably common criterion across tools. Presumably the reasons for this requirement are to reduce quantities of organic material going to landfill, to reduce production of methane during decomposition in anaerobic conditions, and to promote the reuse of the nutrient content of the organic material. However, there may be a range of options that would also satisfy this criterion without necessarily requiring home composting. For example, a well-operated council collection and centralised compost facility could possibly result in an overall higher production of compost of a quality suitable for use. The point is that it is the ends, and not the means, that are of significance. Provision of a composting facility by a builder does not ensure it will either be used in a proper manner to avoid anaerobic decomposition, or at all, by residents (this is acknowledged by the first draft of NABERS to some degree, with only a portion of total possible points for composting awarded if only some composting takes place). The NABERS first draft claims that its 'headings and subheadings provide targets at which designers may aim in their attempts to make more sustainable buildings, but it does not at present attempt to tell them how to design such buildings' [2]. Despite this claim, in the first draft points are awarded for use of on-site composting facilities, rather than the target of avoiding organic waste to landfill.

This then raises another observation – that it may not actually be most appropriate for ratings tools to focus just on individual houses, but instead allow for ratings at a community scale. It is being increasingly observed that many services are more efficient when decentralised, but benefit from economies of scale greater than the individual household. For example, in a recent discussion paper [8], Melbourne Water proposes use of localized stormwater treatment systems such as pollution traps, constructed wetlands and sediment basins. Similarly, 'there is a global trend towards generating power, heat and cooling locally' [9]. Ratings tools need to be able to reflect that community-scale activities may actually be an environmentally-preferable outcome and have the flexibility to accommodate such initiatives.

It can also be noted that many ratings tools focus on the environment, rather than broader sustainability. Taking more of a triple-bottom line approach, newer tools such as the South East Queensland Regional Organisation of Councils draft Sustainable Housing Code and BASIX are reflecting a greater emphasis on social dimensions of housing.

In response to these observations, it becomes apparent that any meaningful development of long-term sustainability criteria for housing needs to give due consideration to the myriad of interpretations of sustainability that exist. This means also looking to interpretations of sustainability in other industry sectors rather than reviewing only existing material within the building and construction sector. For example, Langston and Ding [10] cite Taylor *et al.* (1993) describing a sustainability index system developed for the agricultural sector. Such examples of indices will also be critically examined to consider their applicability in methodology to a housing context as part of this research.

In raising the above issues, the attempt is not to disparage the tools that currently exist. At this early stage in the development of environmental rating tools, each tool provides opportunities for learning about what does and does not work in various circumstances [11]. Those that already exist, even in draft form, are making a significant contribution in the move towards more sustainable housing by raising awareness of the sorts of issues that should be considered.

Conclusion

In order to determine the sustainability performance of the EcoHome over its life cycle, the first stage is to determine its sustainability performance at a point in time, including the sustainability implications of decisions made during design, construction, commissioning and ongoing operation. Use of any one existing environmental rating tool is considered unlikely to provide a comprehensive answer, and it will be necessary to return to more fundamental sustainability requirements before exploring specific performance criteria. The EcoHome at Cairnlea, Melbourne, will be invaluable in this research by providing empirical data that will allow the outcomes of assessment using a range of environmental ratings tools to be compared in depth. The similarities and differences in performance will provide pointers as to how a more all-encompassing framework for determining sustainability, both now and in the longer term, might be developed.

Comparison of environmental rating tools against fundamental sustainability principles and means of achieving them will also provide feedback that can be used to improve those tools.

References

[1] Reardon, C. *et al.* (2002) *Your Home Technical Manual*, Australian Greenhouse Office, Canberra.
[2] Vale, R., Vale, B. and Fay, R. (2001) The national Australian buildings environmental rating system – Draft Final Version, 14 December 2001, Environment Australia, Canberra.
[3] Australian Government Department of Environment and Heritage (2003) National Australian Built Environment Rating System (NABERS) Project, http://www.deh.gov.au/industry/construction/nabers/index.html (last visited 01/12/03)
[4] US Green Building Council (2001) LEED™ Residential – Draft.
[5] South East Queensland Regional Organisation of Councils (2003). Sustainable Housing Code – Draft Version 6 for Comment, SEQROC, Brisbane.
[6] Rao, S., Yates, A., Brownhill, D. and Howard, N. (2000) *EcoHomes – The Environmental Rating for Homes*, British Research Establishment, London.
[7] NSW Department of Infrastructure, Planning and Natural Resources (2003) Building Sustainability Index: making sustainability simple for developers, councils and government agencies, http://www.planning.nsw.gov.au/index1.html (last visited 01/12/03)

[8] Melbourne Water (2002) Discussion paper – Draft sustainable water management strategy, City of Melbourne, Melbourne.

[9] CSIRO Centre for Distributed Energy and Power homepage, http://www.cendep.csiro.au/ (last visited 01/12/03).

[10] Langston, C. and Ding, G. (eds) (2001) *Sustainable Practices in the Built Environment*, 2nd edn, Butterworth Heinemann, Oxford.

[11] Pears, A. (2003) 'Green building rating: implications for design and practice', in *Green Building and Design 2003 Conference Proceedings*, Centre for Design, Melbourne.

20 Development of building environmental assessment tool on a Website and its implementation

N. Yokoo[1], Y. Kawazu[2] and T. Oka[3]

Summary

Since the beginning of the 1990s, building environmental assessment methods have been developed and implemented in many countries to encourage the construction of green buildings and high performance buildings, and to enhance energy conservation buildings.

Some environmental assessment methods and tools have not been used in practice, due to their complex nature, time required and cost. The purpose of this study is to develop and implement a simplified assessment tool on a Website, and to evaluate whether it works or not in practice. The major goal of developing the Green Building Website-based Assessment Tool (GB-WSAT) is to increase the awareness among designers, engineers and building owners about building environmental assessment tools and green buildings.

Introduction

Today, the development of architectural environmental performance assessment methods is the focus of attention worldwide. Although some of the assessment methods are used in real cases, there are also some methods are not.

The reasons for this include:

- the requirements for extensive amounts of data and time
- the requirement that the user learns the methods before implementing them
- the limited awareness of assessment methods in general.

[1] Department of Architecture and Civil Engineering, Utsunomiya University, Japan.
[2] Utsunomiya University, Japan.
[3] Department of Architecture and Civil Engineering, Utsunomiya University, Japan.

The purpose of this study is to develop a Website-based system for the environmental performance assessment of buildings to streamline and popularize environmental assessment methods (for use by practically anyone without the constraints of location or time), then to gather and analyze data, providing Website users with feedback on the findings. In addition, this study strives to compare assessment results with those obtained using other methods (e.g., EcoHomes, LEED, GBTool2002) and identify the objectivity and consistency of the Website-based assessment tool (GB-WSAT).

The need for a simplified environmental performance rating method

According to a questionnaire survey of stakeholders in the architectural community in Japan, the following views of environmental assessment tools were gained:

- Environmental performance assessment tools are not popular in Japan today.
- High expectations are placed on the assessment tool in Japan.
- There is demand for a simpler building environmental performance assessment method that can be implemented during or after the primary design stage, within a few days, and for a total cost of less than 100 000 yen or 500 000 yen, depending on building type and size.

Based on the questionnaire findings listed above, the research discussed here was performed in the knowledge that simpler rating methods are desirable.

Development of a simplified Website-based system for environmental performance assessment

Outline of Website-based assessment tool (GB-WSAT)

GB-WSAT was developed based on GBTool2002, on a Website. Figure 20.1 shows the actual screen of the Website. This study involved the development and Website-based implementation of a program that is capable of generating an assessment with a minimal set of data required for the quantitative GBTool Assessment area R (Resource Consumption) and L (Loading). Table 20.1 shows the assessment criteria of GB-WSAT. Table 20.2 shows the amounts of data required by GBTool and the GB-WSAT. Table 20.3 compares the amounts of data required by GBTool and GB-WSAT to set benchmarks for the assessment criteria R1.2 (annual life-cycle energy during operation).

The common gateway interface (CGI) script and the Perl are used to build the Website for this research (see Figure 20.2). Although the script causes a heavier load on the server, it facilitates a wider range of operations, including not only online computations, but also data storage and updates as well as conditional branching. When the system

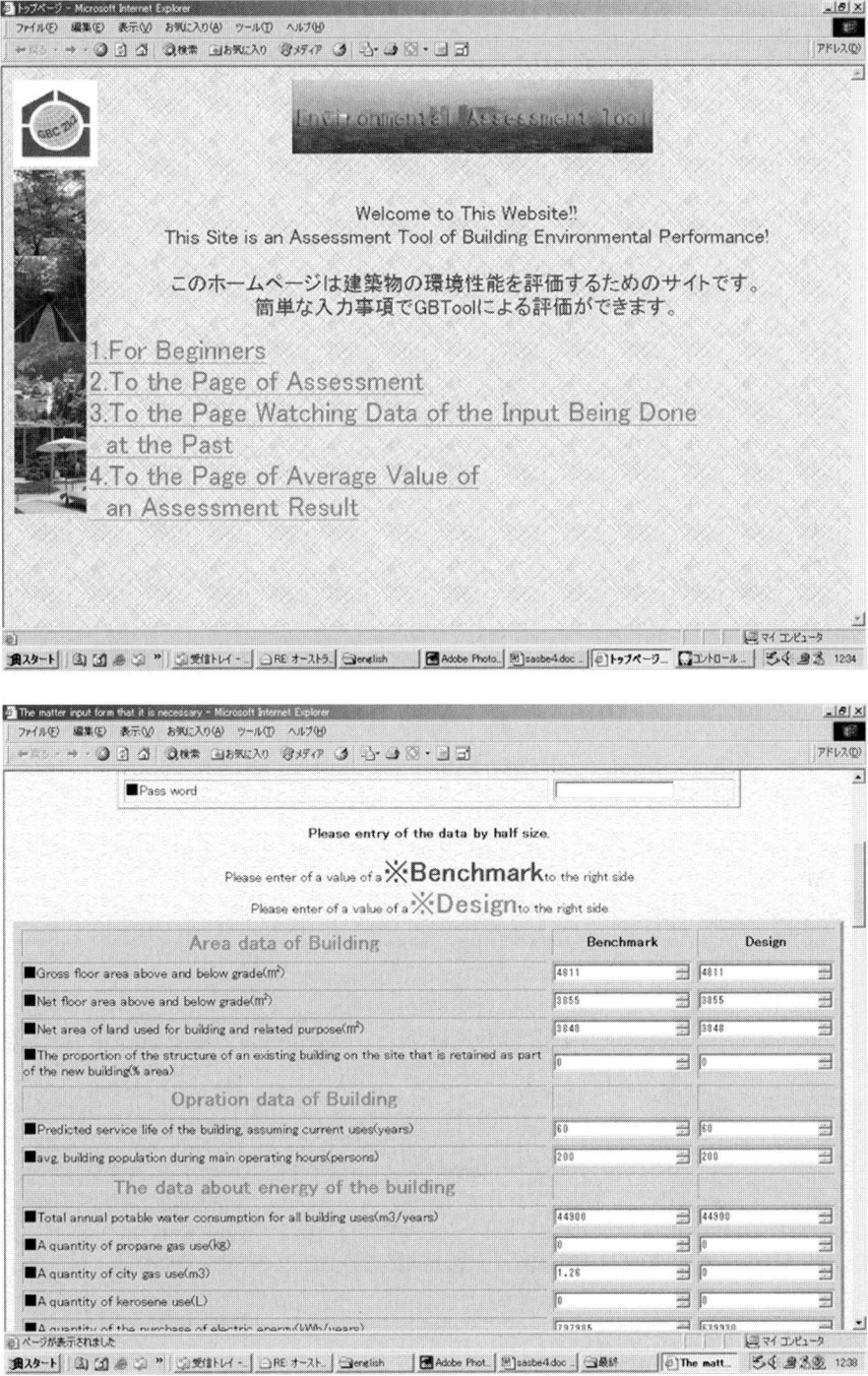

Figure 20.1   Screen shot of the GB-WSAT: (a) Home page, (b) Assessment page, (c) Results page.

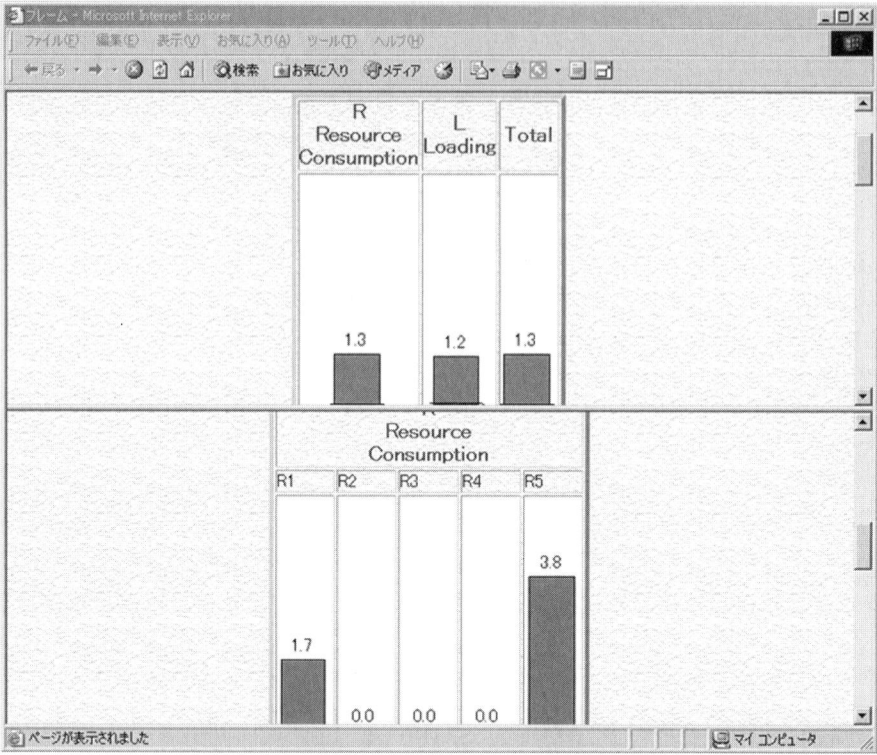

Figure 20.1 (*continued*)

program performs computations, it simultaneously writes the data to a designated file, thus allowing databases to be compiled at a later time (see Figure 20.3).

By avoiding the entry of redundant or difficult-to-understand data into GBTool and utilizing less data overall, GB-WSAT achieves results similar to those achieved with GBTool. Users enter data as required by fields in an online data-entry form, and are able to obtain assessment results automatically at the click of a button. Online graphs facilitate an easier understanding of the results. In addition, if they wish to terminate a data-entry operation, users can save all data entered thus far, and can browse that data later by entering a password and user ID when they resume the data-entry procedure. The system also displays average values for ratings, thus facilitating easy environmental performance comparisons for multiple buildings evaluated on the website.

Comparison of assessment result of other assessment methods for the same building

A case study building was assessed with GB-WSAT and other assessment methods (e.g. EcoHomes, LEED, GBTool2002). Assessment results were compared with each other to discover whether GB-WSAT works or not.

Table 20.1   Assessment criteria in GB-WSAT.

R Resource consumption	L Loading
R1 Net life-cycle primary energy R1.1 Primary energy embodied in materials, annualized over the life-cycle R1.2 Net primary non-renewable energy used for building operations over the life-cycle	**L1 Emission of GHGs from building materials production and operations** L1.1 Embodied emissions of materials, annualized over the life-cycle L1.2 GHG emissions from all energy used for building operations over the life-cycle
R2 Use of land and change in quality of land R2.1 Net area of land used for building and related development purposes	L2 Emission of ozone-depleting substances from building operations L3 Emission leading to acidification from building operations
R3 Net consumption of potable water	**L6 Solid wastes**
R4 Re-use of existing structure or on-site materials and/or recycling of existing materials off-site R4.1 Retention of an existing structure on the site R4.2 Off-site re-use or recycling of steel from existing structure on the site. R4.3 Off-site re-use or recycling of materials and components from existing structure on the site.	L6.1 Avoidance of solid waste from clearance of existing structures on the site L6.2 Avoidance of solid waste resulting from construction process L6.3 Avoidance of solid waste resulting from tenant and occupant operations L6.3.1 Area of central facility provided for sorting and storage of solid wastes L6.3.2 Area of central facility provided for storage of organic wastes L6.3.3 Aggregate area of facilities provided on each floor for storage of solid wastes
R5 Amount and quality of off-site materials used R5.1 Use of salvaged materials from off-site sources R5.2 Recycled content of materials from off-site sources	L6.3.4 Aggregate area of facilities provided on each floor for storage of organic wastes
	L7 Liquid effluents L7.1 Storm water flows disposed of on site L7.2 Sanitary waste water flows disposed of on site
	L9 Environmental impacts on site and adjacent properties L9.2 Reflectance of horizontal building surfaces and hard-surfaced site areas

Table 20.2 Comparison of data collection in GBTool and GB-WSAT for assessment criteria of R1.2.

GBTool	GB-WSAT
Net floor area above and below grade	Net floor area above and below grade
Electricity power base load mix of natural gas	A quantity of the purchase of electric energy
Electricity power base load mix of coal-fired	The electric power company
Electricity power base load mix of oil-fired	A quantity of propane gas use
Electricity power base load mix of nuclear	A quantity of city gas
Combustion and waste losses of natural-gas	A quantity of kerosene use
Combustion and waste losses of coal-fired	
Combustion and waste losses of oil-fired	
Delivery losses of natural gas	
Delivery losses of coal-fired	
.	
.	
.	
Total: 59 items	Total: 6 items

Table 20.3 Outline of a case study building.

Building type	Multi unit residential building
Site area	6014 m^2
Gross floor area	4811 m^2
Net floor area	3854 m^2
Building population during main operating hours	200 person
Design service life of building	60 year
Annual operating energy consumption	2309 GJ/y
Embodied energy of building	40 994 GJ

Main features of introduced techniques

Green roof	PV
Natural ventilation	Biotope
High thermal insulation and air tightness	Skeleton and infill construction method
Heat exchanger	Recycle material
Multi-functional heat pump	Electrical equipment
High floor height	Composting facility
Double-glazing	Low VC emissions materials

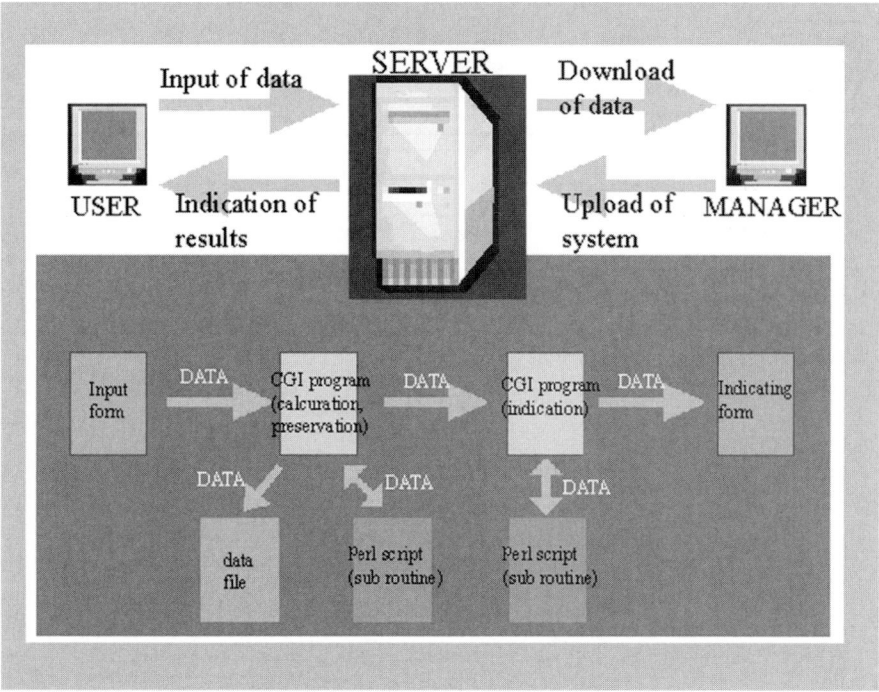

Figure 20.2   Outline of GB-WSAT.

Case study building

Figure 20.4 shows the assessment result of a case study building by using GB-WSAT.

Method of comparison

In order to compare the results of each assessment method, the assessment criteria are re-classified into three assessment areas: Site, Energy and Materials, and assessment points or scores are shown as the ratio of the obtained points to full points in each assessment area:

- site (land use, ecological value, rainwater management, traffic, surrounding environment)
- energy (energy for building construction and operation, CO_2, SO_x, NO_x and CFC emissions, and water consumption)
- materials (recycled materials, re-use of existing structures, management of discarded materials).

Point scores acquired with each method were then compared. Figure 20.5 shows the reclassified assessment results by EcoHomes, LEED, GBTool2002 and GB-WSAT.

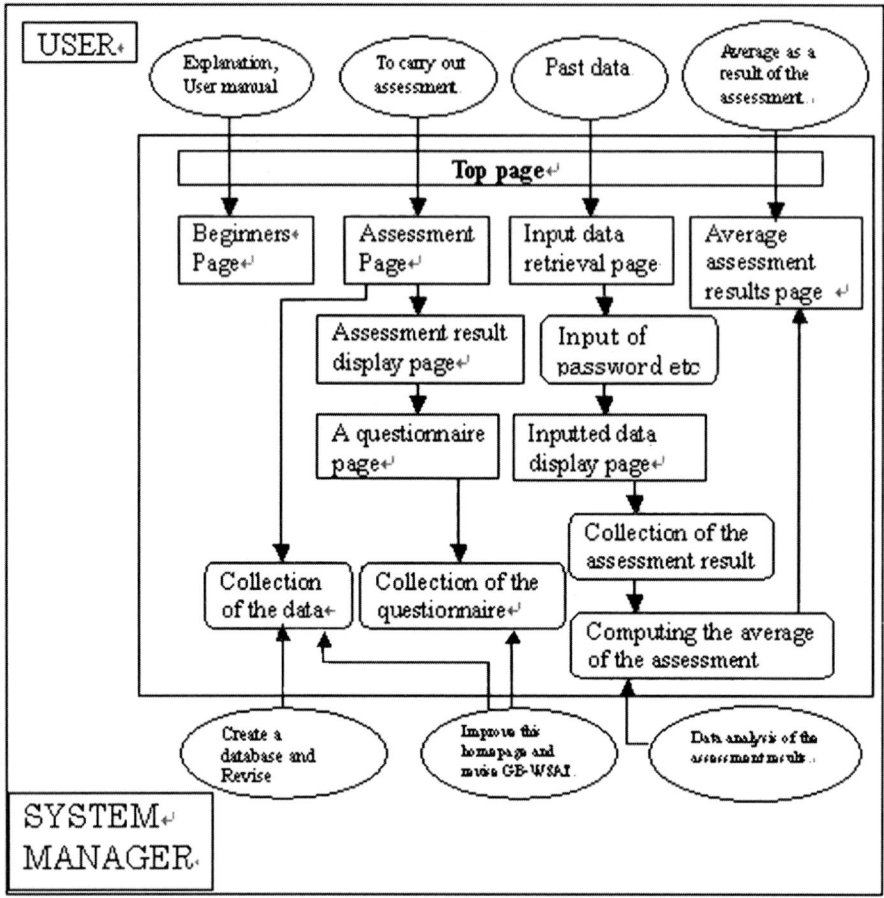

Figure 20.3   Flow chart of GB-WSAT.

Comparison of assessment results

Site

Gaps were highlighted in the assessment scores obtained with each of the assessment methods – EcoHomes, LEED, GBTool, and the GB-WSAT – for the Site field category. This was arguably attributable to the differences in emphasis that each system placed on factors of importance. For example, the EcoHomes provided a high score for the qualitative parameter of change in ecological value, whereas the maximum score possible with LEED in the building site category was comparatively low. Furthermore, in terms of overall score, significant deviations distinguished the 15% score obtained with GB-WSAT from the scores obtained with the other methods. The main reason for this finding is that the method used by the GB-WSAT does not incorporate any qualitative assessment criteria that would allow for a higher score.

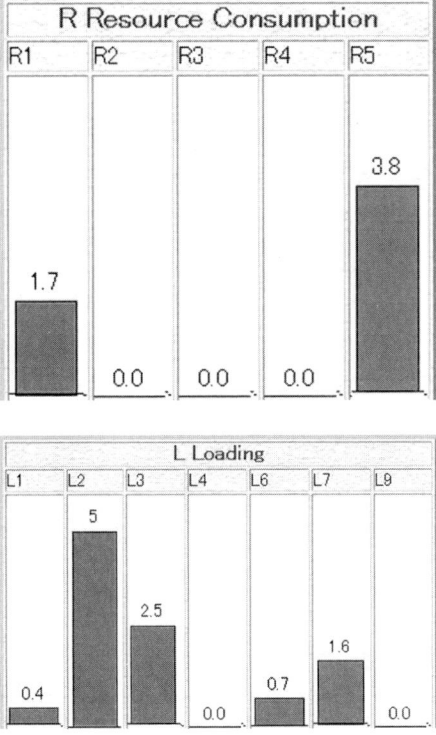

Figure 20.4   Assessment result of a case study building by using GB-WSAT.

Energy

The scores obtained with the compared methods for the Energy field ranged from 35% to 50%, not a significantly large gap. This stemmed from the fact that each of the assessment systems accounted for embodied energy and volume of CO_2 emissions, and otherwise did not differ substantially in terms of the assessment criteria.

Materials

Although LEED, GBTool and the GB-WSAT assessments resulted in a comparatively tight score range of 20–30% for the Materials assessment parameter, the score obtained with the EcoHomes was significantly higher. This was because LEED, GBTool and the GB-WSAT put emphasis on numerous material recycling and re-use assessment criteria, whereas the EcoHomes places emphasis on the use of approved lumber materials and the handling of discarded materials. Hence, although the rated building did not recycle materials to a significant extent, its waste disposal management practices were sound, resulting in the higher EcoHomes score.

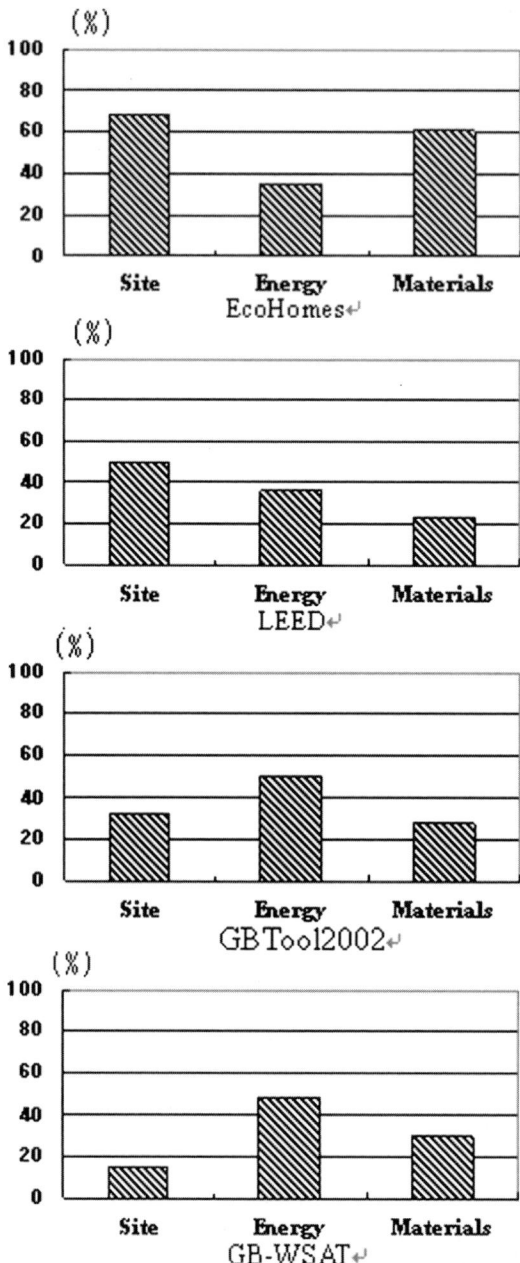

Figure 20.5   Reclassified assessment results for (a) EcoHomes, (b) LEED, (c) GBTool2002, (d) GB-WSAT.

General considerations

Although the GB-WSAT did not return scores for the Energy field that were significantly different from those obtained with the other compared methods, its scores for the Site and Materials fields were significantly different. Even so, the ratings it provided were almost identical to those achieved with the GBTool on which it was based.

Feed back from users on GB-WSAT

Several feedback comments were obtained from the users. Twelve users implemented the GB-WSAT assessment.

The positive aspects of the GB-WSAT were:

- Few data were required for assessment.
- The time needed for the assessment can be shortened.
- The result of the assessment can be compared with the average.

The negative aspects of GB-WSAT were:

- It is difficult to set benchmarks.
- An error message is needed for when users make mistakes.
- A detailed user manual and an explanation of GBTool for beginners is needed.

Conclusions

The following conclusions were drawn on the basis of this study:

(1) The GB-WSAT was able to generate almost identical assessments to the GBTool on which it was based, yet with less data. This facilitated shorter lead times and simplified assessment procedures, both of which were stated objectives of the research.
(2) Comparisons of the GB-WSAT with other assessment methods revealed gaps in the ratings obtained. This was because each environmental performance assessment method placed more or less emphasis on different assessment criteria and rated qualitative criteria differently.

The GB-WSAT developed for this research project enlisted quantitative parameters under the GBTool performance categories R (resource consumption) and L (load). It is presumed that the same ratings as those typically obtained with GBTool will be within reach if the future GB-WSAT incorporates the qualitative parameters in the above categories and 'Indoor environment' and 'Service quality' categories are developed in the future. The objectivity of the GB-WSAT was difficult to ascertain because historical data were referenced for setting GB-WSAT's benchmarks.

Accordingly, the following research topics demand further study:

- The consistency of the GB-WSAT obtained with a future system that incorporates more parameters for assessment.
- Methods for simplifying the inclusion of qualitative parameters for assessment.
- Determination of the objectivity of benchmarks used for assessment with the system devised for the GB-WSAT.

References

[1] Rao, S., Yates, A., Brownhill, D. and Howard, N. (2000) *EcoHomes: The Environmental Rating for Homes*, Building Research Establishment.
[2] Anderson, J. and Howard, N. (2000) *The Green Quide to Housing Specification*, Building Research Establishment.
[3] LEED Green Rating system™ (2000) Version2.0, U.S. Green Building Council.
[4] LEED Reference Guide (2000) Version2.0, U.S. Green Building Council.
[5] Larsson, N., Cole, R. *et al.* (2002) *GBTool2002*, iiSBE.
[6] Cole, R. and Larsson, N. (2002) *GBTool2002 Assessment Manual*, iiSBE.

Part 5
Managing the sustainability knowledge

21 Managing the time factor in sustainability – a model for the impact of the building lifespan on environmental performance

A. A. J. F. van den Dobbelsteen[1] and A. C. van der Linden[1,2]

Summary

Sustainable building has mainly been focused on technological solutions. Until now, as with the space factor, the time factor has played a minor role in environmental performance, mainly due to the lack of models for the implementation of time effects of building designs. This chapter discusses the influence of the lifespan of a building on the real environmental performance of buildings.

Based on financial calculation methods, a mathematical model was developed for the determination of the environmental performance as a function of once-only and annually repeating environmental loads, and taking into account the lifespan and expected service life of the building. This model enables the determinatin of break-even points in environmental comparisons, for instance when a choice needs to be made between the demolition and reuse of a building. The model was theoretically tested and included in environmental assessments of real office buildings and organizations.

Notation

a	Annual environmental load
E_r	Environmental capital remaining
E, e	Total environmental load, total annual environmental load
I, D	Initial environmental load, environmental load at the demolition stage
L, l, L_{ref}	Lifespan of a building, certain moment in the lifespan, reference lifespan
ε	Environmental debit angle

[1] Delft University of Technology, Faculty of Architecture, PO Box 5043, 2600 GA Delft, the Netherlands.
[2] Aacee Bouwen en Milieu, Jan Ligthartplein 39, 3706, 3706 VE, Zeist, the Netherlands.

Introduction

Sustainability in three dimensions: technology, space and time

In the building industry, sustainability has mainly been approached through technical solutions for existing buildings or building designs. The limited progress of environmental performance shows that this focus on technological efficiency alone is not enough [1].

Sustainability has more than one dimension. The space and time factors are also important to the eventual environmental performance but have not often played a great role in sustainable building design. Prolonging the lifespan of objects may contribute to a factor 20 environmental improvement required for sustainable development, but this needs to be integrated into assessment methods to make this visible. So far, few studies have dealt with this methodological problem. Although standard LCA procedures (e.g. Reference 2) prescribe the definition of a reference lifespan for the functional unit or boundary conditions of an assessment; there are no directives for an approach to decisions halfway along the original building lifespan, or in the case of a shortened, prolonged, or varying lifespan. In this chapter we intend to contribute to a better quantitive approach to the time factor in environmental performance.

Mathematical model for lifespan impact on sustainability

The environmental damage to a building consists of once-only and annually repeating environmental loads. An example of once-only environmental damage is the production of waste at the construction or demolition stages. The environmental damage of building materials caused by the pre-construction process of extraction, manufacture, and transport is actually gradual, but can be best allocated to the construction stage. An example of an annual source of environmental damage is the consumption of energy.

The initial, I, annually repeating, a, and demolition, D, environmental loads, which can be determined by means of LCA-based assessment tools, are dispersed over the lifespan of a building, L, as shown in Figure 21.1.

The total environmental load can therefore be characterized as

$$E = I + D + (L \times a) \qquad (21.1)$$

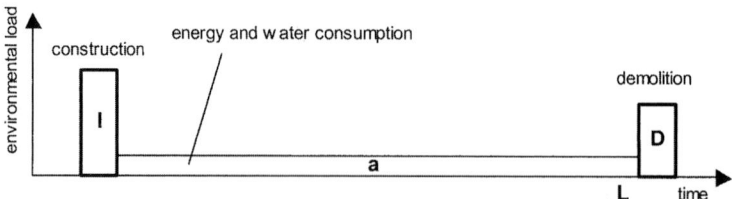

Figure 21.1   Environmental loads during the lifespan of a building.

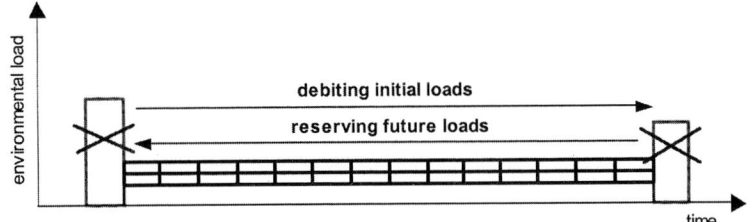

Figure 21.2   Division of environmental loads over the lifespan of a building.

When we want to take into account a varying lifespan of the building, as with financial assessments, one approach is to convert all forms of environmental damage to annual loads. Therefore, the once-only environmental loads need to be equally divided over the expected lifespan, as demonstrated in Figure 21.2.

In this case, The total annual environmental load can therefore be characterized as

$$e = a + (I + D)/L \qquad (21.2)$$

Equation 21.2 clarifies that if the lifespan of a building is prolonged ($L \uparrow$), the total annual environmental load will be reduced ($e \downarrow$). Therefore, in a long-term perspective, much longer than the lifespan of one building, many short lifespans will continuously lead to greater annual loads and overall be less sustainable than a few long lifespans.

When we want to take into account the age of a building, we need to 'debit' or gradually write off the once-only loads over a certain reference lifespan. This is graphically illustrated by Figure 21.3.

In Figure 21.3, the angle ε is called the *environmental debit angle*, a measure of the speed with which the environmental load is debited. It is related to the once-only environmental load and reference lifespan, L_{ref}, and is calculated as follows:

$$\tan \varepsilon = (I + D)/L_{\text{ref}} \qquad (21.3)$$

If the reference lifespan is not yet reached, there will be an environmental load, named the *environmental capital remaining*, still remaining to be debited (see Figure 21.4). It is calculated as follows:

$$E_r(1) = (I + D) - 1 \times \tan \varepsilon = (I + D)(1 - 1/L_{\text{ref}}) \qquad (21.4)$$

In Equation 21.4, $(1 - 1/L_{\text{ref}})$ is the *age correction factor*.

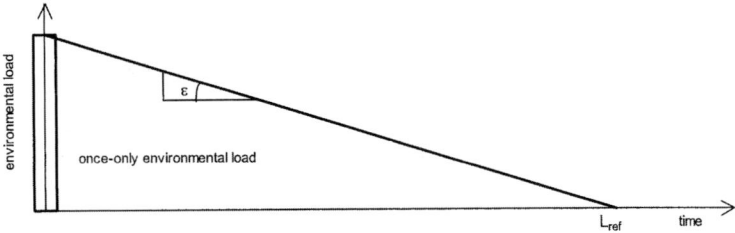

Figure 21.3   Debiting the once-only environmental loads of a building.

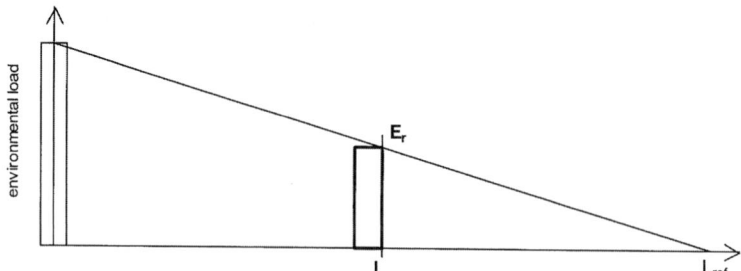

Figure 21.4 The environmental capital remaining to be debited before the reference life-span is reached.

This brief explanation of the mathematical model does not yet make clear how it can be useful to environmental comparisons. Therefore, the following sections will provide theoretical and practical examples of the application of the model.

Theoretical comparison of two building designs

Figure 21.5 illustrates two different building designs. The first has a greater environmental load for the initial and demolition stage and a smaller annually repeating load than the second.

The question is: under which circumstances is which building design more sustainable? The answer can be given when a *break-even point* for the lifespan is determined. The two total environmental loads will therefore be put as equal:

$$E_1 = E_2 \longrightarrow I_1 + D_1 + (L \times a_1) = I_2 + D_2 + (L \times a_2) \longrightarrow L(a_2 - a_1)$$
$$L(a_2 - a_1) = (I_1 - I_2) + (D_1 - D_2) \longrightarrow L = \{(I_1 - I_2) + (D_1 - D_2)\}/(a_2 - a_1)$$
$$L = (\Delta I + \Delta D)/\Delta a \tag{21.5}$$

Equation 21.5 implies that if the lifespan is long $(L > (\Delta I + \Delta D)/\Delta a)$, the environmental load of the second design, E_2, will be greater and, if the lifespan is short $(L < (\Delta I + \Delta D)/\Delta a)$, then the environmental load of the first design, E_1, will be greater. This is logical because the annual load becomes more important in the case of a long lifespan.

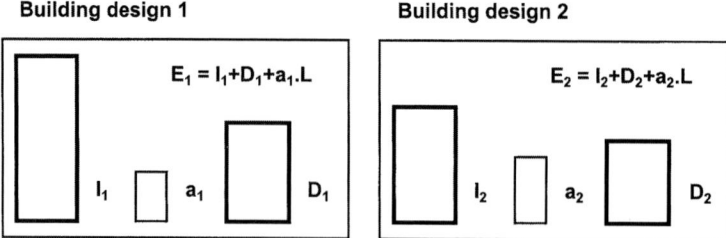

Figure 21.5 Two comparative building designs with differing environmental loads.

Accordingly, a break-even annual load (given a constant lifespan and once-only load) and a break-even once-only load (given a constant lifespan and annual load) can also be determined; however, these are not presented here.

Theoretical comparison of demolition versus reuse

Although a long building lifespan is important, the decision to renovate and reuse a building or to demolish it and construct a new one is not simple. An old building might have its initial environmental load partially debited, it probably also has a less favourable energy consumption pattern than a new building. Since energy is the greatest contributor to the environmental load of buildings [3], demolition and reconstruction will be environmentally preferable in some cases. The question is, however, under which circumstances demolition or renovation will be more favourable. In the following part, we will deduce the break-even points in case a decision needs to be made about an existing building.

First, a reference lifespan, L_{ref} needs to be chosen. When the moment of decision occurs sometime before that lifespan, l, the environmental capital remaining, E_r equals $(I + D)(1 - 1/L_{ref})$, as previously explained. For the demolition solution, the environmental load of the new building ($I_{new} + D_{new}$) needs to be added to E_r.

From the moment L onwards, a new reference lifespan, L^*_{ref}, is taken into account, over which the once-only environmental load at moment L need to be debited. Meanwhile, there is an annually repeating load for both solutions (a_{old}, respectively, a_{new}). Figure 21.6 illustrates the comparison.

The total environmental load of demolition and new construction, E_{new}, and the reuse of the old building, E_{old} equal:

$$E_{new} = E_r + I_{new} + D_{new} + (L^*_{ref} \times a_{new}) \tag{21.6a}$$

$$E_{old} = E_r + (L^*_{ref} \times a_{old}) \tag{21.6b}$$

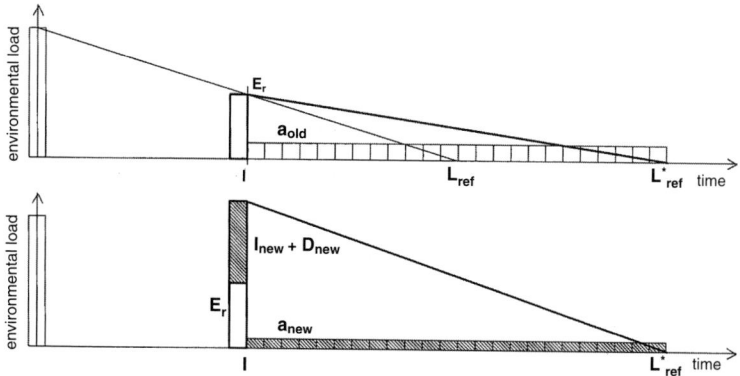

Figure 21.6 Comparative model of reuse versus demolition and new construction of a building.

Note that no environmental load for a possible renovation of the old building is included.

As may be assumed, the annual environmental load of the newly constructed building will be smaller than that for an old building ($a_{new} < a_{old}$). The question is, however, which maximum value of the new annual load (a_{new}) is allowed for a smaller total environmental load for the demolition solution. This can be determined as follows:

$$E_{new} < E_{old} \longrightarrow E_r + I_{new} + D_{new} + (L_{ref} \times a_{new}) < E_r + (L^*_{ref} \times a_{old})$$

$$\longrightarrow a_{old} - a_{new}[= \Delta a] > (I_{new} + D_{new})/L^*_{ref} \qquad (21.7)$$

This outcome provides a clear environmental performance demand to the architect, principal or specialist: the difference in annual environmental load after reconstruction should be greater than the once-only environmental load of the new building divided by the reference lifespan.

The break-even new annual load, $a_{new,be}$, equals $[a_{old} - (I_{new} + D_{new})/L^*_{ref}]$.

Practical application examples

Table 21.1 presents the results of recalculations of the environmental performance (i.e. improvement factors, determined through comparison of the environmental cost with that of a reference project) of seven offices that were originally assessed as if they were newly constructed (the original factor). With the new method, the environmental

Table 21.1 Original and age-corrected environmental performance of different offices.

Government office building					Building materials				
Project	Year of completion				Original			Corrected (2004)	
Specific >	Construction		Renovation		Cost	Factor	Corr.	Cost	Factor
Unit >	yr	%	yr	%	e€			e€	
'Johan de Witt House'	1655	80	1966	20	12 685	1.44	0.10	1252	14.55
Mixed Office	1965	60	1996	40	54 450	1.22	0.65	35 138	1.88
Tax Office	1976	50	1995	50	54 498	1.18	0.75	41 055	1.57
Ministry of SAE	1990	100	–	0	464 221	1.08	0.81	377 566	1.33
Exchequer & Audit	1997	100	–	0	30 042	1.64	0.91	27 238	1.81
Geometry Department	2001	100	–	0	124 588	1.00	0.96	119 604	1.05
Road and Waterworks	2002	100	–	0	66 167	1.65	0.97	64 403	1.69

capital remaining was calculated for the year 2004, as if a decision were to be made about reuse or demolition (the second and third factor). The column with 'corr' gives the age-correction factor previously introduced in Equation 21.4, which is based on a reference lifespan of 75 years and depends on the year of construction or renovation and the percentage of renovated materials in that year (if applicable).

The first office, which is a monument, performed worse when the lifespan was not implemented. As a result of the factor 14.55 in the use of building materials (the once-only loads $I + D$), the building scores factor 1.24 overall (not shown) with respect to a 1990 reference building, which has a more favourable energy consumption pattern. Table 21.1 therefore demonstrates the importance of re-using old buildings, especially when the annual environmental load is not too unfavourable. The Exchequer and Audit office nevertheless performs best, due to a very efficient design. If this building lasts long enough, its performance will improve even more.

In another assessment, we compared different companies that apply an innovative office concept. The main objective was to find the impact of new solutions for office work on space use reduction and other aspects, and consequently, environmental cost. One of the organizations studied was the ICT-company The Vision Web, from Delft, the Netherlands. This company, with 480 employees, uses four small, approximately 100 years old buildings that originally had a residential function but were limitedly renovated for use as office space. Figure 21.7 shows the central office in Delft.

Table 21.2 gives a comparison between two traditional reference projects based on the same function and an equal number of employees. The table shows the year of construction, the proportion of original materials maintained, the year of renovation, the proportion of new materials, the age correction factor and consequential environmental cost. The last columns give the environmental improvement factors with respect to the 1990 or 2000 references. As can be seen, more than factor 20 improvement is achieved through the combination of the reduction of space use and reuse of old buildings. Accounting the lifespan in environmental performance made this visible.

Figure 21.7   Headquarters of the innovative ICT-company The Vision Web, in Delft, Netherlands. Photograph by Andy van den Dobbelsteen.

Table 21.2   Comparison of an office organization and two references.

| Office concept | Environmental performance of building materials | | | | | | Improvement | |
	Constr. yr	Prop. %	Renov. yr	Prop. %	Corr.	Total ke€	1990-ref	2000-ref
1990 traditional reference	1990	100			0.87	117.28	1.00	0.79
2000 traditional reference	2000	100			1.00	92.16	1.27	1.00
The Vision Web						3.89	30.15	23.69

Table 21.2 applies only to the use of building materials. The overall environmental performance is also defined by energy consumption and, especially in the case of a company as The Vision Web, employee travel. The results of each of these aspects are shown in Table 21.3.

As can be seen, compared with building materials, the overall improvement factor is limited. This is due to the great influence of energy consumption and especially employee travel. A modern organization such as The Vision Web relies largely on business travel using private cars. Therefore, the buildings and their age become less significant. A sustainable approach to transport would be more effective.

Nevertheless, the historic buildings in which the company is accommodated enhance the results, which would otherwise have been unfavourable. This would not have been apparent if the building lifespan had been ignored in the assessment.

Accounting the expected service life of a building

So far, buildings have been considered on the basis of a general reference lifespan after the moment of intervention. However, depending on various parameters of quality and functionality, different buildings will eventually have different service lives. The

Table 21.3   Improvement factors of the organization studied.

| Office concept | Materials | | Energy | | Travel | | Overall | |
| | Env. | Impr. | Env. | Impr. | Env. | Impr. | Env. | Impr. |
unit/ref year >	ke€	1990	ke€	1990	ke€	1990	ke€	1990
1990 traditional office concept	116.8	1.00	465.8	1.00	333.8	1.00	916.4	1.00
2000 traditional office concept	92.2	1.27	517.8	0.90	333.8	1.00	943.8	0.97
The Vision Web concept	4.0	29.20	281.2	1.66	398.9	0.84	684.1	1.34

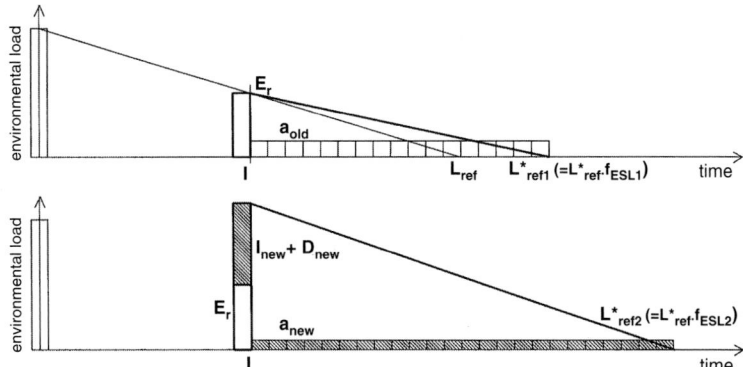

Figure 21.8 Comparison model of reuse versus demolition in the case of differing expected service lives.

comparison of Figure 21.6 would therefore be better represented by Figure 21.8, taking into account a shorter service life for the reuse solution [4].

The eventual lifespan of a building is of course not known beforehand but may be estimated on the basis of experience or calculated by means of aspects influencing the lifespan. An in-between solution is the use of categories for the estimated service life (ESL). Table 21.4 gives an exemplifying, non-oficial overview of ESL-categories and ESL-factors with respect to a reference lifespan of 75 years.

Table 21.4 Categories of estimated service life (ESL) factors [4].

Cat.	ESL*	Description
A	2.00	National monument
B	1.33	Exceptional architecture, flexibility, no functional problems
C	1.07	Exceptional architecture, flexibility, functional problems
D	0.75	Exceptional architecture, no flexibility, no functional problems
E	0.75	Moderate architecture, flexibility, no functional problems
F	0.53	Exceptional architecture, no flexibility, functional problems
G	0.53	Moderate architecture, flexibility, functional problems
H	0.27	Moderate architecture, no functional problems, no flexibility
I	0.13	Moderate architecture, no flexibility, functional problems

* Estimated Service Life factor, reference lifespan $= 75$ years

Table 21.5   Comparison of five accommodation solutions of an existing office in the year 2010, taking into account the age of the old building and expected service life after building [4].

Accommodation solution in 2010	Age and ESL corrected performance				
	Age	Cat.	ESL	Cost	Factor
Reuse without renovation	0.67	G	0.53	166 331	1.01
Demolition and new building traditional	1.00	B	1.33	143 918	1.17
Demolition and new building sustainable	1.00	B	1.33	83 399	2.01
Extensive renovation	0.80	E+	0.85	82 684	2.03
Renovation of building services	0.70	E	0.75	113 042	1.49

As a practical example, Table 21.5 summarizes an environmental comparison executed for accommodation solutions for an existing government office (constructed in 1985) that might be functionally obsolete by the year 2010. The age of the old building and the expected service life after intervention were included in the environmental load of all solutions. In the results given in Table 21.5, energy and water consumption and the favourable influence of new buildings on them were also included in the overall environmental cost.

As can be seen, the traditional demolition alternative leads to the worst environmental performance. The extensive renovation and the sustainable new construction variant lead to a similar performance. If the original building had been older, the sustainable demolition alternative would have led to the best results. If the moment of decision were before 2010, the extensive renovation would have been the most favourable. The limited renovation of building services leads to a performance between the traditional and sustainable new building. It is a better solution than simply reusing the old building. This would not be the result if no environmental capital remaining were involved in the environmental load and if the estimated service life were ignored.

Conclusions and discussion

The age and service life accounting methodology presented in this chapter support an intgegral and just assessment of the environmental performance of buildings. The theory, model and tests discussed form a first exploration of the practical implementation of the time factor in the environmental performance of buildings. The mathematical approach is useful at decision moments in the accommodation process, for instance, when one needs to weigh the reuse of an existing building against demolition and new construction. The model however needs to be more elaborately tested on integrated comparisons of accommodation concepts.

When designing the model, as with financial models, a decision needed to be made about inclusion or exclusion of interest and inflation, influencing both the capitalization and debiting of once-only environmental loads. For simplification, we decided to leave these factors out of the equations. The methodology however offers opportunities for the inclusion of interest and inflation.

Furthermore, building components have differing service lives, requiring separate treatment [5]. Our model for entire buildings however implies multiple replacements for components with a shorter service life than the chosen reference lifespan; however, as presented here, it does not support comparisons between different service life strategies for the building and its separate components. Owing to page limitations, we could not discuss this methodological issue, as well as boundary conditions, design solutions, and functional aspects that enable a long building lifespan. These, and all topics discussed in this chapter, are discussed and clarified in Reference 6.

References

[1] Dobbelsteen, A. A. J. F. van den, Wilde, Th.S. de and Arets, M. J. P. (2002) Sustainability in three dimensions – the importance of improving the use of space and life span beside technological efficiency, Brebbia, C. A., Martin-Duque, J. F. and Wadhwa L. C. (eds) *The Sustainable City II*, WIT Press, Southampton, pp. 305–314.

[2] ISO Environmental Management (1998) Life cycle assessment – goal and scope definition and inventory analysis (ISO 14041:1998(E)), ISO/FDIS.

[3] Dobbelsteen, A. A. J. F. van den, Linden, A. C. van der and Klaase, D. (2002) Sustainability needs more than just smart technology, Anson, M., Ko, J. M. and Lam, E. S. S., *Advances in Building Technology*, Vol. II, Elsevier Science, Oxford, pp. 1501–1508.

[4] Dobbelsteen, A. van den (2004) Reuse of buildings, in *A New Vision on the Building Cycle*. Hendriks, Ch.F., Aeneas, Boxtel, Netherlands.

[5] Erlandsson, M. and Borg, M. (2003) Generic LCA-methodology applicable for buildings, constructions and operation services – today's practice and development needs, *Building and Environment*, **38**, 919–938.

[6] Dobbelsteen, A. van den (2004) *The Sustainable Office – An Exploration of the Potential for Factor 20 Environmental Improvement of Office Accommodation*, Copie Sjop, Delft, Netherlands.

22 Constructability knowledge management in sustainable design

M. H. Pulaski[1] *and M. J. Horman*[1]

Summary

The integration of sustainability considerations into established constructability processes can improve the way sustainable requirements are addressed in projects, and impact the successful delivery of the project. In this chapter, specific constructability practices from the Pentagon Renovation Program are described that have resulted in the successful management of sustainability during project design. They include the use of a full-scale physical model (mock-up), on board reviews and lessons learned from workshops, among others. It is argued that both constructability and sustainability knowledge should be managing in a combined and simultaneous fashion to take advantage of inherent synergistic connections. Four processes and twelve principles for the integration of sustainable design and constructability are outlined.

Introduction

Sustainable design and constructability should be addressed in an integrated fashion on construction projects to maximize the benefits each offers. Traditionally, however, they are not considered to be complimentary in nature. In fact, constructability is often regarded as a constraint to sustainable design and is addressed separately. Efforts to incorporate sustainability in project design focus mainly on schematic design, with emphasis on the design charrette. Constructability efforts are typically dealt with later in the project via periodic design reviews. This separated process can often lead to conflicts between design and construction. Yet when these two initiatives are brought together and addressed simultaneously, a balance is achieved creating common goals and objectives, resulting in significant improvements to project performance.

Sustainable design and constructability have complimentary processes that help to draw the two initiatives together. Pulaski *et al.* (2003) showed that managing the synergy between sustainable design and constructability can lead to significant improvements in project performance and quality. Constructability tools such as workshops, design

---

[1] Department of Architectural Engineering, Pennsylvania State University, University Park, PA, USA.

reviews, mock-ups, 3D/4D Computer Aided Design (CAD) models and feedback/ lessons learned systems provide the means for effectively assimilating sustainable requirements into the design. The use of established constructability mechanisms provides a reliable, well recognized framework for managing sustainability knowledge in building projects.

This chapter examines specific constructability practices that were used to manage sustainable design at the Pentagon renovation. This 20-year, $2.1 billion programme is managing several projects at the Pentagon. One of these is the 600 000 square metre Pentagon building that will become the largest sustainable office building in the US. Specific constructability practices have resulted in the successful management of sustainability during project design. The analysis in this paper distills these practices into four processes and a set of twelve sustainable design and constructability principles in order to enable replication and use on other projects.

Constructability, knowledge management and sustainable design

For constructability and sustainability to be integrated effectively, a methodical knowledge management system is needed. Constructability and sustainable design are integral because they share a common objective of reducing waste through the efficient use of available resources (Pulaski *et al.*, 2003). However, to achieve successful integration and implementation at the project level, a more detailed understanding of the process is necessary.

Managing construction knowledge on projects is complicated. Establishing a structured knowledge management system will help to improve the management of constructability information on projects. Constructability is defined as the 'optimum use of construction *knowledge* in engineering procurement and field operations' (CII, 1986). The definition explicitly states that the management of 'construction knowledge' is essential for effectively implementing constructability. Each project player (contractor, subcontractor, fabricator and supplier) has valuable information to contribute to the project. Moreover, this information needs to be brought to bear at various particular points in time in order for the knowledge to be optimally utilized. Owing to this complexity, a significant amount of research and practical guidance has been developed to help the sourcing and use of the knowledge (CII, 1993; Fischer and Tatum, 1997; Gil *et al.*, 2001; O'Connor *et al.*, 1988). These research efforts, along with many others, improve the management of constructability knowledge on projects. Importantly, these guidelines and practices can also be extended to improve the management of sustainable design knowledge.

A methodical knowledge management system also needs to support the social structures and processes within which knowledge is shared, known as *communities of practice* (Brown and Duguid, 1991). A project team is a community of practice. The usefulness of this feature in knowledge management systems is in supporting and creating processes and activities for individuals to take action and apply their knowledge (Khalfan *et al.*, 2003). Without such processes and activities, valuable information is not usually shared or used. Thus, a constructability knowledge management system must support social structures within which knowledge is shared in order to optimally access and make use of constructability knowledge.

Many practices for managing project constructability (e.g., mock-ups, constructability reviews, integrated project teams, and lessons learned systems) can be modified to assist project teams in managing sustainable design knowledge. There are a growing number of examples where the integration of sustainable design with constructability processes has resulted in mutual benefits to both initiatives. The Pentagon Renovation project provides a number of these examples. Importantly, an analysis of the underlying mechanisms in these events suggests that they are not just random and isolated, but have emerged from the cross-functional management approach adopted at the Pentagon.

Sustainability and constructability knowledge at the Pentagon

The Pentagon is the world's largest office building containing over 600 000 m² of floor space and covering 11 hectares of land. The building contains more than 28 km of corridors and 7754 windows. The building has its own heating and refrigeration plant, water and sewage facilities, police force, fire station, heliport, childcare centre, cafeterias, mini-mall, Metro station, and medical clinic. With a daily population of approaching 25 000 people, it is larger than nine out of ten American towns, and it is more complex than many small cities. It is the nerve centre of the national military establishment. After almost 60 years of operation, the building was in dire need of a thorough renovation.

In 1993, four years before the renovation work began, the Pentagon Environmental Management Committee was formed to advise the Pentagon on environmental issues. The first of five total wedges was scheduled for completion in mid-October 2001. While the September 11 terrorist attacks destroyed a nation's sense of security, it also demolished the first wedge of the Pentagon renovation project. One year later, the damaged area of the building had been completely rebuilt. Meanwhile, the renovation of the remaining portion of the Pentagon (Wedges 2–5) began demolition and reconstruction. The renovation of the Pentagon is currently scheduled for completion in 2010.

The Integrated Sustainable Design and Constructability Team

A vital key to addressing sustainable design and constructability successfully in projects is that a team be empowered to champion these causes. In August of 2001, the Integrated Sustainable Design and Constructability (ISDC) Team was formed within the Pentagon Renovation Program (the owner's managing organization). This team, which draws members from the owner, users, operators, designers, contractors and Penn State researchers, was formed to lead and help manage sustainable design and constructability on the renovation projects. The connection of sustainable design and constructability has helped the Pentagon renovation projects successfully address sustainability and constructability in an integrated manner and led to the development of new project management tools (i.e., continuous value enhancement process (Pulaski and Horman, 2003)). Since the inception of the ISDC Team, the Pentagon currently has four projects registered under the U.S. Green Building Council's LEED™ Rating System.

The formation, functions, and processes of the ISDC Team directly mirror those proposed for state-of-the-art constructability and sustainable design practice. CII's Constructability Implementation Guide states that

> *'the objectives of this (constructability) program shall encourage teamwork, creativity, new ideas, new approaches, and will emphasize total project integration, not optimization of individual parts.'* (CII, 1993)

This purpose closely resembles the objectives of sustainable design as outlined in the newly defined Roadmap for Integrating Sustainable Design into Site Level Operations (Peterson and Dorsey, 2000). Other implementation processes were present in the ISDC Team and these are identified in Table 22.1.

Table 22.1 lists development steps, functions and processes performed by the ISDC Team and their intersection with constructability and sustainability processes. It can be seen that this intersection is extensive. This further reveals the inherent connection between sustainable design and constructability, and strengthens the argument for addressing sustainable design and constructability in an integrated manner.

Constructability knowledge management for sustainable design

The establishment of the Integrated Sustainable Design and Constructability (ISDC) Team is one method of aligning constructability and sustainability processes. Three additional practices are identified and listed in Table 22.2, along with corresponding constructability and sustainability related references. Each practice is introduced and its capability at enhancing sustainable design and constructability is evaluated. For each specific practice identified, a detailed description of the processes that were undertaken on the Pentagon renovation project is outlined. The ways in which sustainable design knowledge was introduced into the processes are described. Finally, supporting literature is referenced to show that the practices are theoretically grounded, and that others have come to similar conclusions about the importance of the processes employed.

Full-scale physical model (mock-Up)

The process

Full-scale physical models and mocks-ups are not typically employed in sustainable design. They are usually a mechanism for complex projects (such as nuclear power

Table 22.1 Comparison of ISDC team processes to established roadmaps.

Constructability Implementation Guide (CII 1993)	Roadmap for Integrating Sustainable Design into Site-Level Operations (Peterson and Dorsey 2000)	ISDC team milestones and deliverables
Identify constructability sponsor/champion	Assign Sustainable Design integration lead/point of contact	Dr Pohlman named Leader for ISDC Team
Develop constructability team. Establish functional support organization and procedures. Assemble key owner team members	Identify Sustainable Design integration team	Identify members and organize regular meeting schedule
Perform self-assessment and identify barriers	Perform Baseline Assessment and Gap Analysis	Submit White House closing the circle award. Compare PenRen with other government agencies
Set goals for constructability effort	Establish Sustainable Design integration goals	Develop schedule, deliverables and team goals
Document lessons learned	Create Sustainable Design plan or manual	Document case studies and develop programme guidebook
Conduct ongoing briefings and awareness seminars at all levels	Provide Sustainable Design training for project	Develop strategy for educating and training construction workforce
Develop constructability procedures and integrate into project activities. Reference to constructability in contract documents	Use existing project design processes to adopt Sustainable Design principles. Integrate into design/build contracts	Incorporate sustainability into standard programme practices
Define constructability objectives and measures	Measure progress/Re-evaluate goals	Establish metrics

Table 22.2   Constructability knowledge management processes for sustainable design at the Pentagon.

Pentagon process	Constructability literature support	Sustainability literature support
Establish Sustainable Design and Constructability Integrated Product Team	(CII, 1993)	(Peterson and Dorsey, 2000)
Full-scale physical model (Mock-up)	(O'Connor *et al.*, 1998)	(Wilson *et al.*, 1998)
On-board review process	(Arditi *et al.*, 2002)	*No supporting literature identified*
Lessons-learned workshop	(CII, 1997); (Arditi *et al.*, 2002)	(Peterson and Dorsey, 2000)

Note: not all steps from CII (1993) and Peterson and Dorsey (2000) are included in this list.

plants) to solve constructability, coordination and scheduling issues. They are expensive and, consequently, are used sparingly. After the mock-up has been constructed, it is typically dismantled and disposed of, creating unnecessary work and waste. At the Pentagon, a 1000 m² mock-up was constructed as part of the actual renovation within the building. Originally this mock-up, now referred to as the Universal Space Plan Laboratory (USP Lab), was used to fine-tune the design details and construction sequence. The construction advantages were substantial as the sequence improvements would impact many times over on this project due to the level of repetition. The mock-up was also used to expedite the design process by building from schematic drawings, testing various configurations, field coordination issues and then producing the construction details from the as-built condition.

Integrating sustainability into the process

After a major portion of the design was complete, the USP Lab was then used in three ways. The first was to test the innovative combinations of materials both from a design-functionality perspective and from a user-acceptance perspective. The second was to test the durability of materials, as many untested products were being incorporated on this project. The third was to educate the project team, visitors and tenants. The materials were installed by the contractor to reveal constructability issues. While installed, the materials were inspected by the PenRen team, the design team, and the building owners and operators. Individuals could observe how the materials performed functionally and aesthetically. This was a key process for obtaining 'buy in' and support from the owners and tenants for new sustainable materials. The USP Lab also served to educate project team members and visitors on the design concept and the sustainability aspects of the project as well as provide a model for tenants to view before moving into the space.

Literature review

The literature revealed similar findings to that of the Pentagon about the use of mock-ups. Mock-ups can be useful tools for both sustainability and constructability. O'Connor *et al.* (1988) found that the use of mock-ups help contractors fine-tune the construction process. Mock-ups also allow designers and clients to understand how a building will look and perform (Wilson *et al.*, 1998). Wilson *et al.* (1998) explicitly identifies building models, three-dimensional CAD models and mock-ups as useful design tools that have been employed by green designers. The mock-up also provides a proven, cost effective solution for managing sustainability knowledge.

On-board review process

The process

The on-board review process and comments session is another method whereby sustainability knowledge can be integrated in the project process and managed effectively. Similar to conventional constructability reviews, on-board reviews typically occur at 35%, 75%, 95% and 100% design completion. All major project stakeholders are involved in the on-board review process. The process begins with the design team issuing drawings to stakeholders for review. Drawings and documents are reviewed and comments are entered into a database. The design-build team enters a response to each comment. The comments are sorted by discipline and discussed in detail during the on-board review meeting. This meeting is held with all the major stakeholders present to discuss all comments individually. During the on-board review process many constructability and sustainability issues are raised, discussed, evaluated and then incorporated into the project where they are best addressed.

Integrating sustainability into the process

'Sustainability' was incorporated into the process by adding it as a defined discipline for review comments. Specific sustainability comments can now be managed in an efficient manner. For example, during one review meeting at 100% design documents, a number of issues were raised concerning the specification of products with low levels of volatile organic compounds (VOCs), which were necessary to achieve a LEED credit. The design team had not detailed the specific levels of VOC for each product in the specifications because they did not want to specify a product that was not available. The reviewer, an expert in sustainable materials, had several comments regarding appropriately low levels of VOCs for each product. After detailed consideration at the on-board review meeting, and subsequent follow-up meetings, the team agreed to use the low levels of VOC identified by the LEED rating system.

Literature review

Review processes are feedback systems that are amongst the most prevalent tools used to achieve high levels of constructability (Arditi *et al.*, 2002). The process allows input from major project stakeholders, such as building operators, maintenance staff and third party industry specialists, who have valuable constructability and sustainability knowledge to contribute throughout the project. The literature search revealed no documentation of using a review process to manage sustainability knowledge. Typically, sustainability efforts are focused heavily on the initial schematic level design effort and diminish as the project progresses. While the early emphasis is important, it is critical that the effort be continued throughout the entire design. The process of having a continuous sustainability review helped to integrate sustainability requirements into the project effectively. This provided a mechanism for individual experts to contribute their relevant sustainability knowledge to the project in an effective and efficient manner.

Lessons-learned workshop

The process

The management of lessons learned is an essential component of any knowledge management system and can substantially contribute to the success of future projects if used effectively. One method typically used in organizations is a lessons-learned workshop conducted on a regular basis. On the Pentagon renovation project, sustainable design and construction lessons-learned workshops have been implemented on a quarterly basis. Issues are discussed, alternative solutions presented and action items assigned. This provides an effective means for transferring lessons learned during this lengthy project. This continuous improvement process has proven to be very effective for the project resulting in small, but significant, process improvements.

Integrating sustainability into the process

The Pentagon renovation project workshops incorporate a focus on issues related to sustainability. One issue addressed at the first sustainable lessons-learned workshop in April 2002 dealt with increasing communication related to current design issues. In response, the distribution list for the design meeting agendas and minutes were expanded to include the additional personnel. By doing so, many more project team members could stay informed on current design issues and the design meetings where their expertise on sustainability and other design issues could best be applied. Additional examples include modifying the contractors' monthly report to include a section on sustainable design and construction, and the inclusion of ISDC team members at the design meetings.

Literature review

Sustainability is new to many project teams, therefore a learning curve is certain to exist and mechanisms such as lessons-learned workshops can help to speed learning and streamline the process of managing sustainable design knowledge. An increasing number of experiences on successful sustainable projects place education as a key

Table 22.3   Sustainable design and constructability principles.

Simplifying trade interfaces
Design for prefabrication and modularization
Simple connection details
Align space dimensions with building materials
Design open floor plans
Use local materials and methods
Design for standard elements and precut items
Maximize straight piping and ductwork runs
Reuse building materials
Design for adaptability
Minimize material use
Ensure proper sizing of materials/components

enabler of these projects. More broadly, a recent study reported that 87% of design firms use some form of lessons-learned feedback system and 68% use brainstorming sessions (Atditi *et al.*, 2002). CII (1997) reports that the most effective (lessons learned) workshops are facilitated meetings, where participants are prepared in advance. These facilitated meetings should be conducted periodically throughout the project, when major issues are current. Identifying lessons learned for sustainable design are critical to project success. Literature on sustainability suggests using brainstorming sessions, benchmarking techniques and lessons learned to identify opportunities for design improvements (Wilson *et al.*, 1998). Brainstorming sessions focus on specific design strategies for recyclability, disassembly, maintenance, energy and water conservation, and hazardous material reduction. In short, lessons-learned workshops and brainstorming sessions are important for managing sustainability knowledge.

Principles for sustainable design and constructability

The integration of sustainable design and constructability initiatives at the Pentagon has resulted in the identification of twelve principles that are beneficial to both initiatives. These broad principles (Table 22.3) have been derived from specific examples on three projects on the Pentagon Reservation. Key personnel from each project were asked to identify specific examples that had positive effects to both constructability and sustainability. Thirty-one examples were identified and grouped into broad categories. These principles capture much of the innovation that is being practised at the Pentagon. They are being used to develop new processes and metrics to help further integrate sustainable design and constructability into the programme.

Future research

The list of four constructability practices represents only a fraction of the potential constructability processes that can be used to enhance and manage sustainable design knowledge. More processes and case study examples should be identified. Additionally, the principles for sustainable design and constructability represent a similarly small fraction of possibilities. Further research is needed to identify and refine more principles. Principles must also be further analyzed to identify at what times or phases of the design process they are most applicable. New sets of project metrics are also needed to identify methods to encourage the integration of sustainable design and constructability at the project level.

Sustainable design knowledge and practices have significant value to add to the constructability of a building. For many years, industry professionals and researchers have striven to develop better ways to integrate design and construction. These have had moderate success. However, sustainable design practices require integrated design and decision making to a much higher level than ever before in projects. This is both a challenge and an opportunity for further improvement. The promise of improvements here is the greatly enhanced use of constructability expertise so as to delivery greater value with fewer resources. Further research is needed to identify specific processes and effective ways to incorporate them in projects.

Conclusion

Sustainability knowledge can be effectively managed by integrating it with constructability processes. An analysis of the Pentagon renovation project revealed four constructability processes to manage sustainability knowledge and twelve principles to support the integration of sustainable design and constructability. An integrated project or organization team was found to be an important success factor for managing sustainable design and constructability knowledge. On the large Pentagon renovation project, constructing a full scale physical mock-up within the building proved to be an effective tool that benefitted both the sustainability and constructability of the project. The integration of sustainable and constructability in a simultaneous manner created opportunities for mutually beneficial processes and decision making to occur. This led to efficiencies and synergies in the project process and resulted in improvements to performance.

References

Arditi, D., Elhassan , A. and Toklu, Y. C. (2002) Constructability analysis in the design firm, *J. Constr. Eng. and Mgmt.*, **128**(2), 117–126.

Brown, J. S. and Duguid, P. (1991) Organizational learning and communities of practice: toward a unified view of working, learning, and innovation, *Org. Science* **2**(1), 40–57.

Construction Industry Institute (CII) (1986) *Constructability – A Primer*, Publication 3-1, Austin, TX.

Construction Industry Institute (CII) (1993) *Constructability Implementation Guide*, Publication 34-1, Austin, TX.

Construction Industry Institute (CII) (1997) *Modeling the Lesson Learned Process*, Publication 123-1, Austin, TX.

Fischer, M. and Tatum, C. B. (1997) Characteristics of design-relevant constructability knowledge, *J. Constr. Eng. and Mgmt.*, **123**(3), 253–260.

Gil, N., Tommelein, I. D., Miles, R. S., Ballard, G. and Kirkendall, R. L. (2001) Leveraging specialty contractor knowledge in design, *J. Eng. Construc. and Arch. Mgmt.*, **8**(5/6), 355–367.

Khalfan, M., Bouchlaghem, N., Anumba, C. and Carrillo, P. (2003) Knowledge management for sustainable construction: the C-SanD Project, *Proceedings of the 2003 Construction Research Congress*, Honolulu, Hawaii.

O'Connor, J. T. and Davis, V. S. (1988) Constructability improvement during field operations, *J. Constr. Eng. and Mgmt.*, **114**(4), 548–564.

Pentagon Renovation Program (2001) Integrated Sustainable Design and Constructability Team milestones and deliverables, discussion paper.

Peterson, L. and Dorsey, J. (2000) *Roadmap for Integrating Sustainable Design into Site-Level Operations*, Pacific Northwest National Laboratory, Publication 13183, Richland, Washington.

Pulaski, M. H. and Horman, M. J. (2003) Continuous value enhancement process, submitted to *J. Construc. Eng. and Mgmt.*

Pulaski, M., Pohlman, T., Horman, M. and Riley, D. (2003) Synergies between sustainable design and constructability at the Pentagon, *Proceedings of the 2003 Construction Research Congress*, Honolulu, Hawaii.

Wilson, A., Uncapher, J., McManigal, L., Lovins, L., Cureton, M. and Browning, W. (1998) *Green Development Integrating Ecology and Real Estate*, John Wiley & Sons.

23 Sustainability, the built environment and the legal system

D. E. Fisher[1]

Summary

The legal system cannot assure sustainability. A sustainable built environment is an outcome of a series of processes many but not all of which are regulated by the legal system. But whether the built environment is sustainable can only be determined after its construction and use. Prior to construction and use, sustainability can be at best only a prediction. The law has traditionally been concerned with the creation of enforceable standards of decision making and behaviour. Standards such as these are sufficiently precise so that it is possible to know in advance how to comply with them. But the nature of sustainability makes this extremely difficult, if not impossible. The law nevertheless has a part to play. The environmental legal system is essentially fragmented and complex. In particular it lacks the necessary integration of procedures and outcomes associated with sustainability. Nevertheless there is a range of liability, regulatory and incentive rules that are enforceable. These rules enable but do not necessarily require the built environment to be sustainable.

The challenge of sustainability

Introduction

The skill of the architect is to use the resources of the environment, including water and energy, to create something that itself becomes part of the environment. The architect does so in accordance with the directions and values of the owner of the land. But now the expectation is that the environment, not only before but also after the skills of the architect are realised, will be neither damaged nor destroyed but positively sustained as a consequence. For this purpose the environment is a potentially all-embracing phenomenon that includes aesthetic, cultural, social and economic as well as ecological

---

[1] Faculty of Law, Queensland University of Technology, Brisbane, Australia; Phillips Fox Lawyers, Brisbane, Australia.

values. So there is another perspective: the interests of the community – local, national and global – in ensuring the satisfaction of these wider expectations.

Sustainability

It is no easy task to ensure that the environment is neither damaged nor destroyed. It is even more difficult to ensure that it is sustained for future generations. Sustainability is a complex concept. It has its critics. But there is little doubt that sustainability has become the driving force of the environmental legal system at international, national and local levels. Ecologically sustainable development – as it is described for the purposes of the Australian environmental legal system – is:

> 'Using, conserving and enhancing the community's resources so that ecological pro-
> cesses, on which life depends, are maintained, and the total quality of life, now and
> in the future, can be increased.' [1]

But what does this mean? First, the resources of the environment used for the purpose of constructing the building must not be exploited to the extent that the environment out of which they are taken is either destroyed or so damaged that any subsequent use is not possible. This applies not only to the element of the environment – the land – upon which the building is constructed but also to all the other resources used in its construction. Second, once the building has been constructed, the resources of the environment required for the continual use of the building after construction must similarly neither destroy nor damage the environment out of which they are taken. This may well be an impossible task. Nevertheless it is the foundation of the concept of sustainability.

The function of the legal system

What part does the legal system play in this? The law is part of the culture of society. But it also stands outside it in the sense that the legal system recognises or creates structures within which society achieves its goals. In traditional terms the function of the law has been to set standards of behaviour and decision-making and to create regimes of liability coupled with sanctions for failure to comply with these standards. This is relatively easy in the case of standards of behaviour that affect other humans. For example, land must not be used in such a way – by re-contouring the landscape, by diverting watercourses or by using premises – so that the use by neighbours of their land is unreasonably affected.

But the law, largely through the influence of legislation, has become much more interventionist. Standards are prescribed for a whole range of circumstances designed to achieve particular outcomes. Many are associated with the environment: for example, controls upon the introduction of polluting matter into the environment; then restrictions upon the use of land and other resources of the environment; more recently, sustainability of the use of resources. In some cases the outcome – sustainable development – has itself become the restriction.

Sustainability of the built environment

And so it has become for the construction of buildings. Thus:

> 'The quest for environmental value in architecture, for a harmonious balance between man and his surroundings, is not new. For centuries, and particularly in domestic and vernacular architecture, people adopted this approach out of necessity. Since the industrial revolution, it has been increasingly abandoned in favour of man's belief in his own omnipotence and ability to draw unrestrainedly on the earth's resources.' [2]

So what has changed over the last 50 years? During this period much of the environmental legal system has been directed at protecting the environment from perceived risks of degradation or damage. It is only in the last 10 years that the direction has been sustainability. There have been many responses to this and one has been 'to approach architecture and urbanism in a way which respects the environment.' [3] But it is respect for the environment in the context of sustainability that constitutes a major challenge not only for planners and architects but also for the legal system with which their projects have to comply.

Sustainability poses problems for all those who have responsibility for the built environment: the owner, the developer, the architect, the landscape architect, the interior designer, the planner as well as public sector regulators. But in practical terms it is the planners and the architects who are expected to solve these problems. And they must do so within the confines of the legal system: in particular, the emerging criterion of sustainability. Sustainability seeks to ensure that the resources of the environment are available for use for future generations. This is a predictive rather than a reactive approach. Whether a building has been sustainably constructed is determined only decades after its construction. Although planning has been concerned to predict the best use of land and resources, the introduction of sustainability has made this function even more difficult to perform.

Those charged with the legal responsibility for ensuring the sustainability of the built environment need to address a number of questions:

- Is the use of the land for the construction of this building the best use of the land?
- What is the best way in which the infrastructural requirements of the building for water, energy and transport can be met?
- What is the best way in which the waste created by the construction and use of the building can be managed?
- What resources of the environment are best used in the construction and use of the building?

The approach of the environmental legal system to these questions has been essentially fragmented in the sense that the standards derive from a number of different institutional sources often acting independently of each other. Sustainability requires the integration of fragmented procedures and the simultaneous consideration of all the issues to enable its achievement as the single outcome of these processes. This is simple to state but difficult to achieve.

The environmental legal system

Its complexity

One of the difficulties in achieving sustainability is the complexity of the environmental legal system itself. It remains essentially disjointed despite attempts at international and national levels to ensure greater integration in terms of procedures and outcomes. The law for the most part approaches independently of each other issues of land use, soil quality, water use, water quality, energy production and consumption, flora and fauna conservation and building standards. The administrative procedures associated with these matters vary: for example, the extent of public participation and the mechanisms for administrative and judicial enforcement. Perhaps even more importantly, each of these elements of the environmental legal system seeks to achieve, either directly or through mechanisms of policy, different and potentially conflicting outcomes.

If these problems are transposed specifically to the built environment, the position is no less complex. To ensure that one building is constructed and used in accordance with the principle of sustainability, the issues that need to be addressed go well beyond the exact boundaries of the land on which the building is to be constructed. The building is part of the wider environment in many senses, for example, part of its wider precincts and curtilage, part of its wider urban community, part of its regional community and part of the national and international communities at large. Thus the criteria for heating and ventilating a building may be expected to reflect the responsibilities of the state for greenhouse gas emissions under international law. The two issues are on the face of it unrelated but the complexity and the interdependency of all of the elements of the environmental legal system render them of practical importance.

Its application to the built environment

The achievement of sustainability of the built environment thus depends upon a whole range of factors. These include:

- building-specific criteria
- site-specific criteria
- criteria for controlling urban development in the relevant area
- strategic criteria for managing the natural resources of the community such as land, forestry, flora, fauna, minerals, water and energy.

For example, a strategic commitment to a 20% reduction over 5 years in the use of non-renewable sources of energy production may require 25% of all new buildings in one suburb to be built in one linked location and for 50% of all new buildings to be fitted with solar collector panels or photovoltaic modules. One option is to leave these difficult decisions to the developers and their architects and engineers. If the interests of the developers – commercial, architectural or aesthetic – coincide with those of the community in using renewable sources of energy, there is no legal problem. But if there is no coincidence of interests, what is the function of the legal system?

Traditionally, the law sets standards with which human behaviour must conform. These are supported by duties, breach of which leads to a liability to criminal or civil proceedings and consequential sanctions. These standards are applied for the most part to circumstances that have arisen and any judicial remedy or sanction is a reactive response to a breach of such a duty. This approach is possible for sustainability. For example, if a development goes ahead and it is proved to be unsustainable in accordance with the relevant criteria, then an appropriate remedy or sanction may be available. But much of environmental law is proactive rather than reactive. One of its original approaches is the prevention of pollution rather than the requirement to take measures to clean up the environment after it has been damaged. The achievement of such an objective is assisted by mechanisms such as environmental impact assessment so that environmental degradation can be anticipated during development planning processes and plans put in place to avoid any foreseeable damage to the environment. But predictions may be unfounded and precautionary measures may be inadequate. Because of its nature, therefore, prevention of damage to the environment cannot be assured.

Nor can sustainable development be assured. The achievement of a sustainable built environment is no more in legal terms than a prediction that certain consequences will come about. The law is devising a range of instruments by means of which it is hoped but cannot be assured that sustainability will be achieved [4]. These include:

- Creating a duty to achieve sustainability.
- Creating a duty to take into consideration sustainability or the need for sustainability in making decisions about developing the resources of the environment.
- Creating a duty to take into consideration in a similar way the principles of sustainability.
- Creating a duty to undertake either generally or in particular sets of circumstances environmental impact assessments, environmental audits, environmental investigations or environmental monitoring to assist in discharging the other duties created by the legal system.
- Creating guidelines as statements of ways in accordance with which the duty to achieve sustainability can be discharged.
- Creating or recognising codes according to which sustainability or particular elements of sustainability can be achieved.
- Providing that compliance with current professional standards constitutes compliance with the duties imposed by the law.
- Creating taxes, charges and other financial disincentives to refrain from activities or practices that are regarded as unsustainable.
- Making available financial incentives such as grants or tax deductions for activities or practices that are regarded as sustainable.

The mechanisms of environmental law

The mechanisms used by the legal system in any particular set of circumstances reflect the political and legal culture of the community as much as the particular circumstances of the built environment whose sustainability is to be achieved. And the manner and the extent to which any of these instruments prescribe enforceable rights and duties are again a matter for the detail of the legal system in question. Let us turn from the issues to the particularities of the environmental legal system.

International norms

Consider some of the provisions of the Draft International Covenant on Environment and Development, [5] which – despite its status as a draft – incorporates many of the most important principles of contemporary international environmental law. Article 1 states that the objective of the covenant is to achieve environmental conservation and sustainable development by establishing integrated rights and obligations. This focuses upon two particular points. The first is that environmental conservation and sustainable development are outcomes to be achieved by whoever is responsible for managing the environment. The second is the mechanism for doing this, namely a system of integrated rights and obligations. It is these three interdependent concepts – integration, rights and obligations – that lie at the heart of the covenant.

The right to development in article 8 is particularly important in the context of sustainability of the built environment. It provides that the exercise of this right entails the obligation to meet the developmental and environmental needs of humanity in a sustainable and equitable manner. The positive nature of this obligation recognizes quite specifically four elements:

- Humanity has environmental needs.
- Humanity has developmental needs.
- These needs must be met in an equitable manner.
- These needs must be met in a sustainable manner.

The covenant goes on to identify an extensive range of obligations together with means for their implementation and ultimately enforcement. Paragraph two of article 13 creates an obligation to ensure that environmental conservation is treated as an integral part of the planning and implementation of activities at all stages and at all levels, giving full and equal consideration to environmental, economic, social and cultural factors. This applies as much to the ongoing use of a building as to the planning for its construction. Environmental conservation is a part of this process. Linked to this is the additional obligation that full and equal consideration must be given to environmental, economic, social and cultural factors.

This approach is taken further by the obligations in article 36 about physical planning. One requires integrated physical planning systems and this specifically includes provisions for infrastructure and town and country planning. Secondly, physical planning specifically acknowledges the relevance of natural systems, natural characteristics, ecological values and ecological constraints.

While the covenant is couched in relatively general terms, its meaning and effect are clear enough. The achievement of sustainability requires integration of procedures and processes as well as integration of the objectives of managing the resources of the environment at large. How is this clear framework extrapolated in practical terms to the detail of a particular legal system? Let us turn to the environmental system of Queensland to seek an answer.

Commonwealth of Australia

In a federation such as Australia, the environmental laws of the Commonwealth have effect within Queensland. There is a substantial body of Commonwealth environmental law and policy that has the potential to impact upon the quality of the built environment. It is always open to the Commonwealth to influence environmental management through policy initiatives, funding mechanisms and taxation measures. The Commonwealth controls aspects of environmental management more directly through strategic as well as operational measures.

The scope of obligations undertaken by the Commonwealth in relation to international concerns about marine pollution, the ozone layer, greenhouse gas emissions, biodiversity, world cultural and natural heritage has prompted the Commonwealth over the last three decades to enact laws directly related to some of these issues. One of the most significant of these is the *Environment Protection and Biodiversity Conservation Act* 1999 (Cth). This regulates at a national level environmental values associated with world heritage areas, endangered flora and fauna, nuclear activities and the marine environment. More recently the Commonwealth has attempted to achieve specific targets for the use of renewable sources of energy for the production of electricity through the *Renewable Energy (Electricity)* Act 2000 (Cth). The Commonwealth continues to exert considerable influence through policy formulation supported by targeted conditional funding arrangements with the states and territories that have primary responsibility for the quality of Australia's environment on a day-to-day operational basis.

Queensland

The environmental legal system of Queensland does not provide for the management of the natural resources of the state at a strategic level. However the way in which interests in land, minerals and water are allocated is specifically governed by statutory regimes that are in some respects strategic. Sustainability and environmental protection are elements of the system for allocating interests in land [6]. While sustainability plays no direct part in making decisions about the development of mineral resources, [7] protecting the environment in mining operations is governed by the *Environmental Protection Act* 1994 (Qld) [8]. Sustainability is fundamental to this Act.

Water management

The *Water Act* 2000 (Qld) is the best example of sustainability in allocating interests in a resource. The purpose of the relevant part of the Act is to advance the sustainable management and efficient use of water and other resources by establishing a system for the planning, allocation and use of water [9]. The notion of sustainable management is expanded by the principles of ecologically sustainable development. One of the principles of ecologically sustainable development with which the economic development of Queensland must be in accordance is the effective integration of long term and short term economic, environmental, social and equitable considerations [10]. This statement of purpose is complemented by a statutory duty linked to the advancement of this purpose [11]. While the precise effect of these provisions remains to be seen, they represent a rigorous and potentially enforceable regime that specifically incorporates by statutory prescription sustainability as the fundamental element of the water management system created by the legislation. Not only that, the sustainable management of water is linked to the sustainable management of other resources [12].

Development control

Queensland has had a system of planning and development control for several decades. Its function is to ensure that land is developed in the best interests of the community. The legislation effectively removes the right to develop land that the owner of the land has under the common law and substitutes for that right the rights that are available under the legislation. The built environment is the primary but not the only focus of the legislation.

Queensland's planning system is governed by the *Integrated Planning Act* 1997 (Qld). The purpose of the Act is to seek to achieve ecological sustainability [13]. This is a balance that integrates three elements:

(1) protection of ecological process and natural systems
(2) economic development
(3) maintenance of the cultural, economic, physical and social wellbeing of people and communities [14].

There is a duty to advance the purpose of the Act [15]. Significantly, section 1.2.3 of the Act indicates how its purpose may be advanced. This includes acknowledging some of the principles of ecologically sustainable development. Two are particularly important for the built environment. These are:

(1) supplying infrastructure in a coordinated, efficient and orderly way, including encouraging urban development where areas of adequate infrastructure exist or can be provided efficiently; and
(2) applying standards of amenity, conservation, energy, health and safety in the built environment that are cost effective and for the public benefit [16].

While the Act is silent on the detail of these standards, there is a clear expectation but perhaps not a legally enforceable duty that these standards should be applied in making decisions about development approval. They relate to the built environment but are linked to wider aspects of managing natural resources, for example, the sustainable use of renewable natural resources and the prudent use of non-renewable natural resources [17].

Decisions about development approval are made with reference to the detailed regulatory regime contained in planning schemes as well as to the purpose of the Act. Planning schemes are expected to achieve the purpose of the Act. There is an expectation if not a duty for a planning scheme to deal with core matters [18]. These are:

- land use and development
- infrastructure
- valuable features [19].

The significance of this is the need to consider the wider context of a particular land use measure. For example, land use and development includes the relationships between various land uses and how mobility between places is facilitated [20]. Similarly, valuable features include resources of ecological significance, areas contributing significantly to amenity, areas or places of cultural heritage significance and resources or areas of economic value [21]. Resources of economic value include water resources and sources of renewable and non-renewable energy [22].

Thus, while the *Water Act* 2000 (Qld) links the management of water resources with the management of other resources including land, the *Integrated Planning Act* 1997 (Qld) links the management of land with the management of water resources and sources of energy. These two statutory regimes require decisions about the built environment to reflect wider and more strategic issues of resource management generally.

Environmental control

Then there is the *Environmental Protection Act* 1994 (Qld). The Act protects the environment from the harmful consequences brought about by the activities of land users and developers. The focus of the Act is the impact of these activities rather than the activities themselves. The object of the Act is the protection of the environment while allowing for ecologically sustainable development [23]. This is achieved in a number of ways including the regulation of environmentally relevant activities [24], the prohibition of activities bringing about unlawful environmental harm [25] and the creation of a general environmental duty to take reasonable and practicable measures to avoid damage to the environment [26]. The object of the Act is supported by a duty to achieve it [27]. This duty is imposed upon all of those who are performing functions and exercising powers under the Act. It applies therefore as much to environmental managers as it does to environmental regulators.

The *Environmental Protection Act* 1994 (Qld) applies to the environment at large and this includes the built environment. It specifically acknowledges ecosystems, natural and physical resources, the qualities and characteristics of locations, places and areas [28]. It extends to the interactions between these and social, economic, aesthetic and cultural conditions [29]. While the focus of this definition may not be the built environment, there is no doubt that the built environment is part of the environment for this purpose. This means that the construction and use of a building are subject to the *Environmental Protection Act* 1994 (Qld) to the extent that the activities of construction and use may have short term and long term environmental impacts.

The object of the Act includes a reference to ecologically sustainable development [30]. So sustainability including sustainability of the built environment is one of the elements of the statutory regime that impacts upon the construction and use of buildings from the point of view of environmental protection. While the construction and use of a building are a site-specific activity, the impacts upon the environment are not so restricted. Once again the potential application of the *Environmental Protection Act* 1994 (Qld) emphasizes the need to consider the built environment in this wider context.

Building control

The *Building Act* 1975 (Qld) provides for the setting of standards for the construction of buildings. Section 4(1)(a) enables a regulation to be made about building work. This is a descriptive provision. There is nothing in the Act to indicate what outcomes are to be achieved by the regulation. So the purpose of these standards is left to the discretion of the regulator. There is no reference – not unexpectedly given the history of the legislation – to sustainability or the principles of sustainability.

The Standard Building Regulation 1993 (Qld) provides for a range of matters some of which are procedural and some of which are substantive. The matters of substance in the regulation suggest that it is all about health and safety. However, amenity and aesthetics may be relevant [31] together with the susceptibility of land to flooding and bush fires [32]. One of the most important provisions is regulation 8, which provides that the Building Code of Australia forms part of and is to be read as one with the regulation. If a matter is dealt with in the Code, any provision of an Australian Standard that also deals with it applies to the extent that it is adopted by the Code [33]. So the standards that apply to any particular building may be scattered across a range of sources.

There is a further structural issue. Considerable flexibility is built into the application of the Standard Building Regulation 1993 (Qld) and the Building Code of Australia. For example, under section 5 of the *Building Act* 1975 (Qld) an application may be made to vary how the regulation applies to the building work. If the application is successful, the regulation as varied by the decision applies to the building work [34]. There are no substantive criteria according to which such a decision is made. These factors render the determination of the specific standards a potentially difficult and complex question.

3.4 Building Code of Australia

Then there is the Building Code of Australia [35]. It is a large and complex instrument. The principal areas of substance with which it deals include:

- structure
- fire resistance
- access and egress
- services and equipment
- health and amenity.

The structure of the code is important. Most parts of the code contain three structural elements:

(1) a statement of the objectives which the provision is designed to achieve
(2) a functional statement indicating how the objectives are achieved and
(3) performance requirements specifying the performance standards necessary to ensure compliance.

The performance requirements are satisfied either by complying with the detailed provisions of the code that are deemed to satisfy these performance requirements or by alternative solutions complying with the performance requirements or shown to be equivalent to the deemed-to-satisfy provisions. The code also refers to standards contained in other instruments. Any such references give to these instruments the same status as the status of the code itself. There is a very wide and extensive range of sources according to which the relevant standards may be determined.

One of the areas with which the code is concerned is energy efficiency [36]. The specific matters for which standards are set include:

- building fabric
- external glazing
- building ceiling
- air movement
- services.

Although the code is a national instrument, individual jurisdictions may vary the way in which it applies to them. An example is the way in which Victoria has provided for energy efficiency – a factor in the sustainability of the built environment. There is a statement of objectives, a functional statement, a performance requirement and a set of deemed-to-satisfy provisions. Paragraph F 6.3 deals with thermal insulation. The requirement is for the building to have an adequate level of thermal performance and this is achieved by having a house energy rating of a certain standard assessed by among others the Sustainable Energy Authority of Victoria. This example shows not only the interaction between the code and issues of sustainability but also the extremely complex set of standards and methodologies included in the code.

The law, sustainability and the built environment

The environmental legal system in Queensland – like others in Australia – is concerned with sustainability of the built environment. However, sustainability of the built environment – if it is achieved at all – is achieved as a consequence of the application of a series of complex and perhaps at first sight unrelated sets of statutory provisions enacted either directly or indirectly for a range of purposes that may in theory be congruent but not necessarily so in practice.

The approaches

There are essentially four ways in which the legal system faces the challenge of sustainability of the built environment. These have been identified as:

(1) strategic planning across the management of natural resources generally
(2) urban planning leading to development control
(3) site-specific criteria whose focus is the impact on the environment of the construction and use of buildings
(4) building-specific criteria that focus upon issues of construction, materials and design some of which are linked to sustainability.

While none of these categories is self-contained, this matrix indicates how the legal system of Queensland approaches sustainability of the built environment. Clearly there is no single objective of sustainability incorporated in the legal system nor is there administrative or procedural integration of decision making. So the system remains essentially characterized as fragmented and disjointed.

This is nothing new for environmental law. Nevertheless it suggests that sustainability of the built environment is emerging as a concept – perhaps no more than that – within the environmental legal system. This will not ensure the outcome of a sustainable built environment. But it certainly renders that outcome less unlikely provided all those involved in making decisions, both in the private sector and the public sector, acknowledge the desirability of sustainability in the first place and are prepared to incorporate it within their processes.

This conclusion avoids the question of enforceability. It has already been suggested that a sustainable built environment is an outcome of a series of processes and that traditionally the law has found it almost impossible to ensure that outcomes are achieved, certainly in advance of circumstances arising reactively when it can become clear that outcomes have not in fact been achieved. The law in other words can determine legality after the event but it cannot ensure proactively legality prior to it eventuating. At least under current doctrines.

The role of law

Does this mean that the law has no part to play at all? Clearly not. Take one example: the Building Code of Australia requires quite specific standards of insulation in a building [37]. The law at the moment does not require the installation of solar collector panels. There may well be financial incentives to install these but there is no duty to do so. That is not to say that a duty to install solar collector panels as an essential requirement for the construction of a new building could not be created. There would moreover be no difficulty in enforcing such a provision. At the other level of abstraction, the law could provide that a building must not be constructed and used unless it is sustainable. And the legislation could go on to indicate how a building can be constructed and used sustainably – along the lines of section 1.2.3 of the *Integrated Planning Act* 1997 (Qld), which indicates how to advance the achievement of ecological sustainability.

Such an approach prescribes how an outcome may be achieved rather than how it must be achieved. Considerable discretion is left to the developer. If the developer supported by relevant professional planning and architectural advice decides to be proactive and innovative in developing the land and designing the building so that sustainability is achieved, then public policy objectives as well as those of the developer will have been achieved.

Is this a realistic scenario? Is it being achieved anywhere? A study into the sustainability of the built environment in Europe suggests positive answers to these questions. It is seen partly as a planning function and partly as a design function. In relation to planning it has been noted:

'In terms of application of sustainable principles to land development and urban planning, strategic aims are broadly the same throughout Europe, as follows:
- equilibrium between urban development and preservation of agricultural land and forest, as well as green spaces for leisure use
- preservation of soils, ecosystems and natural landscapes
- diversity of use in urban areas, with a balance between living and working space
- socially mixed areas (residential and otherwise)
- journey management and control of vehicular traffic
- protection of water and air quality
- reduction of noise pollution
- waste management
- control of natural and technical risks
- protection of exceptional city sites and conservation of our urban heritage.' [38]

Most of these objectives can be achieved in Queensland through the *Environmental Protection Act* 1994 (Qld) and the *Integrated Planning Act* 1997 (Qld). In each case it may require a more dynamic and committed approach to the application of the legislation than perhaps has been the experience so far. It is however suggested that the statutory framework within which these outcomes can be achieved already exists. But much depends upon the imagination and initiatives of the developer and the developer's advisers as well as those of the public sector regulators.

The same is true in relation to the function of design. While elements of design are relevant under the *Integrated Planning Act* 1997 (Qld), the *Building Act* 1975 (Qld) and its related regulations and codes are directly concerned with design. There are often cost aspects to design. The experience in Europe seems reasonably clear:

'A sustainable approach has cost benefits, particularly for developers who retain responsibility for building management, which is generally the case for local authorities. Running costs can be reduced by adopting sustainable measures from the outset:
- integrating energy conservation systems from initial design stage (particularly for social housing and public amenity projects)
- choosing low-maintenance materials, technologies and architectural concepts
- establishing rational systems for rainwater and waste management.' [39]

Achieving outcomes such as these is more problematic for the legal system. Regulating design may well have the effect of stifling the artistic skills of the architect. Similarly with the use of materials. How far should the imagination of the architect be constrained? The same question was asked decades ago about the use of land. Why should the imagination of the owner of the land be circumscribed? It is now acceptable that the use of land should be constrained in the public interest. The issue of sustainability is very much a matter of the public interest. If restrictions on design and use of materials can be justified in the long-term public interest of sustainability, is the analogy complete? This is a political and not a legal question. However, the law has gone someway down the track to restricting the imagination of the architect. Will it continue to do so?

Conclusion

Let us conclude by considering a set of hypothetical circumstances that could quite easily arise in north Queensland. The owner of land on the coastal margin of north east Queensland is proposing a major development that will include the construction of a golf course and other sporting facilities, a hotel complex and an associated residential development of about 1000 houses with appropriate infrastructural facilities. The land is bounded on the landward side partly by a world heritage area, partly by a national park and partly by an Aboriginal community. The land is bounded on the seaward side by the Great Barrier Reef Marine Park.

There are a number of interests likely to be affected by this proposal. The proximity of a world heritage area and the Great Barrier Reef Marine Park – also a world heritage area – triggers the interest of the Commonwealth. The adjacent national park is managed by the state of Queensland. It may contain endangered species protected by the Commonwealth legislation. The indigenous community may have native title, may be claiming native title or may simply have common law interests in the area. Any development impacts upon the environment. So an environmental impact assessment may be required. The land has no infrastructure such as water and electricity. The local government has an interest under the *Integrated Planning Act* 1997 (Qld) and under the *Building Act* 1975 (Qld).

This scenario provides examples of:

- strategic issues relating to the management of the natural resources of the area generally
- strategic and urban planning issues arising under the *Integrated Planning Act* 1997 (Qld)
- the site-specific requirements of the *Environmental Protection Act* 1994 (Qld)
- the building-specific requirements of the *Building Act* 1975 (Qld) and associated regulation and code.

It is within this legal context that the developer together with a group of relevant professional advisers must determine how to go ahead with the project. The standards to be complied with derive from a wide range of sources: some specified in the legislation, the regulations or the codes; some determined ad hoc by regulators; some determined within the limits of discretion afforded to the developer. Sustainability is a part of this scenario. It is a complex set of legal arrangements with which no single professional discipline can ensure compliance. The focus of the architect's skills may be the building specific requirements. But the architects – like the other professionals – cannot perform their functions without accepting a degree of responsibility for the sustainability of the development from all perspectives.

References

[1] Commonwealth of Australia (1992) *National Strategy for Ecologically Sustainable Development*, Commonwealth Government Printer, Canberra.

[2] Gauzin-Müller, D. (2000) *Sustainable Architecture and Urbanism*, Birkhäuser, Basel, Berlin and Boston, p. 12.

[3] Ibid., p. 12.

[4] See e.g. Fisher D. E. (2001) Sustainability – the principle, its implementation and its enforcement, *Environmental and Planning Law Journal*, **18**, 361–368: (2002) The achievement of sustainability – an Australian perspective, *Environmental Law and Management*, **14**, 145–150; (2003) *Australian Environmental Law*, Lawbook Company, Sydney, Chapter 10.

[5] IUCN Commission on Environmental Law (2000) *Draft International Covenant on Environment and Development*, 2nd edn, updated text, IUCN, Gland and Cambridge.

[6] *Land Act* 1994 (Qld), s 4.

[7] *Mineral Resources Act* 1989 (Qld), s 2.

[8] *Environmental Protection Act* 1994 (Qld), s 18.

[9] *Water Act* 2000 (Qld), s 10(1).

[10] *Ibid.*, ss 10(2)(c)(ii) and 11(a).

[11] *Ibid.*, s 12.

[12] *Ibid.*, s 10(1).

[13] *Integrated Planning Act* 1997 (Qld), s1.2.1.

[14] *Ibid.*, s 1.3.1.

[15] *Ibid.*, s 1.2.2.

[16] *Ibid.*, s 1.2.3(1)(d) and (e).

[17] *Ibid.*, s 1.2.3(1)(b).

[18] *Ibid.*, Sched. 1, para 3(1) and (2).

[19] *Ibid.*, Sched. 1, para 4(1).

[20] *Ibid.*, Sched. 1, para 4(2)(a) and (c).

[21] *Ibid.*, Sched. 1, para 4(3).

[22] *Ibid.*, Sched. 1, para 4(3)(d).

[23] *Environmental Protection Act* 1994 (Qld), s 3.

[24] *Ibid.*, ss 426 to 428.

[25] *Ibid.*, ss 436 to 440.

[26] *Ibid.*, s 319.

[27] *Ibid.*, s 5.

[28] *Ibid.*, s 8(a) to (c).

[29] *Ibid.*, s 8(d).

[30] *Ibid.*, s 3.

[31] Standard Building Regulation 1993 (Qld), regs 50 and 51.

[32] *Ibid.*, regs 53 and 55.

[33] *Ibid.*, reg 10.

[34] *Building Act* 1975 (Qld), s 9(2).

[35] Australian Uniform Building Regulations Coordinating Council (1996) *Building Code of Australia*, Commonwealth Government Printer, Canberra.

[36] *Ibid.*, Pt 3.12.

[37] *Ibid.*, para 3.12.1.1.

[38] Gauzin-Müller, D. (2002) *Sustainable Architecture and Urbanism*, Birkhäuser. Basel, Berlin, Boston, p. 39.

[39] *Ibid.*, pp. 39 and 40.

24 Simple and smart SCADA – a survey of user expectations

F. Basiri[1] *and T-G. Malmström*[1]

Summary

The Supervisory Control And Data Acquisition (SCADA) system is used as a front-end for the Building Energy Management System (BEMS). There are many factors that can affect the choice of the SCADA system. In this study, we tried to measure how different categories of people (i.e. buyers and users of these systems) evaluated 21 different factors that could influence the system choice. The study was conducted through a Web-based poll that was e-mailed out to buyers and users in the public and private sectors in Sweden.

The result was evaluated by using statistical methods and significant differences were found in how buyers and users evaluated different factors. In addition, significant differences in answers were found based on the respondents' main fields of competency.

Introduction

Many buildings do not perform well. There is a large potential to create a better indoor climate and to save energy by means of improved design, better construction, proper commissioning, and careful operation and maintenance. Our experience is that operation and maintenance, in particular, are crucial. According to a study performed in Sweden a couple of years ago [1], the main correlating factor between the functional availability (regarding heating, refrigeration and ventilation) of about 200 different buildings was the owning organization. The best was seven times better than that ranked lowest.

Both indoor climate and energy saving in buildings depend on the control strategy and its implementation. Building Energy Management Systems (BEMSs) are now implemented in almost all new commercial buildings (offices, hotels, shopping centres, hospitals, airports, etc.).

The data from a BEMS is usually accessible for plant managers through a software interface in a computer connected to the plant. Through this interface, the managing staff are able to monitor system status, receive alarms, change device parameters such as setpoints, reschedule system operations, create reports, etc. These software interfaces are commonly called SCADA (Supervisory Control And Data Acquisition) applications.

---

[1] Division of Building Services Engineering, KTH (R. Inst. of Techn.), SE-100 44 Stockholm, Sweden.

Table 24.1 Professional role.

Professional role	Frequency
Client	14
Engineer/Technician	20
Other	3

Who is the potential customer or user? Citing the final report of IEA Annex 34 [2]:

'Different users may have very different goals: "quality of service", "energy conservation", "indoor climate quality", "reliability of the system". A hotel manager may only be interested in the comfort of his clients; the service team in a computer centre may be mainly interested in the reliability of the HVAC system; a utility may only be interested in the peak power demanded by its customers...
...It is important to differentiate between the customer who will buy the systems and the end users who will use them, and the service providers who will use the tools to improve the service offered to their clients.'

This chapter will mainly focus on how different factors affect the choice of software interface for a buildings HVAC plant. We focus on BEMS, and have done a survey in Sweden. The result is analysed inter alia through comparing the answers from customers (clients) and end users (engineers/technicians).

The survey

The survey consisted of two parts. The first part was composed of questions concerning:

- How different factors that could affect the choice of a SCADA system were appreciated by the respondents.
- How these factors had affected the choice of the last plant with which the respondents had worked.
- What the respondents thought of the available SCADA and automation systems on the market.

Table 24.2 Main field of competence.

Main field of competence	Frequency
Building engineering	1
Electrical engineering	2
Property management	4
Control-system engineering	9
HVAC engineering	17
Other	4

Table 24.3 Highest level of education.

Highest level of education	Frequency
Senior high school	18
MSc	2
BSc	11
Other	6

- What the respondents thought of the energy-, information- and staff-management policies of their companies.

The second part of the survey consisted of questions about the respondent to help the evaluation of their answers. The survey which was developed for this purpose was then ready to be moved on to a Web-based platform to be filled in over the Internet.

The poll was e-mailed to 180 persons and a total of 37 people responded.

The respondents

We asked the respondents about their professional role, main field of competence, highest level of education, age, and finally the percentage of the building stock in their organization that was equipped with BEMSs.

The respondents were asked to choose the *professional role* that described them best from the options: Client, Contractor, Building services engineer/technician, Consultant and Other.

The people who responded 'Other' stated that they were working as Clients as well as Engineers (see Table 24.1).

As for the *main field of competence*, the respondents could choose from among these options: Architecture, Building engineering, Economics, Electrical engineering, Property management, Structural engineering, Building control-system engineering, HVAC engineering and Other (Table 24.2).

For the *highest level of education*, the available options were: Senior high school, University college (i.e. architecture or MSc. degree), University college (i.e. BSc degree) and Other (Table 24.3).

The age distribution of the respondents is presented in Table 24.4.

Table 24.4 Age.

Age	Frequency
Younger than 30	2
30–40	14
41–50	11
51–60	9
Over 60	1

Table 24.5   Percentage of computerized buildings.

Percentage of computerized buildings	Frequency
20–29%	1
30–39%	2
40–49%	3
50–59%	1
60–69%	5
70–79%	6
80–89%	8
More than 90%	10

We asked the respondents to state the *percentage of their organization's building stock that was equipped with computerised energy management systems*. The answers can be seen in Table 24.5.

Results

We asked the participants how much, in their opinion, each of the 21 factors listed in Table 24.6 should influence the choice of SCADA system. We also asked them how much each of the factors affected the choice of the last system that they had purchased (in the case of clients) or worked with (in the case of other professional roles). Each of the factors could be given a grade from 1 to 6 where a grade of 1 corresponded to 'not at all' and grade 6 was equivalent to 'very much'. The factors are presented in Table 24.6 together with the means from the results. Columns labelled 'Val' represent how the factors were valued and columns labelled 'Last' represent what the participants thought of their last system.

We used statistical factor analysis technique to find out on which of the control variables (i.e. age, highest level of education, main field of competence, professional role, etc.) we should focus, when analysing the results from the poll. We tried two different methods (principal axis factoring and principal component analysis) for variable selection and both methods pointed out the *professional role* and the *main field of competence* to be the primarily variables explaining the variance in the results.

The two largest groups among the fields of competence were HVAC and Control System engineering (Table 24.2). Therefore, the decision was made to continue comparing all the means according to the respondents' professional role (client, engineer/technician) and the main field of competence (HVAC, control system engineering). The result is presented in Figures 24.1 and 24.2. In Figure 24.1, the means for each of the factors are presented in the staples and the standard deviations for the answers are traced in the diagram. Other important factors, apart from the ones in Table 24.6, according to the participants were 'different access levels' and 'password protection'.

Figure 24.2 shows the difference between the means from Figure 24.1 and how the respondents thought that the same factors had affected the choice of the last system

Table 24.6 Factors that could affect the choice of SCADA system.

		Means									
		Total		Clients		Eng./techn.		HVAC spec.		Control system spec.	
Num.	Factor	Val.	Last	Val.	Last	Val.	Last	Val.	Last	Val.	Last
1	User-friendliness	5.59	4.22	5.64	4.50	5.6	4.1	5.35	4.24	5.89	4.78
2	Brand/supplier of the system	3.16	3.70	3.43	3.79	2.9	3.7	2.71	3.76	4.00	4.44
3	Remote access ability (via modem or the Web)	4.86	3.84	5.29	3.71	4.6	3.8	4.65	4.12	4.89	3.78
4	Functions for long-term detection and analysis of plant failure	4.62	3.62	4.71	3.86	4.7	3.5	4.59	3.76	4.56	3.44
5	Connection ability to external alarm-management facilities	5.03	4.05	5.36	4.93	4.8	3.4	4.88	4.00	4.67	3.78
6	Alarms by SMS and/or e-mail	4.70	3.81	5.36	4.36	4.3	3.4	4.71	4.00	3.67	2.67
7	Ability to alter the system graphics (i.e. colours and design of plant schemes)	4.76	3.84	4.29	4.00	5.2	3.9	4.65	4.00	5.67	4.44
8	Ability to alter the application software in the connected field controllers	5.03	3.54	5.00	3.93	5	3.2	4.88	3.71	5.78	3.78
9	Price (for acquisition)	4.57	4.43	5.29	5.14	4	3.8	4.47	4.71	4.67	4.22
10	Ability to simulate the plant operation (i.e. how changes of the application software in connected field controllers affect the plant operation)	4.11	2.51	4.21	2.57	4.1	2.5	4.12	2.65	4.44	2.67

(Continued)

Table 24.6 (Continued).

Num.	Factor	Means									
		Total		Clients		Eng./techn.		HVAC spec.		Control system spec.	
		Val.	Last	Val.	Last	Val.	Last	Val.	Last	Val.	Last
11	Possibility for the tenants to access selected parts of the system and to control their facilities (i.e. monitor system operation, alter set-points, reschedule operation, etc.)	3.54	2.59	3.79	3.07	3.4	2.2	3.71	2.94	3.00	2.78
12	System is module-based	5.24	4.32	5.36	4.71	5.2	4	5.06	4.94	5.56	4.00
13	Price (for maintenance and upgrades)	4.86	3.70	5.07	4.71	4.8	3.1	4.71	3.71	4.78	3.67
14	References for the contractor/supplier	4.84	4.22	5.29	4.93	4.6	3.6	4.94	4.65	4.56	3.33
15	Tools for data-gathering and generation of reports	4.92	3.86	4.36	4.14	5.3	3.7	4.53	3.71	5.33	4.22
16	Swift support from manufacturer/supplier	5.65	4.41	5.79	4.86	5.6	4	5.47	4.76	5.89	3.89
17	Standard interfaces and protocols (i.e. LonWorks, OPC, etc.)	5.49	4.05	5.79	4.86	5.5	3.6	5.35	4.29	5.67	4.11
18	Expandability/Scalability	5.46	4.22	5.43	4.86	5.6	3.8	5.29	4.35	5.44	4.11
19	Backwards compatibility	4.84	3.70	5.29	4.29	4.6	3.4	4.59	3.53	4.89	4.44
20	Ability to simulate the plant operation (i.e. before ant changes of system parameters are made, control how they will affect the indoor climate)	4.00	2.38	4.00	2.43	4	2.3	3.88	2.41	4.00	2.44
21	Logging and trending ability	5.46	4.73	5.64	5.36	5.3	4.2	5.41	4.94	5.78	4.89

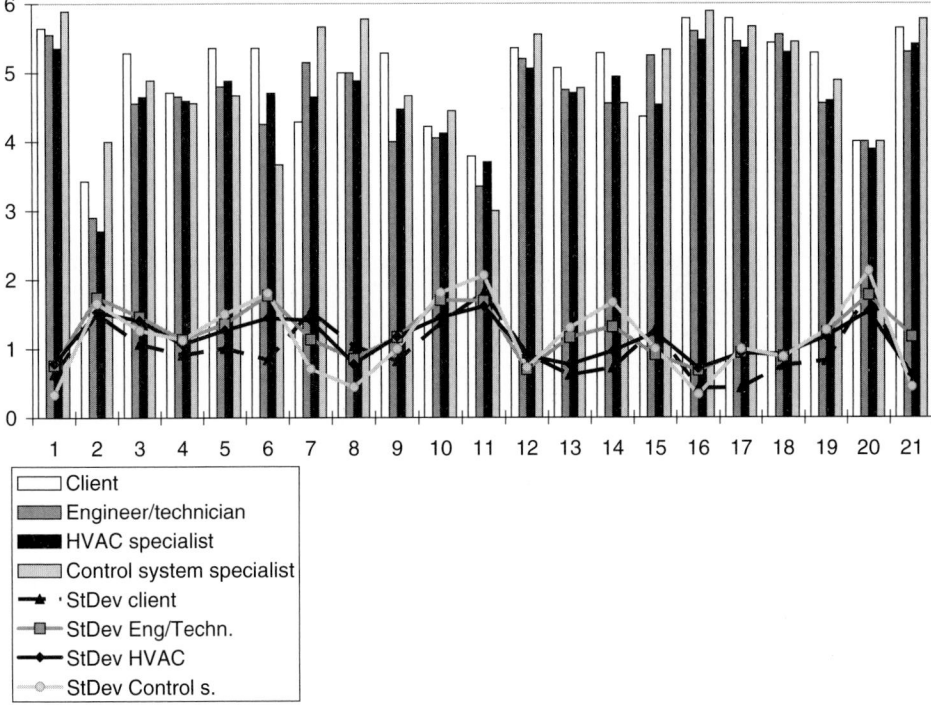

Figure 24.1   How different factors should influence the choice of SCADA system.

that they had worked with. Negative values in Figure 24.2 indicate that a factor had less influence on the choice than it should have.

Statistical Analysis Of Variance (ANOVA) tests show that the mean difference between the answers from clients and engineers/technicians (Figure 24.1), is significant at a 0.05 level (assuming equal variances) for factor 6, factor 9, factor 15 and factor 17 and if equal variance is not assumed, also for factor 14 and factor 19. The answers from HVAC specialists and control system engineers are significant (assuming equal variances) for factor 7 and for factor 8 and if equal variance is not assumed, also for factor 1. A significant difference was also noticed between the answers from the client group and the engineer/technician group about how they believed that the factors in Table 24.6 affected the choice of the last system for factors 5, 9, 13, 14 and 21. The ANOVA test for mean comparison between HVAC specialists and control system specialists however, didn't show any significant differences in this case.

To extract some additional information about how the respondents thought of the functionality of their systems, we asked them some questions about the last system that they had purchased (if clients) or worked with (if other). We asked:

(1)   if the system contained too many functions
(2)   if the system lacked important functions
(3)   if the system was user friendly
(4)   if they were satisfied with their systems.

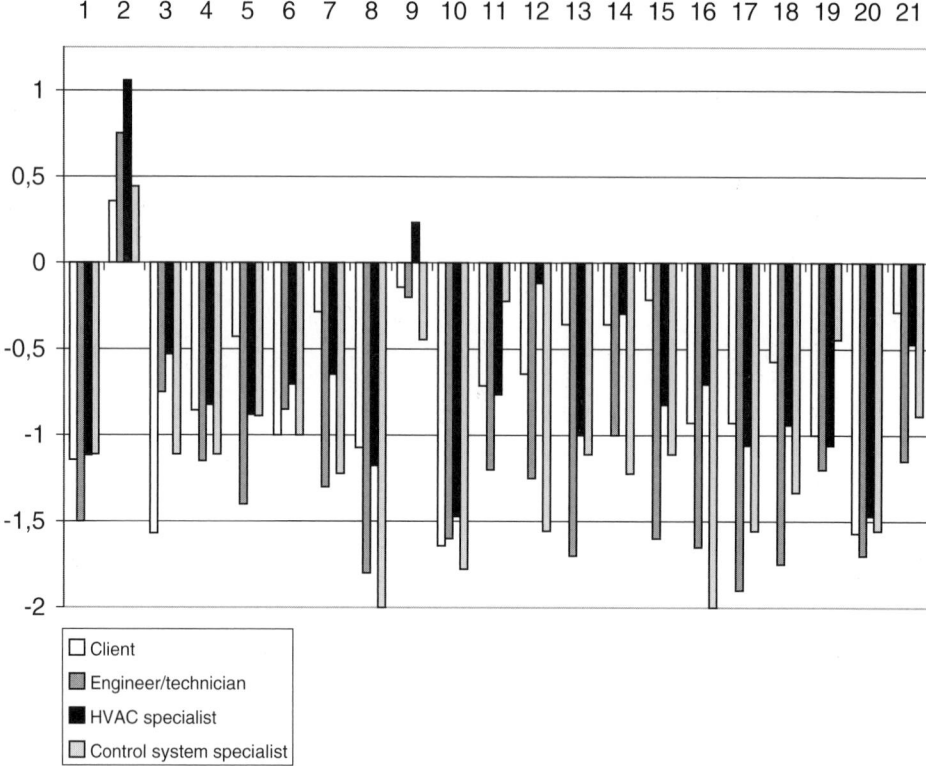

Figure 24.2 The difference between how factors should, and did, affect the choice.

The respondents could grade their answers on a scale ranging from 1 to 6 where 1 corresponded to 'not at all' and 6 corresponded to 'very much so'. The results are shown in Figure 24.3. None of the differences amongst the groups was statistically significant. However, the difference between the clients and the engineers/technicians was almost

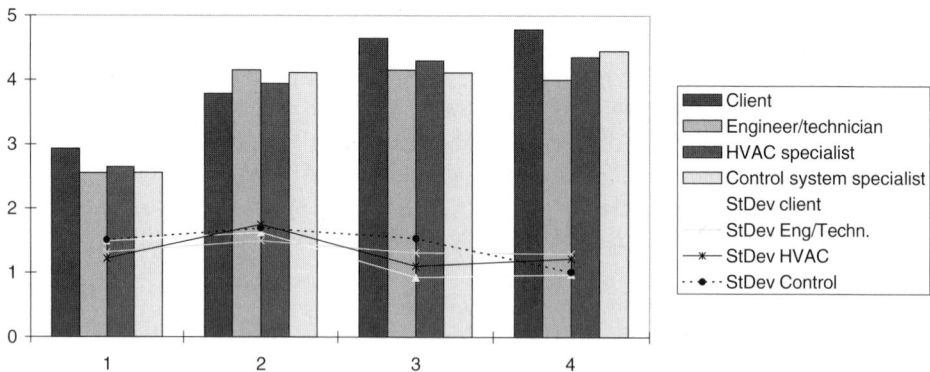

Figure 24.3 User views on the SCADA functionality.

Table 24.7 Energy saving possibilities, potentials and actual situation. Grade 6 is the maximum.

Num.	Statements	Means				
		Total	Clients	Eng./Tech.	HVAC Spec.	Control system spec.
1.1	There are potentials for economic savings through proper commissioning.	5.84	5.86	5.80	5.82	5.89
1.2	It is important to have a strategy about purchase and use of computerized energy management systems (i.e. SCADA systems).	5.46	5.43	5.45	5.29	5.78
1.3	It is important for operators of the HVAC plants to have a proper understanding of the interactions between the building and its HVAC system.	5.68	5.71	5.60	5.65	5.78
1.4	It is important to have methods and/or tools to measure the results from energy management and enhancement efforts.	5.65	5.50	5.70	5.53	6.00
1.5	It is important to work continuously to detect faults and enhance the function of the HVAC plant.	5.68	5.57	5.70	5.47	5.89
1.6	Energy- and resource-saving efforts should be rewarded through a bouns and reward system.	4.95	4.86	4.95	4.59	6.00
2.1	We work actively to make sure that our plants function as they are supposed to.	5.14	5.00	5.15	5.24	5.22
2.2	We have worked out a strategy for purchasing and using computerized energy management systems (i.e. SCADA systems).	3.65	3.79	3.50	3.82	3.67
2.3	In our organization, the operators of the HVAC plants have enough understanding of the interactions between the building and its HVAC plant.	4.08	3.93	4.15	4.06	3.89
2.4	We have tools to measure the results from energy management and enhancement efforts.	3.84	3.71	3.70	4.18	3.56
2.5	We work continuously to detect faults and enhance the function of our HVAC plants.	4.51	4.36	4.50	4.65	4.56
2.6	In our organization, energy- and resource-saving actions are rewarded through a bonus- and reward programme	2.92	3.07	2.60	3.35	2.22

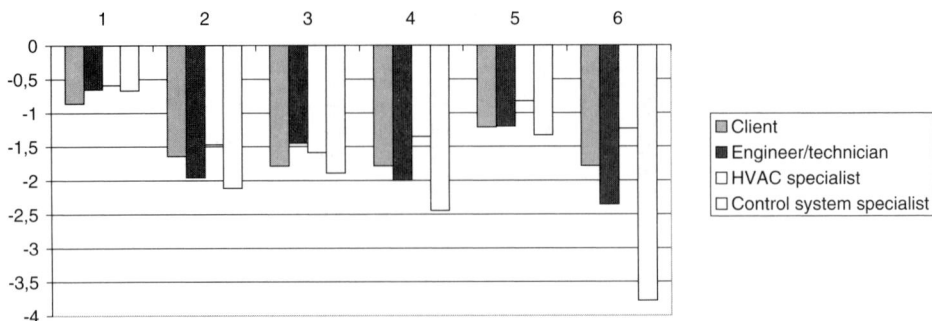

Figure 24.4   Energy saving possibilities, potentials and actual situation (charting the differences).

significant in the case concerning whether or not they were satisfied with the systems. Lastly, the attitude towards energy saving, function improving and the estimate of the respondents' own company's policy in those matters was tested, see Table 24.7 and Figure 24.4.

Conclusions

The number of respondents (37 out of 180) is low, but the method used to evaluate differences between groups, concerning functions of a SCADA system, takes this into account, and the statistical significance is clearly stated in the evaluation reported among the results.

All categories find it most important to understand the function of the HVAC system and to have methods for measuring the results of efforts for achieving better function, and a bonus system for rewarding good results. All categories also thought that there were potentials for economic savings through proper commissioning and found it necessary to work continuously to improve the function of the HVAC plant. Clients and users are generally satisfied with their (latest) systems, consider them user-friendly, but think that they lack some important functions.

Important factors were user-friendliness, swift support, standard interfaces and protocols, expandability and logging/trending ability. Factors that were less important were brand/supplier, simulation possibilities and access possibilities for the tenants. It is interesting to note that simulation capabilities (factors 10 and 20) and tenant access (factor 11) also had the highest variances in the answers.

System operators (engineers/technicians) were generally less interested in functions for alarm management by e-mail and SMS, the acquisition price and standard interfaces and protocols than the buyers (clients), but were more interested in tools for gathering data and for creating reports. Engineers/technicians were generally less satisfied than the clients, and control engineers were less satisfied than HVAC engineers.

Acknowledgement

This study was funded by the Swedish Energy Agency.

References

[1] Myrefelt, S. (1996) *Modell för erfarenhetsåterföring av driftsäkerheten i klimat-och ventilationssystem*, Technical note 41, Building Services Eng., KTH, Stockholm.
[2] Vissier, J-C. and Heinemeier, K. (2001) Customer benefits, user needs, and user interfaces, in *Demonstrating Automated Fault Detection and Diagnosis Methods in Real Buildings*, VTT Technical Research Center of Finland, Espoo.

Part 6
Development processes

25 Sustainable building and construction: contributions by international organizations

W. Bakens[1]

Summary

There is no sector that has such a potentially large contribution to make to achieving sustainable development than the building and construction sector. This sector, however, is very broad and very fragmented in its organization, with many internationally-operating and often non-cooperating organizations. The actual application of the measures taken towards sustainable building and construction, which is expected to make a major impact, requires close cooperation between various professionals, decision makers and other stakeholders. The relevant international organizations have a role to play in accomplishing such non-traditional cooperation. In addition they have an important role to play in presenting the sector as being potentially important as concerns sustainable development and in attracting resources to the sector that up until now have usually been made available to more politically 'sexy' sectors. In support of this, though, there is the need for a consensus-based definition of what the building and construction sector encompass, what the main are issues are, and who are the important stakeholders, here also the international organizations have a role to play. There are some positive developments towards international cooperation that do encompass the sector as a whole, but the decisive step has not yet been taken.

Introduction

Sector contributions to sustainable development

In order to find the position of the built environment and the building and construction sectors properly and to be able to indicate to which of the many international organizations attention should be paid in this chapter, it is important to first define

---

[1] CIB (General Secretary, International Council for Research and Innovation in Building and Construction) Post box 1837, 3000 BV, Rotterdam, The Netherlands.

the sector's possible contributions to achieving sustainable development. These organizations have different natures and include different stakeholders and have different types of potentially opportune measures. These different types of measures include:

(1) *Sustainable construction*   The sustainable production, maintenance and demolition of buildings and constructions (roads and other civil infra-structural works), including, for example, the use of local construction materials that do not require long distance and energy wasting transport, construction technologies that generate less construction waste and require less energy and methods of demolition (or better, deconstruction) that result in more re-useable construction materials.

(2) *Sustainable buildings and built environments*   Contributions by buildings and the built environment to achieving – components of – sustainable development, such as through the design of healthy and less energy-consuming buildings and the planning for urban environments that discourages the use of private motor transport.

(3) *The building and construction industry*   Contributions by the building and construction industry to local and national socio-economic sustainable development, such as through the employment of local labour as opposed to a foreign construction firm that flies in its employees from its home country just for the construction of one project.

(4) *The employers in this industry*   Proper behaviour by the employees in the building and construction industry towards their employees, including the application of the various principles included in social sustainability, such as taking care of safe and healthy work environments, offering opportunities for a positive personal career development to employees, transparent company governance and proper ethical company behaviour.

In most countries – both developed and developing – the building and construction industry has a huge social and economic importance. In most countries the building and construction industry is the single biggest employer in the national economy and its activities involve a very high multiplier effect: each dollar spend on construction may generate up to three dollars of economic activities in other sectors. This makes the above-mentioned types (3) and (4)'s possible contributions by the industry and its employers to achieving sustainable development really important, bearing this in mind, no other industry has more potential than the building and construction industry.

However important, though the possible contributions of types 3 and 4 are, they are not especially unique for the sector and therefore this article will focus on the roles of international organizations and their possible contributions of the above-mentioned types 1 and 2: sustainable construction and sustainable buildings and built environments. These are together called sustainable building and construction.

Definitions

When looking at Sustainable Building and Construction Research and Technology Development (RTD) programmes worldwide and especially at such programmes in developed countries, a relatively strong emphasis can be noticed in areas such as:

- *energy*, including RTD that aims for the reduction of energy consumption and the use of renewable energy sources for buildings, e.g. by using wind and solar energy
- *waste*, including both waste prevention through, for example, designing for re-use and recycling of construction materials and waste management.
- These are methods to measure and thus prevent negative *environmental impacts* caused by construction.

However important these topics are, the potential of sustainable building and construction encompasses much more, but for this we first need a proper definition in order to grasp the full scope of the potential measures that can be taken towards achieving sustainable development, and this requires definitions of both 'sustainability' and 'building and construction'.

Sustainability is defined as including social, economic and environmental sustainability, the so-called triple-bottom line, lately often promoted as 'the' approach toward sustainable development.

In this context *environmental sustainability* often refers to the above-given topics of energy, waste and environmental impact, and is often related to direct and indirect Green House Gas (GHG) emissions, as addressed in, for example, the Kyoto Protocol.

Aspects of *social and economic sustainability* are more or less clearly defined in 'Agenda 21'. It must however be acknowledged that these aspects of sustainable development are far less well understood and applied in the building and construction sectors. We do not have clear performance indicators for social and economic sustainability; we do not know how to measure it and we do not have consensus-based systems to balance between, or to weigh possible trade-offs between, on the one hand environmental sustainability and on the other hand social and economic sustainability. In other words, we talk a lot about the triple-bottom-line approach, but in reality for many people in our sector it is still little more than a buzzword.

If 'sustainability' is in fact difficult to define, 'building and construction' may prove to be even more difficult. To begin with, we won't investigate the different interpretations of the words 'building' and 'construction' in American English, UK English and European English; this will only complicate things and not add real value to a debate that tries to define issues, potential measures and stakeholders.

For many people the building and construction sector or industry – we do not even have a worldwide acknowledged term for this – stands primarily or even only for the conglomerate of construction firms or (general) contractors, possibly including (specialized) sub-contractors.

A somewhat broader definition, as implicitly used by some, also includes the firms that in the traditional organization of projects, supply goods to such contractors: the construction material, component and equipment manufacturers and possibly also the engineering and design professionals.

If we really want to be able to identify potential measures towards achieving sustainable building and construction, and incorporate relevant stakeholders and international

		Phases of the building and construction process				
		Programing and planning	Design and engineering	Construction	Maintenance, management	Demolition, deconstruction
Levels of building and construction	Construction materials, components and technologies					
	(Whole) Buildings and the immediate built environment					
	(The wider) Built environment (including urban issues)					

Figure 25.1   The various 'components' in the broad context of building and construction.

organizations in the debate, we must have a clear understanding of what we are talking about.

In this context the *building and construction* sector is defined as all professionals, firms and organizations (and their representative associations), who contribute to:

- the development, maintenance, management and demolition/deconstruction of buildings and (other construction in) the built environment.

In Figure 25.1 the various 'components' in this broad definition are shown. It concerns a matrix with three different levels of building and construction and with five different phases in the building and construction process as traditionally organized in building and construction projects. The important issue here is that in each separate cell in this matrix different building and construction professionals are at work, different stakeholders are involved and different decision-making processes are crucial. This matrix illustrates both the complexity and the wide scope of the building and construction sector that is to be addressed when defining stakeholders and potential measures towards achieving sustainable building and construction.

Why is it important to have such a wide ranging and complex definition that it deviates from other authoritative definitions as, for example, those incorporated in ISO, CEN and ASTM Standards?

Many of the known potentially important measures toward sustainability, if they are to be successfully applied, require a *cooperation of actors* (professionals, stakeholders and actual decision makers) who traditionally are operating in different cells in the matrix and who often do not even feel that they are incorporated in the same sector or industry. Three simple examples follow, where the so-called actors belong to different cells in the matrix in Figure 25.1:

- An architect (an actor in the cell *Design and engineering* × (Whole) *Buildings and the immediate built environment*) designs a housing project with a specific south-facing facade for a desired reduction in energy consumption. The city-plan, as set up by an urban planner (actor in the cell *Programming and planning* × (The wider) *Built environment*), however does not allow for this. It will require close cooperation between these two professionals and the related decision makers, who work within traditionally unrelated agencies, to address this dilemma successfully.

- A facility manager (actor in the cell *Maintenance and management* × *(Whole) Buildings and the immediate built environment*) wants to introduce a method for controlling the heating of each room in a building, but the construction of the building does not allow for this because of the construction technology that the contractor has chosen (actor in the cell *Construction* × *(Whole) Buildings and the immediate built environment*).
- An owner of a building (actor in the cell *Maintenance and management* × *(Whole) Buildings and the immediate built environment*) wants to encourage the workers in the building to use public transport when coming to work, but the responsible city-planner (actor in the cell *Programming and planning* × *(The wider) Built environment*) does not allow a required adjustment to the zoning regulations to allow for a bus-stop close enough to the building.

These are just three common examples of possible measures towards achieving sustainable building and construction that show the need for different actors to cooperate, while traditionally they do not even feel that they are part of the same industry or sector and therefore why they should cooperate for the same goal.

If the various actors (building and construction professionals, stakeholders and decision-makers) are to be successful in applying such measures that often require non-traditional cooperation, they first need to accept that indeed they are part of the same stakeholder community and that they all belong to the same sector; however wide-ranging and complex its definition.

Programmes in support of achieving sustainable building and construction that address only a single cell in the above-given matrix are of course not in vain. They may motivate or enable people towards applying simple measures – and there are indeed many of those – that do not require such non-traditional cooperation. All the known 'big' measures, from which a substantial break-through may be expected, however, do require such cooperation. It is therefore to be assumed that if 'the sector' aims for the desired break-throughs, it will first need to act as one sector, with professionals and other stakeholders who realize that actually achieving such non-traditional and 'sector-wide' cooperation is conditional for needed break-throughs toward sustainable development in building and construction.

Stakeholders

The above reasoning shows that there are many professionals, decision makers and other stakeholders who all have their own role to play concerning the achievement of – components of – sustainable building and construction. What are the decisive rules for applying more complex and far-reaching measures in building and construction projects? Who should be the primary target for awareness-raising and capacity-building campaigns?

To enable a somewhat easier answer to this question, let's make the simple assumption that in all situations – for all building and construction projects – properly motivated, educated and facilitated professionals are available to do proper planning, design, construction, management/maintenance and deconstruction of buildings and constructions.

This may be over-simplifying things because, for example, in many developing countries such properly educated and facilitated professionals sometimes are actually not available. And also such an assumption ignores the fact that there is often no consensus-based professional opinion on the appropriateness of certain tools and technologies for sustainable building and construction, such as a method of clearly assessing the environmental impact of buildings or a method to define and measure sustainability performance indicators.

Nevertheless, under the assumption that the required knowledge, tools and technologies for achieving sustainable building are available and that the professionals who can apply them in building and construction projects can indeed be involved, who are the key-decision-makers who will demand that such professionals to do a proper job? Who is actually in the driver's seat? The answer to this question will be different according to the segment of the building and construction market and the country. For example key-decision makers on subsidized housing in the Netherlands will be different from key-decision makers on commercial buildings in Japan or on infrastructure projects in Chile.

Ignoring the many differences between countries, however, the decisive decision makers who can incorporate measures towards sustainability in building and construction projects, often are:

- *Local governments*, including local politicians and governmental agencies.
- *Project developers*, who develop building and construction projects on a commercial basis to be sold/handed over to owners; it has to be acknowledged, however, that in many countries, and especially in many developing countries, such commercial project developers do not yet exist.
- *Owners* of buildings and constructions, if the building and construction market in a country is such that they indeed can – directly or indirectly – influence major planning and design decisions. This includes, for example, housing cooperations and governmental building agencies, that in some countries are expected to set the example, and firms that sometimes own a very large building stock.

Such decision makers, however, most often have to work within the framework of:

- National – and often also local – regulatory systems, including codes and standards for building and construction, and this may go as far as to include references in such systems, for example, to energy performance rating systems and to environmental impact assessment systems and related rating and labelling systems,
- and this makes *regulatory government agencies* – either national or local – a fourth potentially decisive type of decision maker as concerns applying sustainability enhancing measures in building and construction projects.

Unfortunately, what we see is that these four often decisive actors: local governments, owners, project developers and agencies, which are responsible for regulatory systems, are most of the time poorly represented in international debates on how to accomplish sustainable building and construction. On the other hand, the various building and construction professionals, such as the urban planners, architects, engineers, contractors, HVAC suppliers and manufacturers, in general have well organized representative international associations that play active roles in such debates, but that unfortunately often have a dominant attitude in such debates that can be characterized as 'our members are able and would like to contribute to sustainable building and construction, but most of the time are not asked to or allowed to do so by the actual decision makers'.

In the meantime the research community keeps on researching and developing all kinds of new concepts, methods and tools and analyses why these do not have the impact that is envisaged.

Of course there are exceptions, but in general the communities of actual decision makers, building and construction professionals and researchers seldom show the willingness and ability to cooperate towards a common goal of achieving sustainable building and construction. They seldom really align their strategic agendas and their international representative organizations are indeed seldom equally represented in international debates. Many of such international debates on sustainable building and construction that aim for international strategic or action oriented agendas are, in fact, either research debates or debates among only one, or at best two, types of professionals.

Joint framework and terminology

If representatives of decision makers, professionals and researchers do have debate together on how best to achieve sustainable building and construction, they often seem to be speaking different languages, based upon different understandings of what sustainable building and construction entails, upon different cultural and educational backgrounds and upon different roles and interests in the building and construction process. They often will have different definitions of what 'the' issues are that should be addressed, including different priorities and different possible solutions to problems.

What is needed is a clear conceptual framework that covers all aspects of sustainable building and construction and that incorporates a clear terminology that is understood and used by all parties concerned: by all types of decision makers, professionals and researchers.

For the English-speaking part of the world there are three publications/projects that could be cornerstones of such a joint framework and terminology. All three aim to cover all aspects of sustainable building and construction in its broadest sense in an integral and internally consistent system. They are:

- *Agenda 21 on Sustainable Construction*
- *Agenda 21 on Sustainable Construction in Developing Countries*
- the CRISP projects on performance indicators for sustainable building and construction.

Agenda **21**
on sustainable construction

CIB Report Publication 237

Figure 25.2   CIB publication – *Agenda 21 on Sustainable Construction.*

Agenda 21 on Sustainable Construction

This publication (Figure 25.2) was produced in English in 2000 by CIB – the International Council for Research and Innovation in Building and Construction – in cooperation with RILEM, CERF, IEA and ISIAQ. It aims to provide a bridge between the statements on sustainable development in 'the' Agenda 21 that are of a general and non-sector oriented nature and the more specific agendas needed for research and action per country, per stakeholder and/or per aspect of sustainable building and construction. The objective behind it is to provide a framework and terminology such that, if used, the detailed national, stakeholder oriented or aspect oriented agendas will be compatible.

It distinguishes social, economic and environmental aspects on three different levels of sustainable building and construction: (1) materials and components, (2) buildings and the micro built environment and (3) urban environments. It provides an indication of possible objectives, barriers, challenges and actions for further development towards sustainable building and construction in the following main domains: (1) products and buildings, (2) resource consumption, (3) process and management, (4) urban development and (5) social, cultural and economic issues.

Since 2000, the English version of the 'Agenda on Sustainable Construction' has been translated into Spanish, Portuguese, Czech and Catalan (and in the immediate future also in Russian). The English and Czech versions can be downloaded from respectively http://www.cibworld.nl/pages/begin/AG21.html and http://www.cibworld.nl/pages/begin/CzechA21.html

The versions in other languages can be obtained through contacting the CIB Secretariat at secretariat@cibworld.nl

Agenda 21 on Sustainable Construction in Developing Countries

After the publication of the 'Agenda 21 on Sustainable Construction' in 2000 it was rather quickly concluded that the incorporated thinking was in fact – be it implicitly – strongly dominated by the average thinking about what could or should be accomplished as common in developed countries. And that barriers and challenges towards sustainable building and construction in developing countries were substantially different, to such an extend that some suggested approaches must even be regarded as infeasible for developing countries because of their special social, economic and also institutional characteristics. CIB consequently started a programme aiming for the production of a special agenda for such developing countries. This resulted in 2002 in the publication of the *Agenda 21 on Sustainable Construction in Developing Countries* (Figure 25.3) by CSIR-BOUTEK, South Africa (with Mrs Chrisna du Plessis as the main author) and with financial contributions by CIB, UNEP-DTIE-IETC, BOUTEK

Figure 25.3   CIB publication – *Agenda 21 on Sustainable Construction in Developing Countries.*

and the South African Construction Industry Development Board (CIDB). A full copy of this publication can be downloaded at http://www.cibworld.nl/pages/begin/Agenda21Book.pdf.

This agenda is not so much a conceptual framework, but is relatively more action-oriented than the *Agenda 21 on Sustainable Construction* and it strongly focuses on short, medium and long-term actions to create the (1) technological, (2) institutional and (3) cultural enablers for achieving sustainable building and construction in developing countries.

CRISP project

CRISP stands for 'Construction and City Related Sustainability Indicators'. It is a project funded by the European Commission, led by VTT, Finland and CSTB, France and involves 22 other European organizations. It aims for the development of an integral system of performance indicators (and related assessment methods) for sustainability for the, broadly defined, building and construction sector. More information on the project, its current status and expected outcome can be found at http://crisp.sctb.fr. The outcome of the project was presented at the conference 'Construction and City Related Sustainability Indicators: Development and Use of Indicators' that took place on June 26–27, 2003 in Sophia Antipolis, France.

It is hoped that the use of the above-mentioned three publications will contribute to a better understanding between the international and national organizations that represent the various decision makers, professionals and other stakeholders who are concerned with the application of measures towards sustainable building and construction, and to a better alignment and compatibility of agendas for action that are being developed with specific foci.

Sustainable development and the building and construction sector

In the above discourse some of the main problems have been identified in moving towards an optimization of the collective contributions by international organizations to achieving sustainable building and construction:

- The social and economic dimensions of sustainable building and construction are given lip-service in the building and construction sector but they are not well understood, let alone appropriately incorporated in integral decision-making processes as related to building and construction projects.
- There is no clear and consensus-based definition of the building and construction sector that encompasses all relevant stakeholders such that they feel that they are part of the same stakeholder community and have a joint responsibility.
- In many international meetings the decisive stakeholders are not well represented. These include: local governments, project-developers, owners and national regulatory agencies.

- If the various building and construction professionals, decision makers and other stakeholders actually meet and discuss options for cooperation, there is often no common conceptual framework and key terminology and therefore no common basis of understanding of the issues to be jointly addressed.

When reading the 'sector problems' above, one may conclude that the situation is difficult enough, but possibly the biggest problem towards actually achieving sustainable building and construction has not even been mentioned: the resource problem.

Those from the building and construction sector who, for example, attended the World Summit on Sustainable Development (WSSD) in 2002 in Johannesburg, will have noticed that amongst the multitude of events that took place as part of or in conjunction with the WSSD, there were almost none that actually focused on sustainable building and construction. In fact there was only one such event: the launching of a new organization the Global Alliance for Building Sustainability (GABS) (see below). Of the thousands of attendees, no more than some tens were actually interested in building and construction related issues.

This reflects the situation in general, especially amongst politicians, but also amongst members of the public in general who, although they do have a real interest in sustainable development, almost none seems to realize and have a real interest for the enormous potential contributions by the building and construction sector to actually achieving sustainable development. Given the magnitude of social and economic impact by the sector in national societies, scales of employment, magnitudes of produced waste (and the implicit potential to reduce this), magnitudes of consumed energy (and again the potential to reduce this) and the long life span of its products, there is in fact no sector worldwide with a larger potential contribution to sustainable development than the building and construction sector. We in the sector all know this, but somehow we are not able to get this message across, for example, to those politicians who decide about national or international RTD budgets in support of sustainable development. Even if one looks at the United Nations Development Program (UNEP) one sees relatively little attention being paid to this, although luckily there are exceptions (see below).

Something similar is happening concerning the European Commission (EC). It may be the single biggest financer of RTD worldwide and in its strategy, for example for the EC 6th Framework Programme, it explicitly presents enabling contributions to achieving sustainable development as one of the most important strategic goals. Nevertheless, it prefers to focus its relatively enormous resources on more high-tech and more 'sexy' sectors.

What is to be concluded is that the building and construction sector has a main communication problem: however potentially important; policy makers, politicians and, in general, decision makers on resources for sustainable development, do not seem to realize the sector's importance and prefer to resource other sectors. And herein lies maybe the biggest challenge for the many international organizations that represent stakeholders in the building and construction sector: to join forces, to jointly develop a strong message about what the sector is capable of and show that it wants to play a very substantial role towards achieving sustainable development and to convince the decision makers on national and international resources that a dollar spend on RTD for sustainable development in building and construction will have a bigger impact than a dollar spend on RTD in any other sector.

Developments towards worldwide sector cooperation

Those who work in the building and construction sector know that its organization is very fragmented: all professionals have their own representative organization and almost all organizations focus on one specific role in the building and construction process (planning, design, construction, facilities management, etc.) and/or on only one level of activities (construction materials and components, whole buildings or the built environment). There is little integral or holistic thinking. This fragmentation is reflected in the scopes and objectives of almost all international organizations in the sector. Most, if not all, say that contributing to sustainable building and construction is one of their prime strategic objectives, but they all go about achieving this in their own way, reflecting their own members' interests and almost never working in cooperation with other such international organizations. Those organizations that do incorporate different roles in the process and different levels often focus on only one aspect of achieving sustainable building and construction, such as energy consumption in or renewable energy for buildings only or waste from construction only.

When looking at initiatives towards a more integrated and all-encompassing approach for sustainable building and construction that have taken place over, say, the last five years, no more than five can be mentioned as being of a major potential importance.

CIB, the International Council for Research and Innovation in Building and Construction

As indicated by its name, CIB has a strong focus on research and innovation, but it does have amongst its members representatives of all professionals and other stakeholders in the building and construction sector and its objective is to stimulate and actively facilitate worldwide exchange and cooperation. This worldwide membership and wide scope is reflected in its broad programme of activities, which includes projects on all aspects of building and construction.

In 1995 CIB decided to begin focusing on the Theme Sustainable Building and Construction. This has resulted in a refocusing of many of its existing expert commissions, the establishment of new ones, international cooperative research projects and conferences and in publications such as the earlier-mentioned *Agendas 21 for Sustainable Construction and for Sustainable Construction in Developing Countries*. It also resulted in the launch of strategic partnerships for joint activities in support of sustainable building and construction with other international organizations, like IEA (the International Energy Agency), ISIAQ (the International Society for Indoor Air Quality and Climate), FIG (the International Federation of Surveyors) and UNEP-DTIE-IETC (the International Environmental Technology Centre that within UNEP is the focal point for sustainable building and construction).

A summary description of the CIB activities that are especially related to sustainable building and construction can be found at http://www.cibworld.nl/pages/begin/Pro2.html

iiSBE, the International Initiative for Sustainable Built Environments

iiSBE was launched as an organization in 2000 with primarily individual experts as members and with an objective: to facilitate and promote the adoption of policies, methods and tools to accelerate the movement toward a global sustainable built environment.

One of its main activities is the management of the Green Building Challenge (GBC) process that aims to develop the theory and practice of environmental performance systems for buildings. Other activities include the establishment of a dedicated R&D database and the joint responsibility with CIB for a series of international Sustainable Building (SB) conferences (see below). More information in iiSBE can be found at http://iisbe.org/

GABS, the Global Alliance for Building Sustainability

GABS was launched in 2002 at the WSSD as a voluntary alliance of individuals and organizations and has the objective of raising awareness for sustainable development in four areas: land, property, construction and development. Although at present it functions as a virtual organization only, it has recognition from the UN as a 'type 2 partnership'. More information on GABS can be found at http://www.earth-summit.net

UNEP-DTIE SBC Forum

In 2003 the UNEP Division for Technology, Industry and Economy launched the Sustainable Building and Construction (SBC) Forum, the objective of which is to facilitate dialogue and the exchange of information among key stakeholders, and with their constituencies, on issues related to sustainability in building and construction. Members of this SBC Platform are the international organizations that represent all possible professionals, decision makers and other stakeholders in achieving sustainable building and construction as broadly defined as in this paper. Information on the SBC Forum can be found at http://www.unep.or.jp/ietc/sbc/index.asp

SB04/05 Conferences

CIB and iiSBE jointly are responsible for a series of international conferences that have developed into 'the' events worldwide, at which all the experts concerned with the various aspects of sustainable building and construction meet on a somewhat regular basis. Recently UNEP-DTIE has also become involved in this conference series.

The next main international Sustainable Building Conference (SB) will take place in 2005 in Tokyo and in 2004 preparatory regional SB conferences will be held in Eastern Europe, Southern Africa, Latin America and Asia (with one covering China and one covering the rest of Asia). It is expected that both the preparatory regional conferences and the next main international conference will attract a magnitude of other events and activities related to sustainable building and construction and will be a focal point for many national and international SB projects.

A positive recent development is that increasingly the above-mentioned organizations and initiatives are beginning to recognize the need to cooperate or at least align their activities as much as possible. At the same time platforms are being created for the other, more focused or specialized international organizations that are to contribute to sustainable building and construction, to getting involved also.

So far, however, most of these potential alignments and cooperations happen on a more-or-less voluntary and somewhat incidental basis and are not really commitment-based. However, they could provide the basis for the next, possibly decisive, step.

26 Some UK experience of gauging progress on introducing sustainable business practices in the construction sector

I. Cooper[1], D. Crowhurst[2], S. Platt[3] and R. Woodall[4]

Summary

In the UK, the Construction Support Unit of the Department of Trade and Industry has sponsored the development of a set of simple self-assessment tools called MaSC (Managing Sustainable Companies) for helping organisations to assess their current progress in introducing sustainable business practices and to plan what steps they need to take next to improve their performance.

This chapter describes the MaSC tools and reports their use by 145 UK construction firms through a series of industry briefings targeted on Managing Directors sponsored by the UK's Construction Industry Training Board conducted during late 2002/early 2003. The paper reports the average 'organisational profile' of these organizations against the six key management areas measured by the MaSC matrix: strategy, responsibility, planning, communication, implementation and auditing. It also records what the 145 construction firms saw as the key external and internal drivers for provoking their organisations to adopt and develop more sustainable business practices. The organisational profiles drawn for these 145 construction firms are compared with those drawn for 45 construction clients and 61 local government planning authorities.

These findings are compared with the progress made using the MaSC tools, over an initial six-month period, by the six test-bed organisations with whom the MaSC team

---

[1] Eclipse Research Consultants, 121 Arbury Road, Cambridge, UK, CB4 2JD.
[2] Centre for Sustainable Construction, Building Research Establishment Ltd, Bucknalls Lane, Garston, Watford, UK, WD25 9XX.
[3] Cambridge Architectural Research Ltd, Unit 6, 25 Gwydir Street, Cambridge, UK, CB1 2LG.
[4] Centre for Sustainable Construction, Building Research Establishment Ltd, Bucknalls Lane, Garston, Watford, UK, WD25 9XX.

worked to develop and evaluate the self-assessment tools. These test-bed organisations ranged from very large multi-national corporations to small companies and spanned both the demand and supply sides of construction.

Sustainability and continuous improvement in UK construction

In the UK, sustainable construction is still primarily a policy imperative driven by the public sector (central and local government) rather than a market requirement demanded by private sector clients. The principles underlying this policy imperative were first set out explicitly by the UK Government [1] in 1999, in *A Better Quality of Life: a Strategy for Sustainable Development in the UK*, as:

- maintaining high and stable levels of economic growth and employment
- prudent use of natural resources
- effective protection of the environment
- social progress that meets the needs of everyone.

These principles were directly applied by the UK Government [2] to construction in 2000, in *Building a Better Quality of Life: a Strategy for More Sustainable Construction*, to mean:

- being more profitable and more competitive
- delivering buildings and structures that provide greater satisfaction, well-being and value to customers and users
- respecting and treating its stakeholders more fairly
- enhancing and better protecting the natural environment
- minimizing its impact on the consumption of energy and natural resources.

In the UK, sustainable construction also has to be seen against the background of multiple initiatives introduced over the last decade in attempts to improve the performance of the country's construction industry. These initiatives, prompted by central government, began with the publication of Sir Michael Latham's *Constructing the Team* in 1994 [3], followed in 1998 by Sir John Egan's *Rethinking Construction* [4] and by the Movement for Innovation (m4i) [5] intended to deliver his recommendations. Subsequently these initiatives were superseded by *Accelerating Change* [6] in 2002 which, in turn, was amalgamated with the parallel but separate Construction Best Practice (CBP) programme [7] to be re-branded as Constructing Excellence [8] in 2003. In response to these initiatives, in 2000 the Construction Clients Panel of the Office of Government Commerce published its action plan [9] for achieving sustainability in the procurement of construction by Central Government. A similar but regional response can be seen in the manual [10] on project management produced for the English Regional Development Agencies. This manual contains specific guidance on sustainability that refers directly to MaSC as an assessment method.

What all these initiatives share is an emphasis on continuous improvement of industry performance. MaSC was grounded in experience gained on behalf of CBP from reviewing how 'leading edge' construction firms undertook a wide range of continuous improvement activities in the UK [11]. This experience was then applied to making business practices in construction firms more sustainable. In this sense, MaSC is an attempt to treat the introduction and development of sustainability in UK construction firms in the same manner as any other continuous improvement initiative – such as partnering, lean construction, or risk management – that they may have tackled over the last decade. The deliberate intention here is to lessen the mystique that surrounds sustainability by sequencing it into a series of small steps, each of which can be easily achieved.

How was MaSC developed?

MaSC [12] was developed collaboratively by three firms based in the UK construction sector:

(1) The Centre for Sustainable Construction, part of the Building Research Establishment Ltd, the UK's leading centre of expertise on buildings, construction, energy, environment, fire and risk
(2) Cambridge Architectural Research Ltd, an independent, multidisciplinary consultancy providing specialist advice and research for the construction industry, design professions and policy-making institutions
(3) Eclipse Research Consultants, a micro-firm specialising in sustainable development and change management.

Funding for the development of MaSC was provided by Partners in Innovation, a collaborative research programme run by the Construction Support Unit of the UK's Department of Trade and Industry. MaSC was originally developed and tested with a range of businesses involved in the construction process – from clients through designers, consultants and contractors, see below. Despite its construction provenance, however, MaSC is made up of a set of generic tools that can be applied to organisations of any size and in any sector of the economy.

What is MaSC?

MaSC is a set of simple self-assessment tools for helping managers to introduce and develop more sustainable business practices in their organisations. MaSC has two principal tools. While these both look deceptively simple, they are, in practice, quite sophisticated. The first tool is the MaSC Matrix, see Figure 26.1. You can use the matrix to assess where you currently think your organisation has got to in introducing sustainable business practices and to identify where you want to get to in the next 12 months.

In filling in the matrix, you are constructing an organisational profile that provides a snapshot of current progress against each of the matrix's column headings. These

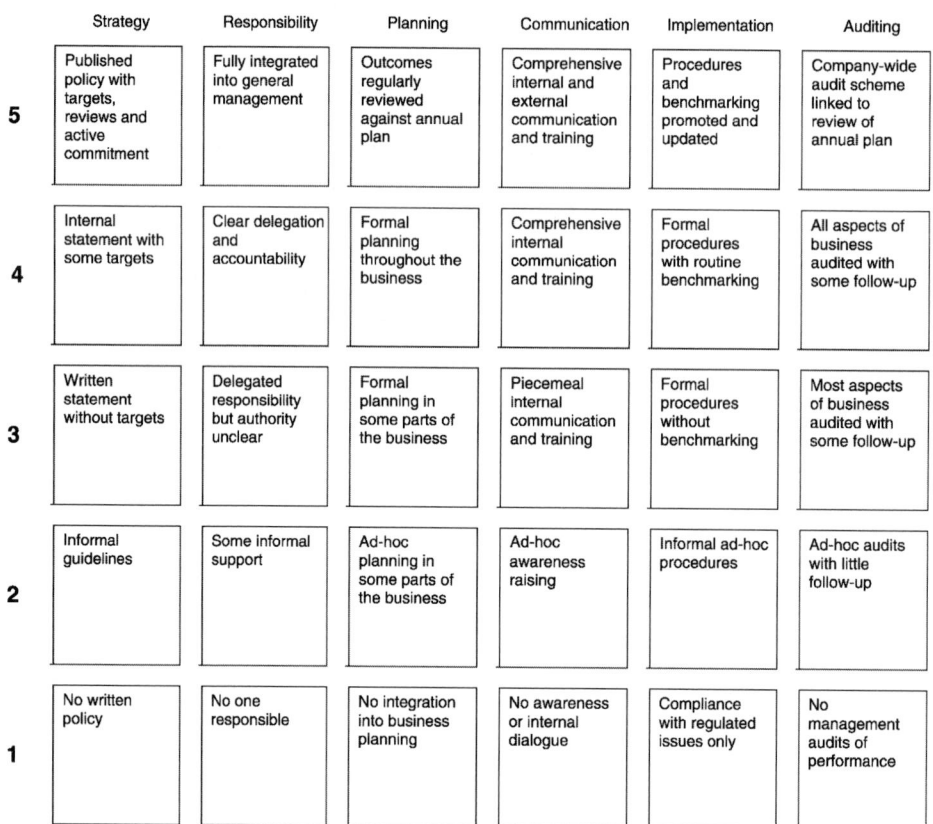

	Strategy	Responsibility	Planning	Communication	Implementation	Auditing
5	Published policy with targets, reviews and active commitment	Fully integrated into general management	Outcomes regularly reviewed against annual plan	Comprehensive internal and external communication and training	Procedures and benchmarking promoted and updated	Company-wide audit scheme linked to review of annual plan
4	Internal statement with some targets	Clear delegation and accountability	Formal planning throughout the business	Comprehensive internal communication and training	Formal procedures with routine benchmarking	All aspects of business audited with some follow-up
3	Written statement without targets	Delegated responsibility but authority unclear	Formal planning in some parts of the business	Piecemeal internal communication and training	Formal procedures without benchmarking	Most aspects of business audited with some follow-up
2	Informal guidelines	Some informal support	Ad-hoc planning in some parts of the business	Ad-hoc awareness raising	Informal ad-hoc procedures	Ad-hoc audits with little follow-up
1	No written policy	No one responsible	No integration into business planning	No awareness or internal dialogue	Compliance with regulated issues only	No management audits of performance

Figure 26.1   The MaSC matrix.

identify six key management areas that you need to address to make your business practices more sustainable:

(1)	*Strategy*	Do you have a published policy in place?
(2)	*Responsibility*	Have you delegated authority for making the performance improvements that you are pursuing and has that been integrated into your general management structure and mechanisms?
(3)	*Planning*	Do you have a plan with specified outcomes that you regularly review?
(4)	*Communications*	Do you communicate performance both inside your organisation and with external stakeholders about how you're attempting to improve?
(5)	*Implementation*	Have you got the procedures in place and the benchmarking capacity to know whether or not you're achieving your aims?
(6)	*Auditing*	Have you got measures in place to review progress regularly and amend your objectives?

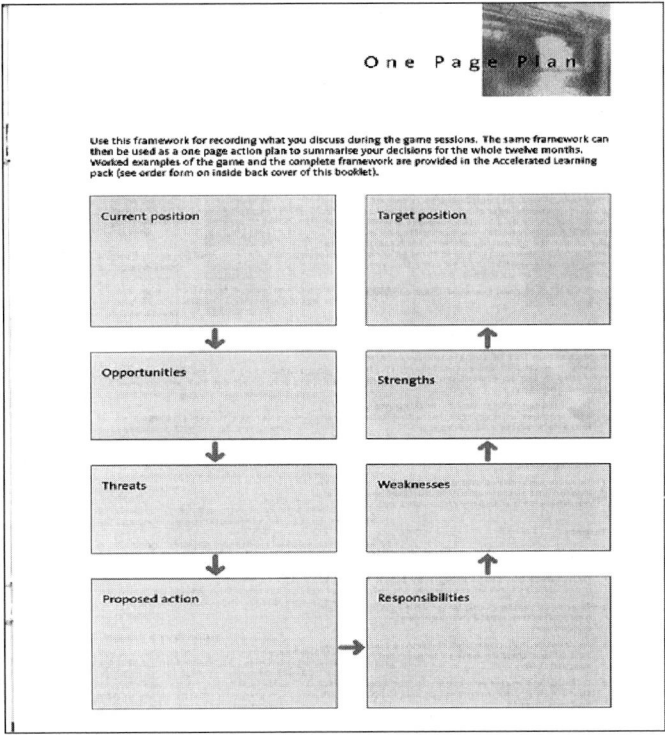

One Page Plan

Use this framework for recording what you discuss during the game sessions. The same framework can then be used as a one page action plan to summarise your decisions for the whole twelve months. Worked examples of the game and the complete framework are provided in the Accelerated Learning pack (see order form on inside back cover of this booklet).

Current position		Target position
Opportunities		Strengths
Threats		Weaknesses
Proposed action		Responsibilities

Figure 26.2 The MaSC One Page Plan.

You fill in the matrix by considering each column in turn and marking a cross where you think your company is now. You then join the crosses to produce your profile. Underlying the matrix are the notions that:

- moving up from level one to level five represents increasingly better practice, and
- such movement should occur in a balanced way in order to avoid making rapid progress in one area while paying insufficient attention to others.

The second tool is the MaSC One Page Plan, see Figure 26.2. This is completed during a Strategic Review Workshop, held to help you to accelerate your understanding of what you need to do to achieve the improvements you are seeking in your organisation. The one page plan records how you are going to improve your performance. You start in the top left-hand corner by identifying what your current position is. You then identify the opportunities that are open to you to make progress and the threats that are likely to prevent you from taking there. Next you specify a proposed set of actions that you can take now to make the performance of your organisation more sustainable. Then you identify who is responsible for implementing these and look at the organisational strengths and weaknesses that are currently confronting you. From this you develop the target position you want to reach in 12 months' time. During the Strategic Review Workshop, you complete a one page plan for each column in the matrix. By the end of

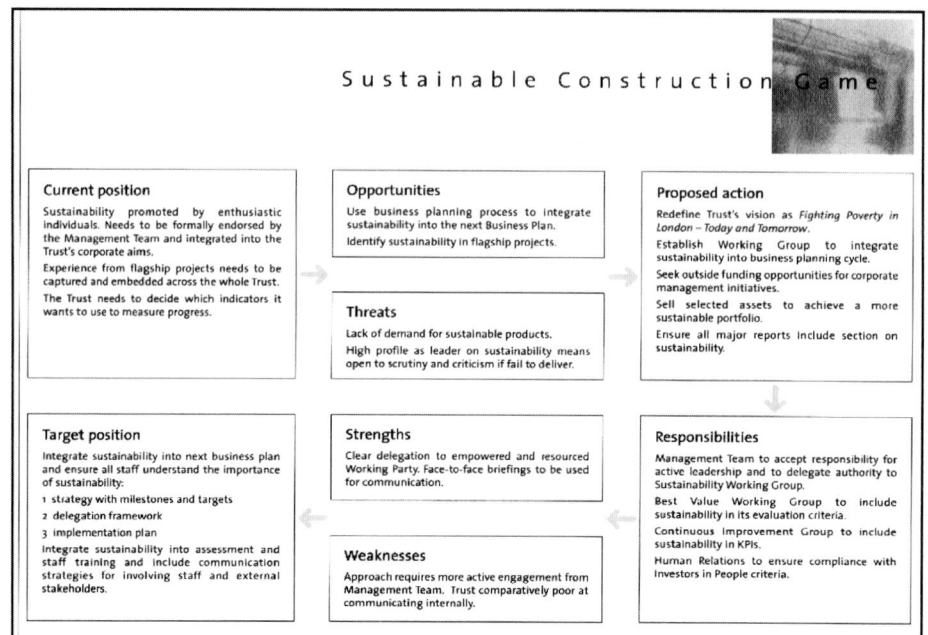

Figure 26.3 Worked example of the One Page Plan for Peabody Trust.

the Review, you are in a position to be able to summarize all of your decision-making into a single one-page plan. You can then hand this to the board of your organisation as a record of your decision-making and as a statement of intent, which you are asking them to endorse. Figure 26.3 shows a worked example of this process for one of the MaSC test-bed organisations – the Peabody Trust, a large provider of social housing in London [13].

Benchmarking current progress on sustainability in UK construction

With sponsorship from the UK's Construction Industry Training Board [14], MaSC was presented at series of five regional briefings to Managing Directors of construction firms, (architects, engineers, contractors and suppliers) held in late 2002 and early 2003 to raise awareness of sustainability as a senior management issue. The 145 self-selecting MDs were asked to draw the current organisational profile for their firms using the MaSC matrix. Figure 26.4 shows the average profile for these construction firms.

Figure 26.4 also shows average profiles drawn by construction clients and Planning Authorities at subsequent workshops on sustainability [15, 16]. Of course, for each of the three groups, these averages hide wide variations. However, taken at face value, they suggest that in the UK, amongst those who chose to attend events focused directly

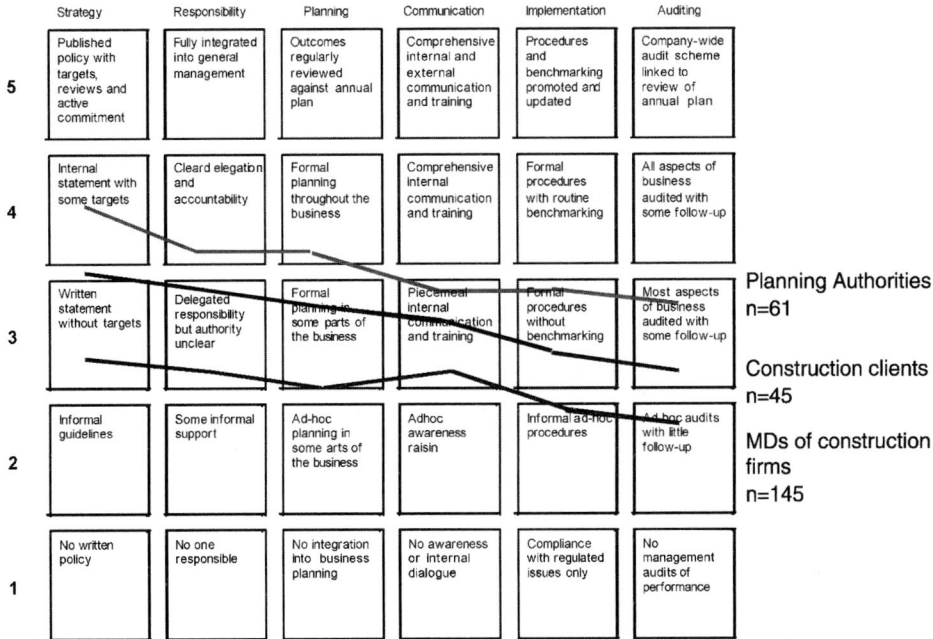

	Strategy	Responsibility	Planning	Communication	Implementation	Auditing
5	Published policy with targets, reviews and active commitment	Fully integrated into general management	Outcomes regularly reviewed against annual plan	Comprehensive internal and external communication and training	Procedures and benchmarking promoted and updated	Company-wide audit scheme linked to review of annual plan
4	Internal statement with some targets	Cleard elegation and accountability	Formal planning throughout the business	Comprehensive internal communication and training	Formal procedures with routine benchmarking	All aspects of business audited with some follow-up
3	Written statement without targets	Delegated responsibilty but authority unclear	Formal planning in some parts of the business	Piecemeal internal communication and training	Formal procedures without benchmarking	Most aspects of business audited with some follow-up
2	Informal guidelines	Some informal support	Ad-hoc planning in some arts of the business	Adhoc awareness raisin	Informal ad-hoc procedures	Ad-hoc audits with little follow-up
1	No written policy	No one responsible	No integration into business planning	No awareness or internal dialogue	Compliance with regulated issues only	No management audits of performance

Planning Authorities n=61

Construction clients n=45

MDs of construction firms n=145

Figure 26.4 Average organisational profiles for construction firms, clients and planning authorities.

on sustainability, clients and regulators currently see themselves as more advanced at making their business practices more sustainable than do the construction firms (whose services they seek or control).

At the CITB Briefings, MDs were asked to identify what they saw as the most significant drivers currently capable of promoting the adoption of sustainable business practices in their organisations. Figure 26.5 presents the drivers they highlighted.

In the UK, MDs see 'legislation' as the primary driver. But direct legislation on sustainability is not imminent. Central government wants the transition to sustainable production and consumption to be delivered by the market, to occur as the result of voluntary activities, not mandatory impositions. 'Increased profitability' (with 're-duced costs' close behind) is the next most frequent driver, indicating that firms want sustainability to be business-led by delivering immediate and direct financial benefits. Next comes another market-based issue 'client demand', particularly in the form of pre-qualification questionnaires for getting on lists of preferred suppliers, closely followed by 'long term survival'. These are then followed by a swathe of drivers – that firms could already be addressing without necessarily having signed up to the broader sustainability agenda of the 'triple bottom line' [17], such as 'reducing risk', 'cutting waste', 'enhancing reputation', 'winning market share' and 'keeping up with competitors'. Indeed drivers specifically associated with social sustainability – 'empowering staff' and 'fostering community relations' – are ranked lowest. Yet MaSC experience with test-bed companies suggests that organisations driven by internal and social

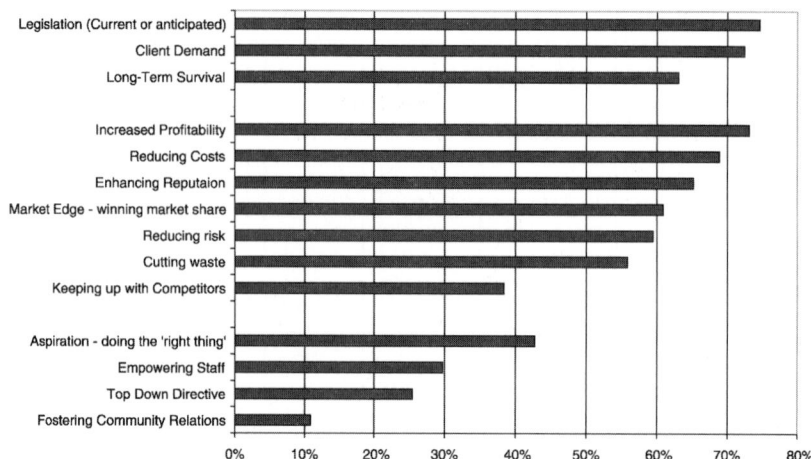

Figure 26.5   Primary drivers for promoting sustainability in UK construction. $n = 145$ MDs.

aspirations are likely to make more progress and have a higher profile on sustainability than those driven solely by economic or environmental considerations.

The performance of test-bed companies

Although the MaSC matrix can be used for benchmarking between organisations, it is primarily intended to encourage internal discussion within a firm about its own current performance and how this can be improved. The matrix and other MaSC tools were developed and tested with six widely differing construction firms:

(1)   Bovis Lend Lease: a large, international contractor
(2)   Crisp & Borst: initially a medium sized contractor subsequently purchased by Vinci, Europe's largest contractor
(3)   Dyer & Butler: a small engineering civil engineering contractor
(4)   Levitt Bernstein: a medium sized architectural practice (but small by construction standards)
(5)   The Peabody Trust: a large social housing provider
(6)   The WSP group: the third largest engineering consultancy in the UK.

In each organisation, a Strategic Review Workshop was undertaken with broadly based groups of staff. At the beginning of the workshop, participants were asked to agree the organisation's current profile and identify where they thought it should be six months from then. At the end of this period, the organisations were revisited and asked to redraw their profiles. Figure 26.6 shows the progress the test-bed companies reported they had made in this very short period.

   As Figure 26.6 shows, although only six months had elapsed, all the organisations had made some progress in just six months – even Crispin & Borst, which lost its

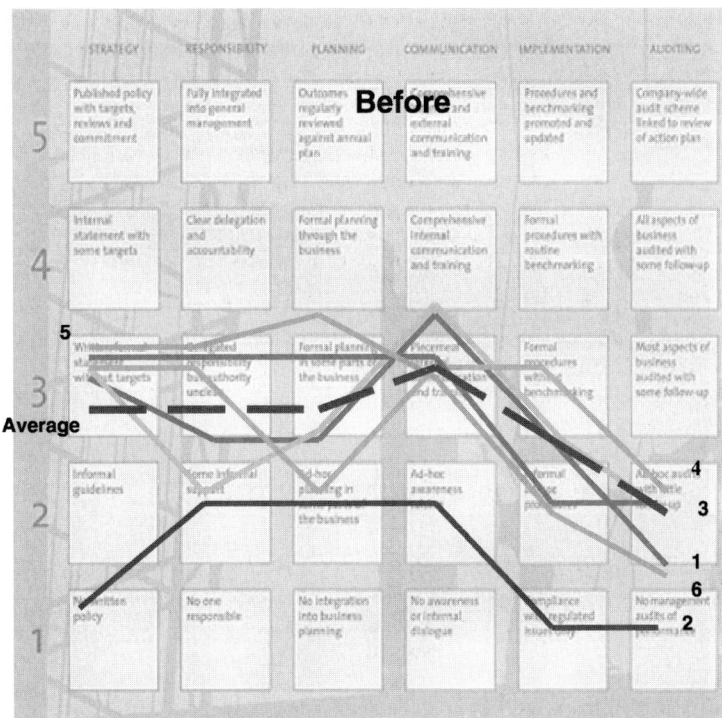

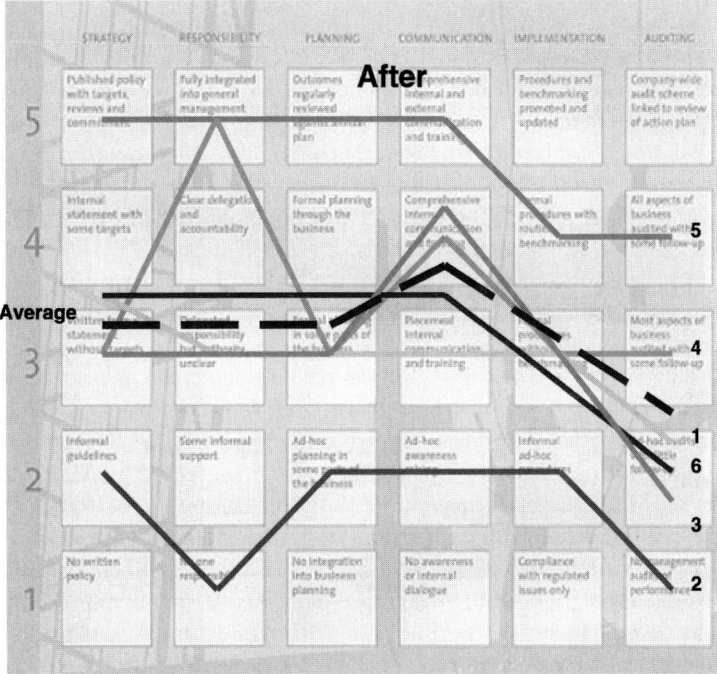

Figure 26.6 Reported progress of test-bed companies in six months (numbering as in list on p. 296).

Environmental Manager during this period. The Peabody Trust made most progress. It treated sustainability as the central focus of its next annual business plan and developed a detailed sustainability strategy containing wide ranging and highly aspirational targets:

- by 2010
    - reduce waste by 50% on sites, offices and estates
    - improve energy efficiency of stock by 20%
    - produce clean electricity for 700 homes
    - substitute cars in fleet by environmentally friendly vehicles
    - an ecology plan for all estates to maximise plants and wildlife
    - a social and economic plan for 40% of estates
    - a community regeneration programme for up to 12 000 residents
    - a rolling programme for investing in staff
- by 2020
    - a carbon-neutral, zero-waste business.

Its Chief Executive summed up the benefit of engagement in MaSC to the Trust as, 'MaSC has shown that, despite the mystique that surrounds it, sustainability is achievable, we leapt a considerable number of months if not years through engaging in the process.'

Conclusions

The introduction of sustainability into construction in the UK remains a policy-driven issue, rather than a response to market demand. Given the current lack of client demand, MDs of construction firms see legislation as the primary driver for sustainability, but such legislation is not imminent. Central government is treating the transformation to both sustainable production and consumption in the UK as a voluntary initiative. MDs also want sustainability to be immediately business-friendly, delivering both increased profitability and reducing costs. Many of the drivers for sustainability that they cite – such as reducing risk, cutting waste, keeping up with competitors – address only economic and environmental sustainability. Few firms are currently addressing the third component of the 'triple bottom line' – social sustainability. Yet MaSC experience to date suggests that those that do address all three components – because they are driven by internal aspirations rather than external factors – are likely to make more progress and have a higher profile than those prompted by just economic or environmental considerations.

The use of the MaSC matrix for inter-organisational comparisons indicates that, even amongst the 'leading edge' and 'early adopters' (defined here as those who choose to attend events on sustainability), there is still much progress to be made. Although they are the advanced guard of the industry, such construction firms are typically less than half way on the journey to current best practice. Advanced guard clients present themselves as marginally further ahead but slightly behind Planning Authorities that regulate them both.

References

[1] Department of Environment, Transport and the Regions (1999) *Building a Better Quality of Life: A Strategy for Sustainable Development in the UK*, DETR, London.

[2] Department of Environment, Transport and the Regions (2000) *Building a Better Quality of Life: A Strategy for Sustainable Construction*, DETR, London.

[3] Latham, Sir Michael (1994) *Constructing The Team*, final report of the government/industry review of procurement and contractual arrangements in the UK construction industry, HMSO, London.

[4] Egan, Sir John (1998) *Rethinking Construction*, available from website www.rethinkingconstruction.org

[5] Movement for Innovation, www.m4i.org

[6] Strategic Forum for Construction (2002) *Accelerating Change*, Rethinking Construction, London, www.rethinkingconstruction.org

[7] Construction Best Practice, www.cbpp.org.uk

[8] Constructing Excellence, http://cbp.idnet.net/

[9] Government Construction Clients Panel (2000) *Achieving Sustainability in Construction Procurement: Sustainability Action Plan*, Constructing the Best Government Client, Office of Government Commerce, London.

[10] Bovis Lend Lease Consulting (2002) *Project Management Manual for English Partnerships and the Regional Development Agencies*, BLLC, London.

[11] Construction Best Practice, *Best Practice Profiles series*, http://cbp.idnet.net/resourcecentre/publications/profiles.jsp

[12] Managing Sustainable Companies, www.bre.co.uk/masc, contact masc@bre.co.uk

[13] Peabody Trust, www.peabody.org.uk

[14] Construction Industry Training Board, www.citb.org.uk

[15] Centre for Facilities Management and Sodexho (2003) Sustainability and Asset Management, Breakfast Seminar, Portcullis House, London, 2/2/03.

[16] Go-East (Government Office for the Eastern Region) (2003) Rural communities need affordable housing too, Conference, Robinson College, Cambridge, 19/3/02.

[17] Elkington, J. (1998) *Cannibals with Forks: The Triple Bottom line of the 21st Century Business*, Capstone, Oxford.

27 Developing a sustainable development approach for buildings and construction processes

J. T. Gibberd[1]

Summary

This chapter argues that buildings have an important role in supporting sustainable development in developing countries. It explores the relationship between buildings and sustainable development and suggests how this should influence buildings and construction processes in developing countries. The chapter describes work being carried out that aims to support a sustainable development approach in building and construction processes. In particular it describes lessons learnt, and the tools and approaches developed as part of Thuba Makote, a pilot school construction programme and the Sustainable Building Assessment Tool (SBAT), a project that is developing a framework that aims to integrate sustainable development considerations into the briefing and design processes of buildings.

Introduction

Sustainability can be described as a state in which humankind is living within the carrying capacity of the earth. This means that the earth has the capacity to accommodate the needs of existing populations in a sustainable way and is therefore also able to provide for future generations (World Commission on the Environment and Development, 1987). However, as we are currently in an unsustainable state as a result of humankind having exceeded the carrying capacity of the earth, we must make a strong and concerted shift in direction in order to return to within that carrying capacity (Loh, 2000). This concerted, and integrated, action and change of direction can be referred to as *sustainable development*.

A review of literature on sustainability suggests that sustainability can be described in terms of social, economic and environmental states that are required in order for

---

[1] CSIR, Division of Building and Construction Technology.

overall sustainability to be achieved (Gibberd, 2003). The World Summit on Sustainable Development Plan of Implementation provides a range of sustainable development objectives that should be aimed for in order to achieve sustainability (United Nations, 2000). This plan is complex, but the key objectives relevant to buildings and construction can be extracted and summarised. These are listed below. It is therefore possible to describe sustainability in terms of *states* and sustainable development in terms of *objectives* within the environmental, economic and social arenas.

Environmental sustainability state: Robust, vibrant, productive and diverse biophysical systems that are able to provide resources and conditions necessary for existing and future populations on an ongoing and steady basis.

Environmental sustainable development objectives

- *Size, productivity and biodiversity:* Ensure that development conserves or increases the size, biodiversity and productivity of the biophysical environment.
- *Resource management:* Ensure that development supports the management of the biophysical environment.
- *Resource use:* Ensure that development minimizes the use or support of environmentally damaging resource extraction and processing practices.
- *Waste and pollution:* Ensure that development manages the production of waste and pollution in order to minimize damage to the biophysical environment.
- *Water:* Ensure that development manages the extraction, consumption and disposal of water in order not to adversely affect the biophysical environment.
- *Energy:* Ensure that development manages the production and consumption of energy in order not to adversely affect the biophysical environment.

Economic sustainability state: Responsive structures, systems and technologies able to accommodate change and ensure that limited resources are used and maintained as efficiently and effectively as possible to provide for the needs of existing and future populations without damaging the biophysical environment.

Economic sustainable development objectives

- *Employment and self-employment:* Ensure that development supports increased access to employment and supports self-employment and the development of small enterprises.
- *Efficiency and effectiveness:* Ensure that development (including technology specified) is designed and managed to be highly efficient and effective, achieving high productivity levels with few resources.
- *Indigenous knowledge and technology:* Ensure that development takes into account and, where appropriate, draws on indigenous knowledge and technology.

- *Sustainable accounting:* Ensure that development is based on a scientific approach that takes into account, and is informed by, social, environmental and economic impacts.
- *An enabling environment:* Develop an enabling environment for sustainable development including the development of transparent, equitable and supportive policies, processes and forward planning.
- *Small-scale, local and diverse economies:* Ensure that development supports the creation of small scale, local and diverse economies.

Social sustainability state: Safe, happy, healthy, fulfilled, educated societies that have organizational structures and innovative capacity that enables limited resources to be shared equitably and in ways that ensure that existing and future populations needs are met.

Social sustainable development objectives

- *Access:* Ensure that development supports increased access to land, adequate shelter, finance, information, public services, technology and communications where this is needed.
- *Education:* Ensure that development improves levels of education and awareness, including awareness about sustainable development.
- *Inclusive:* Ensure that development processes, and benefits, are inclusive.
- *Health, safety and security:* Ensure that development considers human rights and supports improved health, safety and security.
- *Participation:* Ensure that development supports interaction, partnerships and involves, and is influenced by, the people that it affects.

This description provides simple definitions for sustainability and sustainable development. A useful aspect of the definition is that it provides an ultimate state that must be strived for as well as a set of actions or objectives that, if addressed and implemented, will lead towards this state. The list of sustainable development objectives can be related to buildings and building processes in order to develop an understanding of the relationships that exist. In order to do this, buildings, in a similar way, need to be categorised and described in terms of a set of key elements. For instance, buildings can be described in terms of building elements and building lifecycle stages.

Building elements

- *Location:* This described the location of the building.
- *Site:* This describes the site and landscaping in which the building is located.
- *Size and shape:* This describes the size and shape of the building as a whole.
- *Building envelope:* This describes the physical envelope of the building.

- *Internal space:* This describes the space enclosed by the building envelope.
- *Furniture and fittings:* This describes equipment, furniture and fittings located within the internal space.
- *Services:* This describes services in the building such as water, electricity and telephone.
- *Materials and components:* This describes the materials and components used in the construction of the building.

Building life cycle stages

- *Briefing:* This stage starts with the decision to develop a building and includes initial conceptualisation of the requirements of the building.
- *Design:* This stage includes the development of the design of the building through to tender documentation.
- *Construction:* This stage refers to the construction of the building and ends at handover to an owner or users, on completion.
- *Operation:* This describes the stage where the building is in normal use and ends when a building is refurbished or demolished.
- *Refurbishment/Demolition:* This describes the stage when the building is refurbished for further use or deconstructed/demolished.

Both of these descriptions are somewhat simplistic, however they are useful in that they begin to allow the relationships between sustainable development and buildings to be plotted. For instance, this can be described in the form of a matrix seen in Figure 27.1.

This matrix enables a set of objectives for buildings to be generated. For instance the following building and construction sustainable development objectives can be inferred from the sustainable development objective of 'employment':

- *Location:* Locate the building where employment is needed.
- *Site:* Design site to support job creation and self-employment.
- *Size and form:* Design building size and form to support job creation and self-employment.
- *Building envelope:* Design building envelope to support job creation and self-employment.
- *Furniture and fittings:* Specify furniture and fittings that support job creation and self-employment.
- *Materials and components:* Specify materials and components that support job creation and self-employment.
- *Construction:* Use labour intensive construction processes.
- *Operation:* Use labour intensive facilities management and maintenance processes.
- *Refurbish/Demolish:* Use labour intensive construction processes.

The number and complexity of objectives generated by this matrix indicates that, in order for these to be applied, it is important to have an effective *assessment framework*

Design Elements

	Brief	Design	Location	Site	Size and Form	Envelope	Internal Space	Furniture and Fittings	Services	Materials & Components	Construction	Operation	Demolition / Refurbishment
Size, productivity & biodiversity			●	●	●	●	●	●	●	●	●	●	●
Resource management								●	●	●	●	●	●
Resource use			●	●				●	●	●	●	●	●
Waste and pollution			●	●			●	●	●	●	●	●	●
Water			●	●	●	●		●	●	●	●	●	●
Energy			●	●	●	●	●	●	●	●	●	●	●
Employment and self employment	●	●	●	●	●	●	●	●	●	●	●	●	●
Efficiency and effectiveness	●	●	●	●	●	●	●	●	●	●	●	●	●
Indigenous knowledge	●	●						●	●	●	●	●	●
Accounting	●	●									●	●	●
Enabling environment	●	●									●	●	●
Small-scale, local, diverse	●	●	●	●	●	●	●	●	●	●	●	●	●
Access	●	●	●	●	●		●		●			●	
Education	●	●	●	●				●	●	●		●	●
Inclusiveness	●	●	●	●		●	●	●	●	●	●	●	●
Health, safety and security	●	●	●	●	●	●	●	●	●	●	●	●	●
Participation	●	●	●	●	●	●	●	●	●		●	●	●

Figure 27.1 Matrix for generating building and construction sustainable development objectives.

in which objectives can be prioritised. In addition, to evaluate progress towards the achievement of objectives, linked performance indicators are required. An effective *structured approach* is also required in order to ensure that objectives identified are considered, and actively addressed, by the appropriate parties at the appropriate stages in the building's lifecycle.

An assessment framework

The assessment framework should consist of an overarching goal, objectives and indicators. The overarching goal is to ensure that buildings and construction processes support sustainable development. The objectives describe a range of actions and approaches that should be taken to support this goal. These objectives should be defined through a structured approach (described below) and refer to local sustainable development priorities and opportunities. Indicators are used to measure progress towards achieving objectives set. The assessment framework aims to ensure that the correct sustainable development objectives are set in terms of the state of knowledge and technology, the context, the project, and the stakeholders. It aims to make sustainable development directly relevant to buildings and construction through breaking this down into easily implementable steps, which can be integrated into buildings and construction processes.

A structured approach

A structured approach should aim to make sure that the assessment framework is used to maximum effect. It does this in the following ways. It ensures that the framework is based on accurate up to date and relevant information. It ensures that key project stakeholders gain some understanding of sustainable development and are able, through the assessment framework, to discuss and agree on building and construction objectives that will support sustainable development. Finally, it ensures that once an assessment framework has been populated and agreed this will be used to influence the development of buildings and construction processes.

It is suggested that an assessment framework and structured approach is an effective way of beginning to integrate sustainable development into buildings and construction processes. This has been reflected in the work carried out by CSIR. Here an assessment framework, called the Sustainable Building Assessment Tool (SBAT), and a structured approach consisting of nine stages with associated tools, is being developed.

The Sustainable Building Tool (SBAT)

The tool was developed to relate strongly to the context of a developing country and is designed to support sustainable development. This is reflected in the assessment criteria and how the tool is to be used.

The tool describes fifteen sets of objectives that should be aimed for in buildings. It suggests that the extents to which these objectives are achieved in buildings provide a simple, yet reasonably effective, measure of the level of support for sustainable development. Objectives are arranged under the headings of Environmental, Economic and Social, and are as follows:

Social

- Occupant comfort
- Inclusive Environments
- Access to facilities
- Participation and control
- Education, Health and safety

Economic

- Local economy
- Efficiency of use
- Adaptability and flexibility
- Ongoing costs
- Capital costs

Environmental

- Water
- Energy
- Waste
- Site
- Materials and components.

These objectives were established through a process of describing, and understanding, buildings in terms of their relationship to social, economic and environmental systems (Gibberd 2001). Different environmental and economic and social systems have different levels of sustainability and the approach used to develop the SBAT aimed to assess not only the performance of buildings in terms of support for sustainable development but also assess the extent of the building's contribution to supporting, and developing, a wider set of more sustainable systems around it. The SBAT interface can be seen in Figure 27.2.

An important part of developing the SBAT was consideration of how this could become part of, and influence, normal design, construction and building management processes. This led to the development of a nine-stage structured approach based on the typical life cycle of a building.

Structured approach

The structured approach has been developed to ensure that an assessment framework, like the SBAT, can be used to maximum effect. It is designed to ensure that at particular

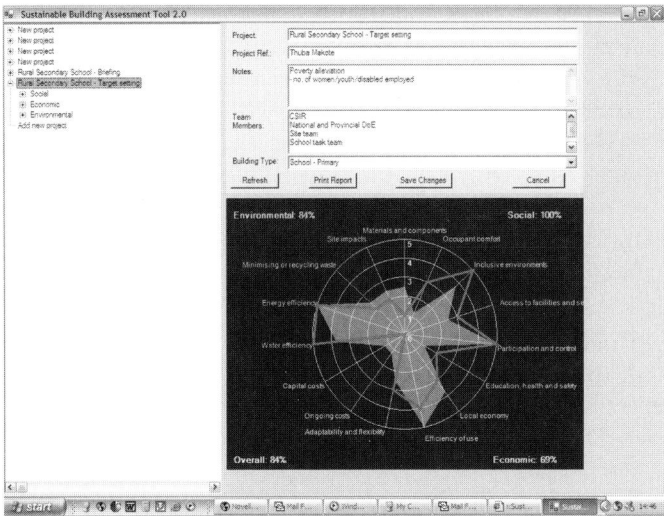

Figure 27.2   The SBAT interface.

stages targets are set and agreed by key stakeholders (during the briefing, site analysis and target setting stages) and that these are then used to guide design decisions and the selection of procurement and construction options. However, in addition to the SBAT (which sets the high-level objectives or targets for the building), additional tools are required at different stages of a building's life cycle in order to support the achievement of these objectives. This is because there are different role players and processes at the different stages. A variety of tools are being developed for each of these different stages, these are listed and described in more detail below. This is also summarized in Figure 27.3.

The first of these stages is *Briefing*. During this stage a detailed picture of the building users and building users' organization is developed. This includes an understanding of the work and lifestyles of building users and an understanding of the trends affecting the organization as well as possible future scenarios. It also aims to establish the level of commitment to sustainable development within the client. This can be prepared through a briefing workshop that uses the SBAT. This helps to make sustainable

	Stage	**Tools**
1	Briefing	Briefing document (SBAT)
2	Site Analysis	Site Analysis Report
3	Target Setting	Target Setting Report (SBAT)
4	Design	SBAT Design Review, Design for Sustainability Methodologies
5	Design development	SBAT Detailed Design Review Report
6	Construction	Construction Monitoring Indicators and Guidelines
7	Handover	Building User Manual, Facilities Management Training
8	Operation	Facilities Management Systems, User Awareness
9	Reuse/refurbish/recycle	As stages 1-6

Figure 27.3   Nine stage structured approach and related tools.

development an *explicit, practical issue,* allowing discussion and agreement. The target-setting component of the SBAT enables a 'target footprint' of 'required performance' to be developed for the building. This footprint can be developed and agreed by the stakeholders as part of the workshop and issued to the design team as part of the briefing documents. This is a useful reference document as it helps establish the level of commitment by the client to sustainable development at the outset of the project and enables the client and the design team to monitor design development against an explicit performance brief. This stage aims to ensure that there is a shared understanding about sustainable development and a strong commitment to addressing this amongst all stakeholders in the project.

Often, the development of a new building offers the opportunity to make parallel changes in the organizations, communities and people that will use the buildings and that exist around them. For instance, an organization may want to become more efficient, or government may want to use the construction process of a public building to create employment within a community. Where this requirement or desire exists, it should be made explicit in the brief. The briefing document therefore may contain a description of both 'hard' outputs such as buildings as well as 'soft' outcomes such as increased local employment, organizational change and increased capacity. It is suggested that *briefing documents* for projects that aim to support sustainable development are increasingly likely to describe sets of mutually supportive soft and hard outputs that aim to ensure that limited resources are used for a maximum, long term effect. This type of briefing document was used in the Thuba Makote project, where beneficial social and economic impacts such as job creation and training as well as buildings were required.

The second stage is *Site Analysis.* This stage investigates the site in terms of Social, Economic and Environmental aspects in order to establish the context in terms of problems to be addressed and potential resources that can be used. This approach was used by the CSIR in the Thuba Makote project to analyse school sites and assist in the development of project briefs. This aimed to ensure that, where possible, the project helped address local social, environmental and economic problems, using local resources where available and appropriate. The output of this stage is a *Site Analysis Report,* which is used to inform the development of the brief and the target setting document.

The third step is the *Target Setting Stage.* This stage is used to develop detailed sustainable development performance targets for the building. This draws on a range of information. It uses outputs from the briefing workshop to establish the level of commitment by the client to sustainable development and the capacity and understanding within this area by the design team. The site analysis is used to provide a description of the local context in terms of problems and resources. Finally benchmark performance figures for similar buildings in similar contexts are required. All of this information is then used to develop a detailed set of achievable but challenging performance targets for the building. This document provides a detailed reference document that is used to guide many aspects of the projects such as the design, selection of procurement method, type of construction process and facilities management policies. This type of document was developed by the CSIR with Arup to support the development of a sustainable hotel by the Ritz Carlton at Spier (Perkins *et al.,* 2001).

This approach requires the design to be tested against the target document during the *Design Development* stage. This enables the performance of different designs and strategies to be evaluated and the best options chosen. The SBAT can be used at this stage to support the rapid evaluation of different decisions as it enables the 'performance' of an approach to captured readily.

As currently developed, the Structured Approach does not include extensive monitoring of site and *construction* processes. It suggests that sustainable development issues should be addressed in tender documents and covered by the contract and a detailed briefing to the contractor. This enables the contractor to understand the requirements of the project and provides recourse if these requirements are not met. Simple *Construction Monitoring Indicators and Guidelines* have been developed within this area, including tender documentation and site checklists. They also enable information on a number of indicators such as construction training, employment of disabled people and women, and labour intensity to be captured, and reported on.

During and after the design development, it is suggested that the design team prepare potential users and managers of the building for *handover* and occupation of the building. This preparation should occur through briefings on the buildings and its systems, the development of a *building user manual*, and building user induction training and material. This enables the building and its systems to be understood fully and managed effectively by it new users.

Over the life of the building it is important to ensure that performance of the building and its systems are maintained at a high level. This requires systems for logging and reporting information on a range of building performance indicators, such as water and energy consumption. It also requires an appropriate level of awareness amongst building users who are able to improve performance through their actions, for instance by switching off lights. This can be supported through readily accessible information on current and past building performance. An approach using an intranet web page to deliver this information to building users is being explored in a commercial office building.

Finally, at the end of its useful life the structured approach suggests that, where possible, the building is refurbished and reused. Where this is not possible, as much of the building as possible should be recycled or reused. Ideally, this process of deconstruction would be assisted through reference to the building user manual, which would provide the information required to enable the process to happen easily with a minimum of waste.

Conclusion

The chapter suggests that integrating sustainable development into buildings and construction will requires a significant change of approach. Work carried out to date suggests that an approach can be developed through the use of an assessment framework and a structured approach. The assessment framework enables people involved or affected in the development of buildings to discuss, agree sustainable development objectives and monitor progress towards their achievement on the project. The structured approach ensures that actions required at each of the building lifecycle stages to achieve sustainable development objectives are allocated to and implemented by the responsible parties.

Developing and implementing sustainable development objectives in buildings is particularly important in developing countries where there may be considerable social and economic problems such as low/poor levels of health, education and employment and limited economic resources. The approach developed, it is suggested, should ensure that there is maximum beneficial social and economic impact for the investment rather than merely concentrating on the more conventional approach of minimizing environmental impact. Buildings and construction therefore cannot be seen in isolation from users and local communities and should be responsive to local needs and opportunities. They should be seen as systems that require both physical aspects (such as the buildings and site) and non-physical aspects (such as use of the building and management systems) to work together in an efficient and integrated way to achieve sustainable development objectives.

References

[1] Gibberd, J. (2001) *Building Sustainability: How Buildings can Support Sustainability in Developing Countries*, Continental Shift 2001, IFI International Conference, 11–14 September 2001, Johannesburg.

[2] Gibberd, J. (2003) Integrating sustainable development into briefing and design processes of buildings in developing countries: an assessment tool, unpublished thesis.

[3] Loh, J. (ed.) (2000) *The Living Planet Report 2000*, WWF, Gland, Switzerland, p.1.

[4] Perkins, A., Gibberd, J., Campbell, A. and Oliver, N. (2001) Ritz Carlton at Spier, sustainable construction brief, unpublished report.

[5] United Nations (2002) World Summit on Sustainable Development, Plan of Implementation, advance unedited text, http://www.johannesburgsummit.org/index.html

[6] World Commission on the Environment and Development (1987) *Our Common Future*, Oxford University Press, Oxford.

28   Adoption of Local Agenda 21: local councils' views on sustainability initiatives

S. Khan[1] and B. Bajracharya[2]

Summary

Having endorsed the policy of Ecologically Sustainable Development (ESD) followed by the ratification of the Earth Summit (1992) charter, Australia has embraced the principles of Local Agenda 21 (LA21). Our local councils are, consequently, required to include LA21 aims into local strategies and planning [1]. This implies that local councils need to build up effective partnerships with the local community to establish strategies promoting an integrated and holistic treatment of the social, economic and environmental aspects of planning, in line with ESD objectives.

The chapter reports on preliminary findings of a currently on-going study of three local councils in New South Wales, to determine the factors affecting the adoption of LA21. It reviews strategic policy documents to examine the councils' corporate commitment to sustainability and reports on responses of planners working in local councils towards ESD principles, LA21 and community involvement requirements that it entails.

Introduction

The 1992 Rio Earth Summit developed Agenda 21, which outlines the objectives and actions that can be taken at local, national and international levels to make the required transitions towards sustainability to take us into the 21st century. Chapter 28 of LA21 focuses on the role of local government agencies in promoting sustainable development by working with local communities to achieve a local action plan [2]. By ratifying the charter, Australia has also embraced the principles of LA21 and is therefore required to include its aims into local strategies and planning [1]. As a stated objective of LA21:

'(b)y 1996, most local authorities in each country should have undertaken a consultative process with their populations and achieved a consensus on 'a local Agenda 21' for the community.' [3]

[1] School of Construction, Property and Planning, University of Western Sydney, Sydney, Australia.
[2] School of Design and Built Environment, Queensland University of Technology, Brisbane, Australia.

In 1992 the commonwealth government also adopted an Ecologically Sustainable Development (ESD) policy [4].

This chapter is an attempt to understand how local councils in New South Wales have implemented LA21 and sustainability at a local level. The main objective of the paper is to develop an understanding of the factors that influence the ways that different councils adopt LA21. Such understanding is a prerequisite to removing obstacles that may impede the universal adoption of LA21 by local councils in Australia. The paper thus sets out to identify the characteristics of individual local planning authorities that could shape their approach towards promoting community participation in line with LA21 principles.

The chapter begins with a discussion of the main elements of LA21, highlighting the potential benefits from its implementation. It then briefly outlines the adoption of LA21 in the Australian context. It focuses on three local councils in western Sydney, reviews their strategic planning documents and presents the viewpoints of their planning staff to identify key factors influencing the extent of adoption of LA21 in these councils. Lastly, it draws conclusions from the findings and identifies further areas of research based on the case studies.

The main elements of LA21

The basic elements of LA21 include the creation of a community *vision* that brings together the aspirations of all stakeholders; the establishment of a *partnership* between local authorities, communities, and businesses; the engagement in a community-based, inclusive process of *issue analysis*; the preparation of *action plans* based on formalized objectives; and processes installed for the *implementation, monitoring, evaluation* and *feedback* [5]. Commenting on its adoption in Europe, Raemaekers similarly notes that LA21 'tries to be bottom-up, inclusive, participative, and open to scrutiny' [6].

Review of the literature suggests that there are many benefits to a council that implements a LA21 [7]. First, it provides an enhanced opportunity to meet community needs, stay relevant to them and contribute to community cohesion. Second, it provides stronger communication and cooperation between different stakeholders to coordinate joint planning and action for sustainability. Lastly, effective policy integration and long term planning can realise cost savings to both the council and the community.

The adoption of Local Agenda 21

In the face of growing environmentalism, local governments in many parts of the world have sought to improve their environmental performance. The focus has shifted from one-off measures or a piecemeal approach to one that incorporates environmental protection and environmentalist values across all functions and activities of local government. Further, local governments have assumed the responsibility of promoting environmental protection among the various actors in their community [8].

While Australia hasn't quite met its commitment to adopt LA21 by 1996, the Federal government, supported by a number of State level initiatives, has adopted a number of policies supporting sustainable development. These are reflected in various legislative changes such as the enforcement of the Local Government Act 1993 in NSW and its recent amendments, which consequently filter into the policy documents of local councils in NSW. Since 1997, councils in NSW are required to adopt corporate plans and management plans that address their commitment to ecologically sustainable development (ESD) principles. Councils and their staff are expected to adopt sustainable development principles in carrying out all of their responsibilities. They are also required to prepare an annual State of Environment (SoE) report and ensure that its findings feed back into management processes [9].

While there has not been any comprehensive study to gauge the extent and level of adoption of LA21 in Australia, there is reason to assume that its adoption by councils is generally quite limited. Even in the case of the ESD requirements that are legislated, a recent review suggested only partial implementation of ESD principles by relevant Federal government departments and agencies [10].

The fact that local authorities in Australia seem to lag behind their European counterparts in matters relating to planning for sustainable development in general and the adoption of LA21 in particular, are causes for concern. We need to be mindful of the fact that complacency at the local government level in the implementation of LA21 could result in social inequity and a general failure to achieve the goals of sustainable development. This makes it imperative to identify any hurdles that may retard the progress in the implementation of LA21. As Davidson [11] suggests, one of the impediments to the adoption of LA21 could be the attitude of the councils towards community consultation. This could play a significant part in determining the achievement or otherwise of a council's pursuit of LA21 objectives.

A study of three western Sydney local councils

The study focused on three adjoining local councils of varying sizes in the western Sydney region, viz. Blacktown, Holroyd and Penrith. Blacktown (pop. 254 817, area 247 km^2) [12] is the largest of the three councils with much pressure for development. Holroyd (pop. 80 000, area about 40 km^2) [13] is a small council and serves as one of Sydney's established industrial areas. Penrith (pop. 178 361, area 407 km^2) [14] represents a council on Sydney's outer fringe. With reference to distance from Sydney's CBD, Holroyd is the closest (25 km) followed by Blacktown (35 km) while Penrith is the farthest away (54 km). Blacktown, a rapid growth area, charges its residents one of the highest rates in New South Wales. While Penrith is also experiencing development pressure, it is relatively lower than that in Blacktown. Of the three, Holroyd seems to experience the lowest development pressure.

A content analysis of the three councils' strategic planning documents was carried out to ascertain their commitment to ESD. Descriptors used by each council in their publications and corporate statements to describe themselves were compared. These publications and corporate statements were accessed from each council's official website.

An analysis of the descriptors used by the three councils in public documents posted on their official websites reveals that Blacktown Council describes the local government area (LGA) as 'an urban growth area', and emphasizes its 'central locality' (www.blacktown.nsw.gov.au). While using phrases like 'the most populous city' in NSW, the 'third-largest in Australia' and 'eighth-fastest growing City in Australia', it seeks to project itself as an area of 'sustained and rapid' growth. In further pushing its credentials as an area with growth potential, it describes itself as an area with the 'largest quantity of zoned and serviced industrial and commercial land throughout NSW'. Holroyd Council also seems to follow the general direction of Blacktown's approach in describing its LGA, describing it as being 'one of Sydney's most established industrial areas'. It seeks to market its LGA as one that offers 'competitively priced sites close to major markets and a variety of skilled local workforce' (www.holroyd.nsw.gov.au). Penrith Council, however, seems to have a different focus. It describes its LGA as one that is 'distinguished by its natural setting' and as 'a place with a distinct character and identity' [14]. While Penrith Council also seems to market its LGA as 'the capital of outer western Sydney', its focus seems to lie on promoting the LGA as an area 'fortunate to contain much of the natural environment which remains in the Sydney Basin'. In further stressing its rural and natural environment as an asset while compromising on its claim to be an urban centre, it describes its LGA as 'a City where the harmony of urban and rural qualities give the city a relaxed yet cosmopolitan lifestyle'.

A comparison of the three councils' corporate mission statements contained in public policy documents posted on their official websites show subtle variations within their focus. Blacktown Council's mission statement stresses the Council's resolve to provide its community with 'the best living and working environment through commitment to service'. In its vision statement, the council envisages Blacktown to be 'a vibrant, healthy and safe city' and 'a city of excellence'. While the vision is in line with a sustainable future, the focus seems to be on ensuring or promoting the vibrant character of the LGA. Holroyd Council, according to its mission statement, seeks to 'satisfy the reasonable needs of (the) community and efficiently and effectively manage the community assets'. The council's vision statement stresses its 'commitment to an enhanced quality of life in partnership with (its) community'. It is worth noting the stress on the need to develop a partnership with the community in seeking quality of life objectives. The Penrith Council, meanwhile, envisions 'a prosperous region with a harmony of urban and rural qualities and a strong commitment to environmental protection and enhancement'. It envisions a region that 'would offer both the cosmopolitan and cultural lifestyles of a mature city and the casual character of a rural community'. This is in line with its promotion of the LGA as one presenting a balance between urban and rural character.

To summarize, a content analysis of the mission and vision statements of the three councils suggest that Blacktown projects itself as a vibrant, large and growing city seeking to attract investment; Holroyd envisions itself as a city striving for enhanced quality of life as it seeks to continue to attract industry; while Penrith refers to its natural environment and seeks to offer a balance between a mature city and rural community.

Besides the content analysis of public policy documents, semi-structured interviews were carried out to understand the attitude of planners in these councils toward issues relating to ESD and LA21, soliciting the council planners' opinions regarding community participation, urban expansion and the need for balancing development and

Table 28.1 How councils describe themselves: descriptors/statements adopted by the councils.

Issues	Council		
	Blacktown	Holroyd	Penrith
Community consultation in environmental protection initiatives	Scepticism based on two main grounds: first, the projects are nebulous, and second, they is often council-driven.	It is essential – but overcoming the initial inertia within the community is a major task.	Community is aware – but can be confused at times. Needs to be helped to sort out its priorities.
View towards urban expansion	The community is often interested in urban expansion as they can benefit financially from land rezoned from rural to residential.	The community is very concerned about environmental issues such as air, noise and water pollution.	A small but active minority favours expansion while the silent majority wishes to maintain the rural character and lifestyle.
Stance regarding the balance between development and conservation	The council seems to consider itself 'pro-development'. This is explained by the presence of a very high proportion of new residential releases.	The council seems to consider itself conservative and anti-development.	The council sees itself as favouring development fractionally over conservation. The mix of both urban and rural areas explains this.

conservation. As indicated in Table 28.1, Blacktown planners were rather sceptical of councils that promote themselves as ones that involve the community. Holroyd planners complained of the community's inertia while in Penrith the planners stressed the need for councils to take up an educational role. Views on urban expansion similarly varied greatly across councils. According to their council planners, Blacktown wishes to be seen as pro-development council. At Holroyd, planners felt that their council is conservative, reflecting the presence of a large segment of the elderly in its population. At Penrith, planners saw the council as having a balanced view towards environmental protection and development, reflecting the presence of both urban and rural areas within its LGA.

Findings

Based on our analysis of the three councils' strategic planning documents and interviews with their planners, the following five factors appear to influence the extent of adoption of LA21 and sustainable development principles by local councils. These factors are:

(1) characteristics of the councils and their location
(2) local political dynamics
(3) community participation and consultation mechanisms
(4) community support for sustainability
(5) commitment through innovative policies and programs
(6) the role of the state government.

These, however, are only preliminary findings, mainly reflecting councils' views as elicited from their publications and the opinion of their planning staff. The next phase of our on-going research will seek to verify these findings and the assumptions and perceptions upon which they are based, against the communities' perspective.

Characteristics of the councils and their location

Councils' characteristics in terms of location and their roles in the metropolitan region have an important influence on their nature and extent of involvement in implementing sustainability ideas and programmes. Penrith, located close to Blue Mountains national park and farther away from Sydney, seems to represent a council with a much stronger commitment to the protection and enhancement of the environment than councils such as Blacktown and Holroyd, which have stronger development pressures due to their proximity to Sydney. Penrith council's vision statement thus clearly emphasizes the objective of harmony between urban and rural qualities of its area. Both Blacktown's and Holroyd's vision statements, on the other hand, focus more on economy and community than the natural environment. Thus we could conclude that the nature of sustainability envisioned in a council's vision is reflective of its physical and locational characteristics.

Local political dynamics

It is important to understand the local political dynamics within the councils as politicians, the development industry and the community play an important role in facilitating or constraining the implementation of LA21. In the case of Holroyd, the current mayor's strong interest in integrating sustainability into council processes has played an important role in its recent attempts in this direction. Some councils, such as Blacktown, seem to be driven more by a pro-growth coalition of politicians, development industry and large-scale landowners than other councils. In areas where there is a stronger community support to maintain the rural character and life-style, such as Penrith, councils are more keen to protect the natural environment.

Community participation and consultation mechanisms

Community participation and consultation are important mechanisms for implementing Local Agenda 21. As a part of community consultation, Holroyd city undertakes community surveys every four years to understand the residents' viewpoint about council activities (telephone surveys of 800 households across city areas in 1994 and 1998). These surveys are used to develop the council's social plans and management plans and suffice the State requirements of engaging the community. However, the utility of these surveys in actually engaging the community in a meaningful dialogue may be limited. These surveys helped the council identify 900 issues and concerns regarding the state of the environment, which fed into the preparation of the brief for the SoE Report. When the council tried to engage the community through follow-up public meetings, the extremely poor turn-out forced the council to shelve the initiative. Only recently, at the current mayor's insistence, has there been an attempt to revive the process of community meetings and inviting the community on to steering committees.

In the case of Penrith, there was extensive consultation for the Rural Land Survey with attendance of more than 800 people in five rounds of workshops to discuss community aspirations about the long term use of rural land. This study also used five mail-outs to send 25 000 letters to the community. Another consultation for Recreation and Cultural Needs Study undertook a telephone survey of 600 respondents. City wide mail-outs were sent to 60 000 households as part of PLANS 2002, and the community was invited to have its say at seven workshops at various neighbourhood locations. A web page was also set up. The council planners regard their involvement in the Rural Land Survey as a major learning experience. It seems to have served as a pilot project in community consultation. The council became much more aware of the various communities that exist and even learnt of the way the communities delineated their locality which at times was different from that perceived by council staff. Incidentally, the council also learnt about the ineffectiveness of the internet as a medium for dialogue with the community as yet.

Community support for sustainability

The planners in both Penrith and Blacktown are focused on and committed to what they perceive to be desirable to the local residents. In the case of Penrith, a perceived stronger community support to maintain the rural character and life-style seems to motivate the council to protect the natural environment and promote sustainability. They are thus more focused on serving what they see as the local residents' interest rather than on attracting investment from outside. In Blacktown, planners seem to believe that land-owners are inclined to capitalise on potential land value gains resulting from development activity, rather than being overly concerned with ecological sustainability. In their opinion, it is often the State agencies rather than the community that drive the sustainability concerns. This would account for the diametrically opposed views expressed by the Blacktown and Penrith planners on the preferred role of the council in soliciting community involvement referred to in the earlier section.

Planners at Penrith Council believe that the wide support for sustainability in the community is well articulated. To explain why it is so, they point to the fact that most of the elected Councillors in Penrith reside locally and are easily accessible to the community. Another suggested reason refers to the absence of any deep rifts or conflict within the elected Councillors of different persuasions. The implication is that the community can easily get its views across to the Councillors who, in the absence of any major conflicts with other Councillors, spend their energy on solving the community concerns rather than confronting each other and defending political agendas. This situation provides an incentive to find out the community's views on various issues, such as sustainability.

Holroyd seems to be in the initial stages of adopting LA21. The planning staff are reported to be going all out to engage the community in various ways. A series of public meetings and community events including community barbeques are on-going. In their search for effective means of increasing community interest towards setting up steering committees, they seem to be concentrating on community concerns for sustainability. However, at this early stage the extent of community support for sustainability cannot be ascertained.

Commitment through innovative programs and policies

A quick look at the three councils' Websites for lists of council policies, programmes and projects dealing with sustainable development and community partnership initiatives reveal numerous such projects. Each council has undertaken initiatives dealing with environmental and social issues at various levels. Councils' programmes, policies and projects could be considered good indicators of their commitment to sustainability. Each of the councils demonstrates some innovative attempts at leading by example on sustainability. For example, Blacktown city council has a number of programmes such as Community Pride Movement, Environmental Support Program, and Cities for Climate Protection Program. Community Pride Movement in Blacktown is a proactive programme seeking to involve local communities to improve the quality of the local environment by focusing on graffiti reduction as well as improving community facilities, landscaping and community signage. Likewise, the Environmental Auditing Program is a partnership approach to environmental protection in collaboration with a number of government departments (such as EPA) and industrial owners to audit all industrial precincts within the council.

Holroyd has prepared an annual Environmental Management Plan 2002/2006, and has put in place a number of strategies to protect environmentally sensitive areas and contribute to the area's sustainable development. Some of the strategies being implemented include the preparation of SoE reports, education to primary school children under Council's Local Environmental Awareness Program, energy assessment procedure for new major developments, industrial auditing of council areas and preparation of management plans for environmentally sensitive areas.

While Penrith Council is clearly more inclined towards promoting sustainability through environmental protection and natural heritage conservation, this fact is not apparent from comparing its programmes with the other two. The programmes are generally developed in response to State government directives and so tend to be similar across the councils. Hence, while the existence of certain programmes and policies proves that all three councils have adopted sustainability principles to some extent, it does not necessarily reflect the degree of a council's commitment or otherwise to the same.

Strong role of state government

Sustainability agenda in the councils is driven by top down international, national and state government agenda rather than bottom up pressure from the community. Statutory requirements within State planning frameworks such as the requirement for preparing an SoE Report is having a major influence on how councils take up sustainability issues. For example, Blacktown city council initiated an environmental auditing of its industrial premises under the new Environmental Operations Act 1997, which came into force in NSW.

While the State planning framework may force a council to move some paces towards implementing the agenda, that may not be sufficient to create a sustainability oriented community culture. It requires initiative from council staff to develop community awareness of sustainability principles embodied in the agenda through education, leading by practice and demonstration projects. Planners at Penrith Council are well aware that they have gone beyond meeting the statutory requirements of community consultation in their major strategic planning initiatives such as the Rural Lands Study and Residential Land Review. They clearly recognised the need to establish a dialogue with the community to be more important than meeting the statutory minimum imposed by the State legislation.

Conclusions

To make the objectives of sustainable development a reality, there is a need for both bottom up (local/community) and top down (state/federal) initiatives for LA21 implementation. At the local level, there is clearly a strong need to increase awareness and commitment to sustainability through financial support for community initiatives, rather than merely relying on a top down approach to implementing sustainability. More attention needs to be given to developing action plans of the councils that demonstrate a clear commitment of resources and manpower for promoting sustainable development.

In addition to initiatives at the local level, there also needs be a greater focus on creating supportive governance structures and consultation mechanisms to implement LA21 at the State level. The opportunity to learn from successes and failures of different councils in New South Wales, for example, in implementing local agenda should be capitalized upon. State-wide and even nation-wide database of community and council initiatives in sustainable development and their effectiveness would be useful in setting up examples of best practices for similar councils to consider.

This study provides a springboard for further studies into how the State's planning system can influence a council's adoption of LA21. A study of the interface between the State government institutions and policies and the local council's strategic planning activities could explore the possibilities regarding the strengthening of the supportive role State planning frameworks play in the promotion of the agenda. Another area for further research is a comparative study of councils operating within different State planning systems to better understand the role of State and regional planning systems in influencing sustainability outcomes.

References

[1] Greene, D. (1994) Managing for the future: a local government guide to Local Agenda 21, http://www.ea.gov.au/esd/local/agenda21 (accessed June 2003).
[2] UNCED (1992) *Agenda 21: UN Conference on Environment and Development*, United Nations Commission for Environment and Development Secretariat, Geneva.
[3] Agenda 21, *Chapter 28 (1992): Local Authorities' Initiatives in Support of Agenda 21* (as adopted by the Plenary in Rio de Janeiro, on June 14, 1992).
[4] Commonwealth of Australia (1992) *National Strategy for Ecologically Sustainable Development*, Australian Government Publishing Service, Canberra.
[5] Rushmoor Borough Council, October 1996, www.iclei.org (accessed June 2003).
[6] Raemaekers, J. (2000) Planning for sustainable development, in *Introduction to Planning Practice*, P. Allmendinger *et al.* (eds), Wiley, London.
[7] Whyalla City Council (2000) *Whyalla's Local Environment Plan, A Local Agenda 21 Initiative*, May 2000.
[8] Penrith City Council (2002) *Sustainability Review of Council's Processes and Activities*, Consultants Brief, Penrith City Council.
[9] Conacher, A. and Conacher, J. (2000) *Environmental Planning and Management in Australia*, Oxford University Press, Melbourne.
[10] Productivity Commission, (2000) *Implementation of ESD by the Commonwealth Departments and Agencies*, Inquiry Report (February 15, 2000), AusInfo, Canberra.
[11] Davidson, S. (2002) Participation in planning: meeting the challenge, *New Planner*, March 2002, 24–25.
[12] Blacktown City Council (2000) *Blacktown City social plan*, p.1, www.blacktown.com.au, (accessed June 2003).
[13] Holroyd City Council www.holroyd.com.au (accessed June 2003); ABS (2000), CDATA.
[14] Penrith City Council (2003) *Draft Management Plan 2003–2004*, Part A – Summary, p.7, Penrith.

29 Sustainability assessment considering asset and building life cycles

D. G. Jones[1], K. Lyon Reid[2] and D. G. Gilbert[3]

Summary

This chapter discusses development of eco-decision support tools and databases designed to facilitate corporate management in sustainable development and construction processes. Such tool developer's alignment of their products with best practice asset management frameworks is critical if they are to be integrated into corporate management systems. Real world asset management issues include discontinuities in work sequencing, multi-layered communications across stakeholders and between policy, procurement and construction development procedures and timelines. CASE studies are shown from two new Government buildings developed to achieve star ratings for Ecologically Sustainable Fitout of Office Accommodation using asset management decision support tools based on Life Cycle Assessment (LCA) methods and databases. Using these two studies a comparative assessment is drawn of the application of Australian ecodesign tools, based on quantitative and qualitative LCA. Such tools and assessments are discussed to facilitate eco-efficient building design as well as fitout and refurbishment. The scope of work of the qualitative tool is based upon an asset management framework covering policy, planning, design and documentation, procurement, demolition, construction and pre-occupancy assessments. The quantitative tool exploits more construction detail over a shorter scope of work from acquisition, manufacture and supply of building products and systems through to construction and building operation. Results of a gap analysis are presented to illustrate potential outcomes from each tool and that potential uptake of either tool type can best be appreciated in the wider context of sustainable asset and construction management.

---

[1] Queensland Department of Public Works (DPW) Building Division, Built Environment Research, Brisbane, Australia.
[2] Queensland Department of Public Works, Government Office Accommodation, Brisbane, Australia.
[3] Queensland DPW Building Division, Director Built Environment Research, Brisbane, Australia.

Background

In the Australian built environment, there is depletion of natural reserves of freshwater, clean air, naturally productive land and pollution of urban air to an extent that it can be detrimental to the health of both human communities and natural ecosystems. The global scale of habitat deterioration and destruction, climate change and depletion of natural resources elicited responses from the United Nations with the Montreal Protocol on Ozone Depletion in 1987, the Rio Convention on Biological Diversity in 1992 and the Kyoto Protocol on Climate Change in 1997 [1]. Subsequently the Council of Australian Governments' National Strategy for Ecologically Sustainable Development (NS ESD) defined ESD as using, conserving, and enhancing community resources so that ecological processes on which life depends are maintained and the total quality of life now and in the future can be increased [2]. A further national commitment to ESD endorsed in 1996 was the National Greenhouse Response Strategy (NGRS) [3]. Responding to the NS ESD, the national 'CGI-97 Directions Forum' recognized that the total buildings' share of escalating global environmental deterioration was significant and that incorporation of ESD into asset management practices would support such national and global initiatives [4–7].

In 1997 this Forum launched the Built Environment Protocol on issues of:

- ecologically sustainable procurement in building construction, fitout and use
- communication and education to promote improved building products and practices
- minimized building resource use, wastage and emissions to air, land and water
- improved environmental health performance of buildings/products and services [8].

Introduction

The Forum considered that the key to ensuring a sustainable relationship between buildings, occupants and the environment was to improve communication and technical development over all phases of planning, design, procurement, construction, operation and disposition [1]. In support of the NGRS, studies commissioned by the Department of Public Works revealed that Queensland buildings greenhouse gas emission (GGE) share was 22% in 1995 as shown in Figure 29.1, a column chart of GGE by source [3]. The study showed that residential and commercial operations dominated the building sector share as depicted in Figure 29.2, a column chart of GGE by building type and phase for 1999. It also found commercial building cooling produced 28%, ventilation 22% and lighting 21% of GGE as depicted in Figure 29.3, a pie chart of GGE by end use in 1999.

The objective of this chapter is to describe eco-decision making and tools for strategic asset management (SAM) to significantly enhance the performance of occupants and reduce local and global environmental impacts, especially greenhouse gas emissions (GGE) attributed to building operations. In this respect, fitout of new buildings implemented in Queensland, in a regional centre, Cairns, and a capital city, Brisbane, is discussed.

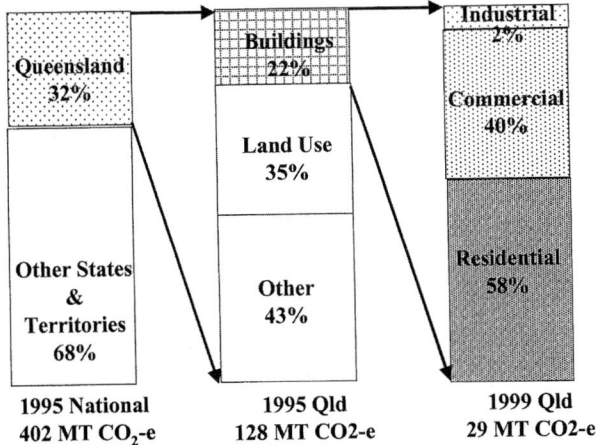

Figure 29.1   Queensland and state building share GGE by source.

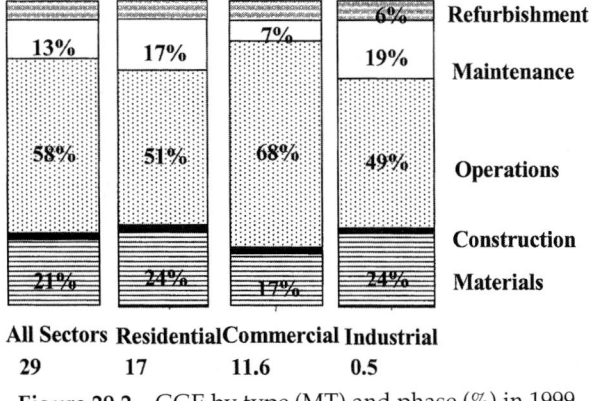

Figure 29.2   GGE by type (MT) and phase (%) in 1999.

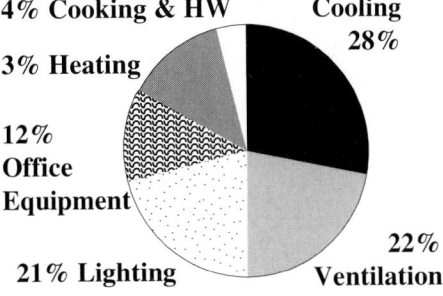

Figure 29.3   Commercial GGE shares in 1999.

Table 29.1   ESD considerations for conserving resources and reducing emissions.

Conserving resources of

Community	Energy	Materials/Land	Water
Natural heritage	Passive solar design	Materials durability	Efficiency in use
Cultural heritage	Efficiency in use	Waste avoidance	Waste avoidance
Built heritage	Waste avoidance	Reliant on renewables	Waste management
Access and safety	Conserve sources	Conserve sources	

Reducing emissions to

Air indoors	Air outdoors	Materials/Land	Water
Volatile organics	Ozone depleting gases	Construction waste	Effluent reduction
Ventilation	Greenhouse gases	Recycled materials	Waste treatment
Indoor air quality	Airborne particulates	Avoid toxic waste	Potable water

For any community, key outcomes include a sustainable source of supply and healthy levels of emissions. Because all supply comes from community, air, land and water sources and all emissions go to community, air, land and water sinks, such resource supply and emission sinks are essential to consider. Therefore, as summarized in Table 29.1, ESD considerations relate to conserving resources of and reducing emissions to these sources and sinks [6, 7].

Eco-decision support tools

Recently, Environment Australia commissioned the Royal Melbourne Institute of Technology (RMIT) Centre for Design to study leading eco-decision making tools and in particular those used for Life Cycle Assessment (LCA) [9]. Such tools were classed as being for building materials and components, modelling, design, computer aided design (CAD), product guides or assessment schemes. This study reported that users felt that their objectives were not being met. This may be compounded when, as other studies have reported, tools are not linked explicitly to asset management systems to effect their application [4, 5, 7, 9].

A matrix of stakeholders versus eco-decision tools coverage, adapted from an emergent International Standard for building sustainability and given in Table 29.2, illustrates, for example, eco-decision making tools apply to few rather than many life cycle phases [10, 11]. To address these deficiencies in life cycle coverage and linkage to asset management, the Department of Public Works has developed a suite of eco-decision making tools.

Asset management eco-decision support tools

Ecologically Sustainable Strategic Asset Management (ESSAM) decision support tools were developed to address stakeholder needs to assess buildings, agencies and

Table 29.2 Stakeholders versus eco-decision tool scope.

Stakeholders/coverage	Strategic plan and concept design	Detailed design and construction	Fitout, operate and maintain	Renewal, disposition
Investor, Developer, Client, Manager, Provider	Eco-asset management, e.g. ESSAM [1]			
Client, Owner, Provider, Designer, Supplier		Eco-design, e.g. LCADesign [11]		
Investor, Owner, Manager, Occupant		Eco-rate existing building, e.g. LEED [12]		
Owner, Manager, Operator, Tenant, Broker			Eco-operation, e.g. NABERS [13]	

Table 29.3 ESD considerations across the asset life cycle.

Phase/Flow	Planning	Design/Procure	Construct	Operate	Disposition
Strategies to conserve sources of					
Energy	Renewables	Passive	Comfort	Efficienct	Embodied
Water	Catchment	Harvesting	Catchment	Low use cycle	Groundwater
Material	Renewables	Interoperable	Disassembly	Recycling	Recover, reuse
Community	Welfare	Educative	Amenity	Interactive	Heritage and habitat
Strategies to protect sinks of					
Air	IAQ	Breeze flow	Dust and noise	Fresh air	Toxic exposure
Water	Potable quality	Clean delivery	Zero effluent	Low effluent	Contaminants
Material	Re-use existing	Zero toxicity	Zero waste	Reparability	OH&S
Community	Habitat	Equity	Refuges	WH&S	Ecosystem

suppliers against criteria to show compliance and achievement over whole of life. They provide planners, designers, contractors and end-users with the policy instruments, guidelines, checklists and rating systems to do this work [1, 7]. In each phase of the work there are whole of life considerations and strategies to apply related to essential sources and sinks as summarized in Table 29.3. The ESSAM Guideline for Ecologically Sustainable Fitout of Office Accommodation (GESFOA), for example, provides guidance for conserving resources and minimizing emissions over the full asset life cycle.

Fitout is the process of designing and building a physical workplace environment within, or in association with, a building's structure, envelope and services and stresses that it is dynamic and continuous and includes consideration of the impact of the fitout in use as well as its disposal [1]. The GESFOA is published on the web as well as on CD ROM as a suite of software applications and tools that are interactive via a wizard to facilitate integrated decision-making and reporting from planning to disposition and renewal in a continuous loop [1].

This wizard, designed as a fixed rim wheel rotating inside a tread, depicted in Figure 29.4, shows a ratings wheel hub driving aims via timely strategies between each spoke, in a phase by phase segmented rim with recommended Energy, Water, Community and Land tactics located by clicking on that region of the tread. As shown in Figure 29.5, to enhance communications and implementation this Guideline also exploits icons that act as signposts across a range of written formats that stakeholders typically require such as those listed below. The guideline comprises a range of tools including:

- a presentation in PowerPoint and on video to provide an overview and introduction
- a wizard of a wheel to guide users over the project lifecycle and relate to the issues
- a star-rating system to provide for performance assessment and easier appraisal
- an Excel spreadsheet to act as an eco-calculator and facilitate performance reporting
- checklists for tracking effort at each phase of work to facilitate effective responses
- a project planner for resource allocation and auditing progress against milestones
- benchmarking material and a trouble-shooter to address problems.

ESD best practice databases

ESSAM relates to large government estates where best and standard practice varies according to the building type, climate, region, occupancy rates and services delivered [1, 4, 5]. Table 29.4 lists examples from an ESSAM database of best practice case studies in energy, water, emissions, ecosystems and biodiversity developed to provide relevant performance benchmarks. This supports the planning, design and reporting that is reliant on project specific data, from a base case, to show improved:

- energy use per unit floor space for service delivery and occupancy rate
- water use per unit of floor space for service delivery and occupancy rate
- recycling rate per unit of material used, reused and disposed.

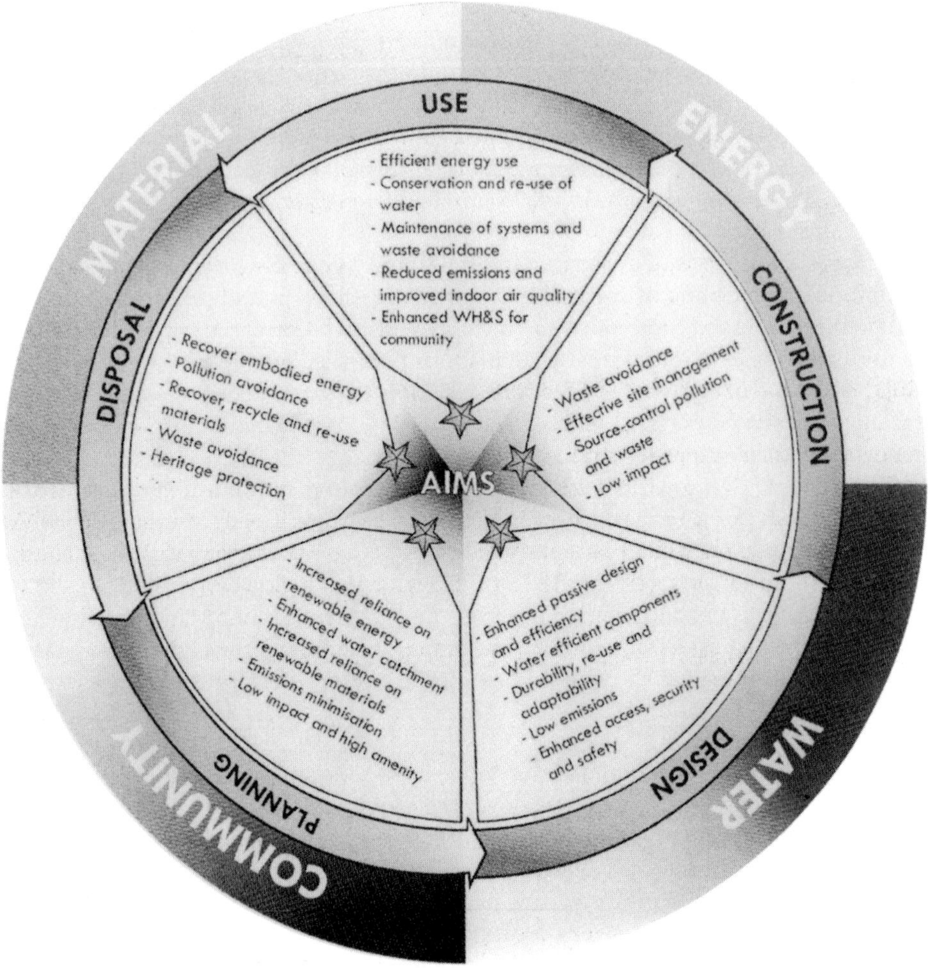

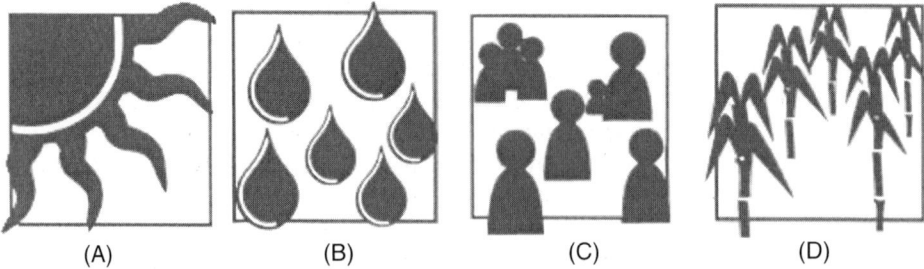

Figure 29.4 A 5 star rated sustainable outcome is awarded for implementing activities on the tread to achieve the given aims at each phase of the project.

Figure 29.5 Icons for energy, water community and land sources and sinks.

Table 29.4 ESD initiatives in buildings.

Issue	Best Australian	Best International
Enegy	Perth Solar Energy Centre	US EPA Research Facility
Water	John Morony Centre Windsor	Boston Park Plaza; Office Complex Ontario
Emissions	Buxton School; SEDA Office Fitout	SA Office Complex, Ontario, Canada
Ecosystem	Thurgoona, Charles Strut University	Boeing Corp., United Parcel HQ

Implementation in Brisbane and Cairns

An extract from the project planning Guideline adopted for both Cairns and Brisbane building fitout is shown in Table 29.5, strategies to conserve community resources and heritage, illustrates management procedures involved in ESSAM and the GESFOA. Checklists, as shown in an extract in Table 29.6 pre-occupancy assessment phase checklist, were also used to track work and facilitate communications in both projects.

Rating ESSAM initiatives using the GESFOA

Each guideline action item is also considered in checklists because both are needed to implement and rate strategies across the issues as well as throughout all fitout phases.

Table 29.5 Strategies to conserve community resources and heritage.

Objective:	To protect natural heritage, habitat and biodiversity
Strategy:	*Enhance reliance on sustainable pracitice*

Recommendation	Implementation
Protect endangered species	Specify certified plantation timbers with origins recorded. Avoid use of tropical hardwood unless it is recycled.
Promote cleaner production	Specify alternative materials to viny/lowest PVC content in floor covering; cabling; conduit, ducting; and furniture components and Venetian blinds.
Select natural fabric	Specify organic fabrics with high flame resistance and low toxicity qualities.
Consider pest management	Specify/check on effective sealing of all service penetrations. Design out spaces for pest/vermin nesting. Include a pest protocol in Handover Manual.
Promote responsible cleaning	Include manufacturers' recommendations/direction for use of natural cleaning products in Handover Manual.

Table 29.6   Pre-occupancy phase checklist.

Consideration	Implementation activity
Awareness and Education	
Inform on ESD issues	Improved outcomes/lessons are reported to clients/occupants.
Maintenance and cleaning	
Cleaning practice	Low impact cleaning recommendations for all elements are on record.
Electrical/Data communications	
Electro magenetic field	EMF fields are checked near electrical risers and all adjacent sub-boards.
Ventilation air handing/conditioning	
Quality fresh air	Opportunties for increasing fresh air volumes are identified.
Lighting	
Efficient energy use	Light switches are labelled and show areas of control/function.
Hydraulic services	
Water pressure	All outlets are checked to ensure there is a minimum water splash.
Mechanical services	
Base building services	The expected life of mechanical plant/fittings is on record.
Records management	
Enhance operations	ESD information formatted for ready reference was distributed/reviewed.

ESSAM uses a range of such approaches to rate performance, to support decision-making, to provide tools for the client and project teams to summarize and analyse progress and discern the level of performance required from the start of project planning. Ratings can be used to prioritise responses, focus the efforts of professionals in areas of concern, compare, verify and analyse strengths and weaknesses and plan for an overall outcome, as well as to differentiate approaches taken by organisations and suppliers.

They also have the potential to drive improvements in overall performance, return on investment and sustainable technology. Both the Brisbane and Cairns office fitout were commissioned to apply the GESFOA to achieve a four star rating. A combined numerical and 'star' rating system was developed to differentiate approaches required of the many agencies and suppliers in fitout. Each item has a numerical value depending on its contribution and importance and users tally numerical scores to a total score of between 1 and 300. To achieve a star rating as shown in Table 29.7 an aggregated total is compared against five performance levels that establish the overall star rating.

Table 29.7   Star rating table.

Rating	Score	Logo	Description
★	0–99	Eco-Starter	Encouragement award for getting started
★★	100–149	Eco-Improver	Improvement award for applied effort
★★★	150–199	Eco-On-track	Fast track award for improved outcomes
★★★★	200–249	Eco-Leader	Leadership award for achievement
★★★★★	250–300	Eco-Best-fit	Best practice award for ESD Outcomes

Reporting on performance

Reporting on project outcomes is considered critical, so ESSAM tools, databases and rating systems provide for auditing and accreditation of Agencies, Suppliers, Contractors, Projects and Project Teams to facilitate:

- benchmarking performance, improved project timing and communication
- professional initiatives, product specification and design for ESD solutions
- conserving resources of and reducing emissions to community, air, land and water.

An example of an improvement report is shown in Figure 29.6 an eco-star model diagram of current, best and ESD fitout practice. Current fitout practice is shown as the outer dashed circle, compared with contemporary best practice shown in solid line.

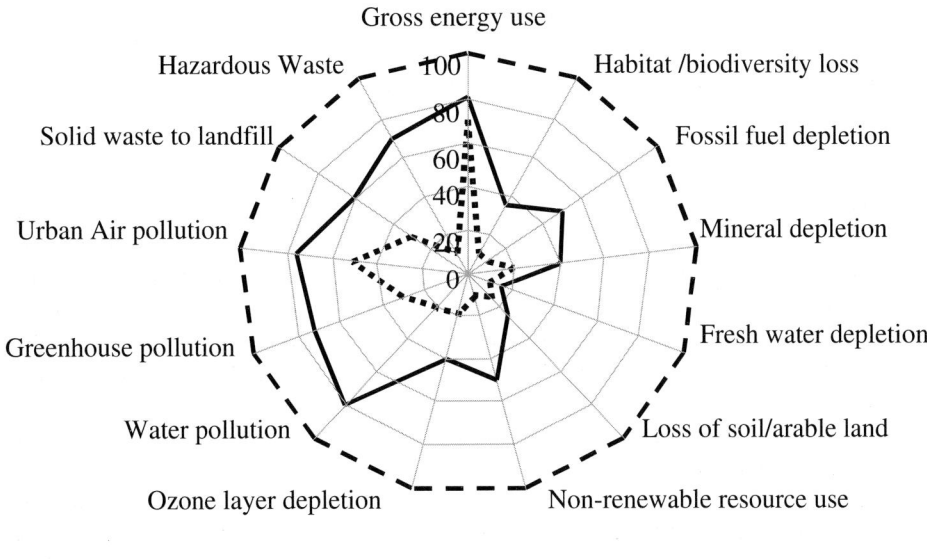

Figure 29.6   Model building fitout eco-star diagram.

The inner dotted line depicts the improvement required for ecologically sustainable performance. ESSAM outcomes can be reported in such categories or as specified in the GESFOA.

Status of implementation in Brisbane and Cairns

From project inception, the Environment Protection Agency, a Cairns Government Office Building tenant, was committed to implementing the GESFOA in their new Offices. In keeping with the $17.5M project's local industry participation plan it was also committed to be a commercially viable regional office building and to the engagement of local firms' innovative architectural design/engineering. The design employed passive design features such as orientation, insulation, shading, double-glazed windows and wide eaves combined with high efficiency lighting and air conditioning.

The building achieved a four and half star eco-rated fitout and was also the first new, air conditioned office building to be formally awarded a five star energy rating from the Australian Building Greenhouse Rating Scheme based on one year occupied operations. Occupants consider it a very effective workplace with 40% lower energy consumption than comparable buildings, saving 360 tons of greenhouse gases annually.

By comparison, the original design of new Government Office Accommodation in Brisbane suffered delays prior to proceeding to construction. In the interim an enhanced commitment to ESD and new ESSAM considerations were implemented in building, services and fitout detailed design. From fitout project inception, the tenant, Queensland Treasury, was committed to implementing the GESFOA and checklists. Now that it is nearing completion, they show that it is on track to achieving a four and half star ESD fitout.

Conclusion

This chapter discussed the development of eco-decision support tools designed to facilitate sustainable development via corporate asset management. Many available tools exploit construction detail over one or two phases, but according to one comprehensive national study the objectives of their use may not always be met.

Asset management requires eco-decision making tools to cover planning to disposition and renewal in a continuous loop and unless such tools are called for in asset management they remain excluded from such decision-making. The ESSAM tools that were described here were for planning, design and documentation, procurement, demolition, construction and pre-occupancy assessments.

The chapter describes the GESFOA, as a suite of interactive applications featuring a wizard to guide users over the project lifecycle decision-making and reporting processes and relate to issues concerning community, energy, materials and water sources/sinks. The applications described included those applications that are designed to focus effort, provide for an easier appraisal and facilitate effective responses. GESFOA tools to facilitate implementation included presentations and checklists to assist communications and tracking, a trouble-shooter to addresses potential problems and a rating system for self-assessment/auditing and performance reporting.

Case studies of two new Government office building projects achieving highly rate ecologically sustainable fitout were provided including the Cairns building project, the first awarded a five star Australian Greenhouse Building Rating following one full year of occupied operation. In this chapter the authors have presented the case that implementation of eco-decision support tools designed to facilitate strategic asset management have achieved new industry benchmarks for best practice.

References

[1] Department of Public Works, (2000) Ecologically Sustainable Office Fitout Guideline, Queensland, Australia, http://www.build.qld.gov.au/aps/ApsDocs/ESDMasterDocument.pdf

[2] Council of Australian Governments (1992) National Strategy for Ecologically Sustainable Development, Department of Industry Science and Tourism, Canberra, Australia.

[3] Energy Efficient Strategies and Faculty Of Science And Technology, Deakin University: LCA of Queensland Building Energy Consumption and Greenhouse Gas Emissions for Building Division, Built Environment Research Unit Department of Public Works (2001).

[4] Gilbert, D., Jones, D., Greenaway, C., Williams, N., Ball, M., Stockwell, B. and Morawska, L. (1998) Directions for a healthier and sustainable built environment: The built environment protocol – an Australian experience, *Proceedings of International Building and the Environment in Asia Conference*, February 1998, Singapore.

[5] Gilbert, D. L. and Jones, D. G. (2000) Healthy Sustainable Buildings: A Queensland perspective, in *Proceedings of the International CIB Conference: Shaping the Sustainable Millennium*, Queensland University of Technology.

[6] Levin, H. (1997) Systematic evaluation and assessment of buildings environmental performance, in *Proceedings of the National Community, Government and Industry CGI-97 Directions Forum*, Brisbane, Australia.

[7] Barton, R. T., Jones, D. G. and Gilbert, D. (2002) Strategic asset management incorporating ESD, *Journal of Facilities Management*, Henry Stewart Publications, London.

[8] Department of Public Works (1997) The built environment protocol, Queensland Government, Australia, http://www.build.qld.gov.au/research/p_bestforce/nocd/theprotocol.html (accessed 05.06. 2003).

[9] Royal Melbourne Institute of Technology (RMIT) Centre for Design, Greening the Building Life Cycle, http://buildlca.rmit.edu.au/menu8.html (accessed September 2003).

[10] ISO (2002) ICO TC59, ISO AW121931 Buildings and constructed assets – Sustainability in building constructions – Framework for assessment of environmental performance of buildings.

[11] Tucker, S. N., Ambrose, M. D., Johnston, D. R., Seo, S., Newton, P. W. and Jones, D. G. (2003) LCADesign: An integrated approach to automatic eco-efficiency assessment of commercial buildings, *Proceedings of the International CIB W078 Conference*, New Zealand.

[12] US Green Building Council (2001) *LEED™ Rating System Version 2.0 Draft*, March, US Green Building Council, San Francisco, CA.

[13] Vale, R., Vale, B. and Fay, R. (2001) The National Australian Buildings Environmental Rating Scheme, Environment Australia, http://www.ea.gov.au/industry/waste/construction/pubs/final-draft.pdf

30   Energy efficiency uptake within the project house building industry

M. D. Ambrose[1], *S. N. Tucker*[1], *A. E. Delsante*[1] *and D. R. Johnston*[1]

Summary

Within Australia, project houses, that is, houses selected from a builder's standard design collection, constitute the majority of new houses built. Many of these houses are built within large planned estates, which are growing in popularity. This paper looks at the impact of energy efficiency education directed towards project house builders and the opportunities that exist for developers to promote energy efficiency concepts within their housing estates.

The chapter reports on the findings of a project undertaken at a large estate development in South East Queensland that included surveying house builders and designers and the running of an educational workshop for them and energy assessment of houses both prior to the workshop and post workshop.

Perceived barriers to the uptake of energy efficiency principles and common misconceptions are explored, such as the additional costs involved in achieving high energy efficiency. Problems associated with the actual delivery of information to builders and developers are also explored. Finally, possible initiatives that could be implemented are discussed, including the effectiveness of demonstration housing projects.

Introduction

The uptake of energy efficient design of residential houses constructed in Australia has, for many years, been limited within the building industry. Uptake has usually been restricted to those States where variations within the Building Code of Australia (BCA) require minimum energy efficiency standards. However, the current housing provision version of the BCA (amendment twelve) has seen the introduction of a new subsection (3.12) to deal specifically with energy efficiency. This focus on a minimum standard for energy efficiency for housing has been included to bring about reductions in energy use and consequently a reduction in greenhouse gas emissions Australia

---

[1] CSIRO Manufacturing and Infrastructure Technology, Melbourne, Australia.

wide. It represents a significant shift in focus and, although not all States and Territories are currently subject to the new provisions, it is planned that they soon will be. Consequently, builders and designers will be required to comply with the new regulations and be able to implement appropriate modifications to their designs to ensure compliance.

In states where previously there have been no statewide energy efficiency regulations, such as Queensland, there will be a need to educate and inform builders, designers and the house buying public to the new requirements and explain the benefits and cost implications behind the changes. There are many possible methods for undertaking this education, some of which are discussed in this paper, with each having benefits and drawbacks. However, their goal should be the same, that is, the effective implementation of energy efficient design into the residential housing market.

Current energy efficient design in South East Queensland

South East Queensland's climate is one of the mildest climate zones in Australia, with an annual mean maximum of 25.5°C, a mean minimum of 15.7°C and a mean of 7.5 hours of sunshine daily [1]. However, despite this mild climate, houses that are being constructed are rating very poorly in terms of energy efficiency. This is particularly true of project houses, that is, houses that are selected from a builder's standard design collection.

Within Australia a star based rating system is used to assess the relative energy efficiency of residential construction. The number of stars achieved is determined through a computer based thermal modelling program that determines average annual energy usage required to maintain a house within a particular thermal comfort range, over a set time period and in a particular climate zone. The resulting energy total, in $MJ/m^2/annum$, determines into which star band a house falls. For example, a house in South East Queensland would achieve 1 star if it consumed between 226 and 290 $MJ/m^2/annum$. To achieve the maximum possible 5 stars, a house would need to consume no more than 60 $MJ/m^2/annum$. A recent study, commissioned by the Australian Greenhouse Office (AGO), into the energy efficiency of houses in South East Queensland found that 91% achieved only a 1 star rating or worse [2]. Figure 30.1 shows the distribution of star ratings achieved by the houses rated. Only one house achieved a 3 star rating, while an environmental builder committed to energy efficiency in his houses constructed the two that achieved 4 star ratings.

Such results indicate serious problems within the building industry in regard to the application of energy efficient design concepts. The reasons behind these problems are no doubt many and varied, but most likely include lack of understanding, perceived cost and marketability. The AGO study found that a 3.5 star rated house should be considered good practice in the South East Queensland climate and be relatively easy to achieve at minimal additional cost.

Potential for energy efficiency in planned estates

The rising popularity of large-scale planned estates provides an opportunity to encourage energy efficient design through the developers. Many of these estates already

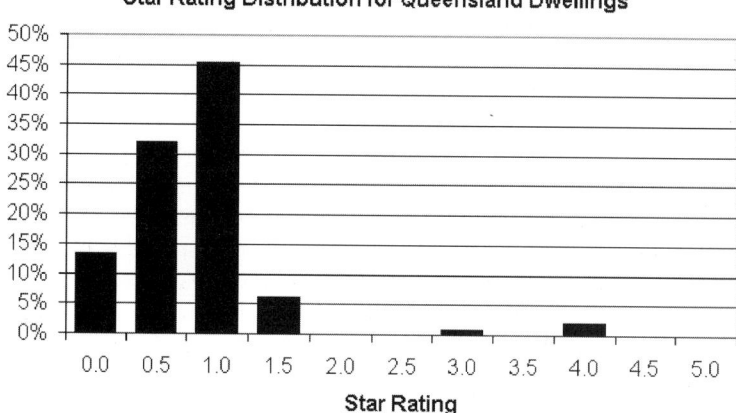

Figure 30.1   Star rating distribution for houses in south east Queensland.

employ covenants on the houses built to include such issues as street appeal, roof colours, lighting, etc. This same idea could be applied to energy efficient design with new houses being required to achieve a certain energy efficiency rating. Alternatively, a less ambitious approach can be through the use of educational devices aimed at both builders and buyers. The running of workshops (for both builders and house owners), providing educational leaflets and brochures and concept display houses are all possible devices that could be employed to promote and encourage the building of energy efficient houses. Such concepts are particularly important in those areas of Australia where building regulations do not cover energy efficiency, such as many areas of South East Queensland.

The AGO study looked at one of these estates over a period of time to see if certain educational strategies would translate into improved energy efficiency in the houses built. The estate selected was a 1000 hectare site 25 kilometres north of Brisbane that is still currently under construction. The 15 year development will ultimately contain 8500 houses, a town centre, schools and a business park. An initial review and rating was carried out on fourteen houses constructed within the stage one display village. The results showed that 79% of the houses were rated at or below 1 star with an average energy consumption value of 313 MJ/m².

These results were then compared with twenty houses built immediately after the opening of the first display village. These houses showed a slight improvement with an average energy consumption of 305 MJ/m², still well below what would be considered good practice. At this stage no educational programmes had been put in place, the idea being to see if house buyers had any influence on the level of performance being achieved on constructed houses. These results would indicate that despite the slight improvement, what builders are constructing in the display villages is what house buyers are purchasing for themselves. This is hardly surprising, as most house buyers will look to the designer/builder to suggest what construction related elements should be included in the building.

Following these initial studies a workshop was held for all interested builders on current techniques for energy assessment and the various costs for implementing

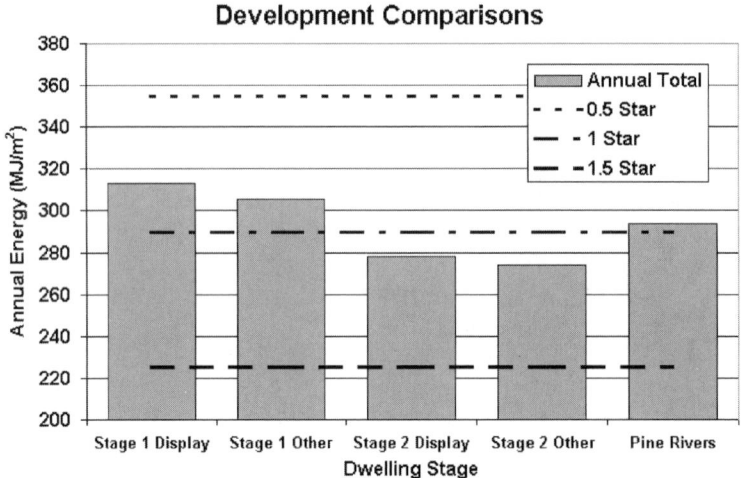

Figure 30.2   Comparison of average energy performance of five groups of houses.

systems into houses. The outcome of the workshop is discussed later, but following the workshop further analysis was done on the second stage display village and the subsequent houses built after the opening of this village. Results again showed a very slight improvement in each step, with houses within the stage two display village having an average energy consumption of 278 MJ/m² and the immediately following houses having an average of 274 MJ/m². In addition to analysing the houses being constructed within the estate development, a random sample of newly constructed houses that had been built in the surrounding municipal area of Pine Rivers were assessed. These houses had an average energy consumption of 293 MJ/m², which closely mirrored the results obtained within the estate development. A summary of the results is shown in Figure 30.2.

These results show that there was a 12.5% improvement in the average energy efficiency of houses within the estate development during its first couple of years of development. However, the houses still fall well below the 3.5 star rating that is considered good practice in this climate zone, with the stage two houses still only averaging a 1.5 star rating.

Educational workshops

One of the often considered impediments to the uptake of energy efficient design principles is a lack of understanding of the concepts by the house building industry. However, two recent surveys of house builders in South East Queensland suggest that builders do understand the concepts behind energy efficiency in their climate with respondents in both surveys indicating the top five energy saving features as:

(1)   cross-ventilation
(2)   roof insulation
(3)   window shading

(4)   landscaping

(5)   wall insulation.

These responses demonstrate a good understanding of the issues; yet there is little evidence to suggest that they are being incorporated into actual houses. A reason for this may be in the relatively low response rate to the surveys with 88 responses in the first survey and only 47 in the second, despite surveys being sent to all members in South East Queensland of the Building Designers Association of Queensland (BDAQ) and the Housing Industry Association (HIA). In many respects, the low response rate is an indication of the low level of importance placed on energy efficiency issues by designers and builders, rather than a low level of understanding.

The survey indicated that respondents saw workshops as one of the main sources of information, along with trade journals, their industry colleagues and, increasingly, the Internet. Educational workshops can serve an important role in disseminating information to designers and builders, but it is important that the workshops focus on the areas where information is needed. The reasons why energy efficient design should be adopted, rather than how it is adopted, would seem to be an important focus area. Workshops carried out as part of the AGO study discovered a keen interest by attendees in energy efficient design but, similar to the surveys, low attendance was a worrying indicator. Consequent follow up on why so few attended found that their absence was not due to lack of knowledge about the workshops, but more to the timing. It was found that many builders are not inclined to attend workshops (even if they are free) if it interrupts their work. Thus it is important to schedule such workshops after working hours as well as providing an informal, yet attractive, venue.

Concept houses

The traditional display village has been an essential part of the project house builders' marketing strategy. The ability to show potential clients the type of houses available is a powerful and effective technique. Likewise, the use of an energy efficient display house or concept house can be an effective means of educating builders and the general house buying public on the benefits of energy efficient design.

The estate development embarked on establishing a GreenSmart village as part of its stage two display village. This involved three houses being built by separate builders using environmental guidelines developed by the HIA under their GreenSmart program. Interest in the project (a first for Queensland), was very strong with significant media coverage, helped by the fact that the houses were also being built as part of a charity fund raising exercise.

Feedback from the builders six months after the houses were completed has been very positive with one of the builders claiming such a rise in demand that he has had to shift to larger offices and is now fully booked for the next year. Indeed, one of the houses received the GreenSmart building of the year award (Figure 30.3) and the builder received the Professional of the Year Award.

Such successes are encouraging for the future long term take up of energy efficient design and indicate a need in the marketplace for such designs. At this stage, any

Figure 30.3   Award winning GreenSmart concept house.

increased uptake in the building of energy efficient houses has yet to be evaluated at the estate development.

Cost of energy efficient designs

Survey results indicate that the greatest single barrier to the adoption of energy efficient design principles by builders and designers is the perception that it is a costly exercise that will not be accepted by the client. Two recent surveys of designers and builders found that more than 90% believed it was more expensive than a standard new house, with 39% of respondents in the first survey believing the cost exceeded 10%, although the number dropped to 33% in the second survey (Figure 30.4).

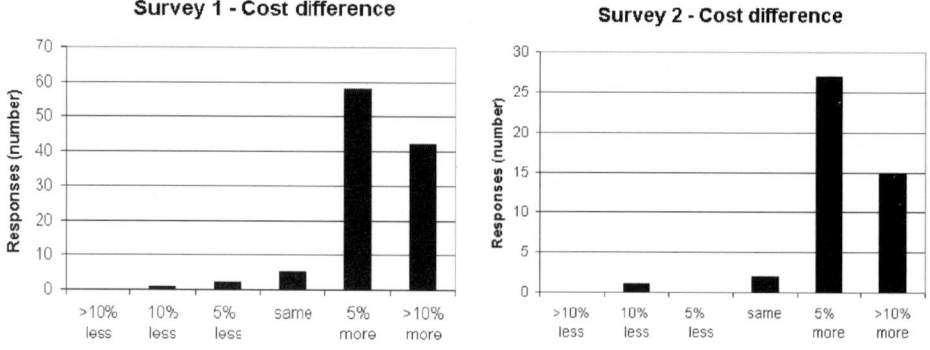

Figure 30.4   Perceived cost difference between standard and greenhouse/energy efficient design.

Table 30.1 Cost of improvements for raising houses to 3.5 stars

Cost range	Cost range (%)	Average cost ($)
$0–1000	0%	
$1000–2000	45%	$1669
$2000–3000	15%	$2487
$3000–4000	15%	$3514
$4000–5000	15%	$4532
>$5000	10%	$6842
Total	**100%**	**$3015**

To quantify the actual cost, the AGO study looked at improving the houses that had been assessed from within the estate development and the surrounding Pine Rivers Shire that had, on average, a 1.5 star rating. The aim was to improve the houses to a 3.5 star rating through the most cost effective methods. The houses intended orientation was left unchanged, as was the floor plan. The results are listed in Table 30.1 and show that for 45% of houses, the cost of improvements was less than $2000, with an average cost of $1669. Only 10% of houses required in excess of $5000 to improve their rating to 3.5 stars. Overall, the average cost for improvement was $3015, which represents a 2% increase in cost on a $150 000 house. This is much less than what the vast majority of builders and designers believed was the case. In reality, the cost could be even less, as optimising orientation and plan layout are usually cost neutral and can deliver significant energy savings.

It is interesting to note also that the award winning concept house was constructed to a much higher standard than a 3.5 star rating, achieving in excess of 5 stars and incorporating a range of additional environmental measures not considered by the rating systems currently used to check compliance. These included water efficiency measures (water efficient fittings, water tanks, solar hot water, etc.) and energy efficient appliances and lighting. Incorporating all these additional environmental measures

Table 30.2 Improvements used to raise sample houses to 3.5 stars.

Improvement	% Used
Ceiling insulation: add R2.5	75%
Ceiling insulation: add R3.0	20%
Roof insulation: add reflective foil lining under roof	35%
Wall insulation: add reflective foil lining foil	50%
Wall insulation: add R1.5 batts	25%
Awnings on west windows	40%
Awnings on north windows	15%
Pergola on living area west windows (shade cloth)	5%
Glazing: add tinting	20%

and achieving the highest energy efficiency rating possible, cost an additional $7774 on the builder's normal inclusions [3]. This casts doubts over certain industry body's claim that it would cost $10 000 extra to obtain a five star rating [4].

It is interesting to note also the types of improvements that were needed to bring the houses up to the 3.5 star rating. Table 30.2 shows that insulation was the key component to improving efficiency, whether the insulation be located in the ceiling or the walls. This simple and cost effective solution is the reason that the overall cost for the improvements is so low, as no expensive solutions, such as double glazing, were required.

Conclusion

Energy efficient design of residential houses should by regarded as the norm rather than the exception. Amongst project house development this is especially important, as these constitute the majority of houses that are currently built in Australia. In Australia's milder climate areas, such as South East Queensland, the opportunities for cost effective, energy efficient design measures are particular good, yet at present it appears that this is not being realised, with the majority of houses currently being constructed rating poorly at around 1.5 stars.

The new energy efficiency regulations within the Building Code of Australia provide a much needed catalyst to drive energy efficient design. However, it is important that builders, designers and the house buying public understand the reasons behind energy efficiency, the methods for implementing and the cost. This will help ensure a smooth and positive uptake of the measures. Educational workshops, concept houses and appropriate rating tools can all play an important part in this process and, when organised and encouraged by large scale residential development organisations, the transition can be easier still.

It has been shown that many of the design changes needed are minimal in cost, with an average increase of only 2% on the cost of a standard house. Of course, the savings gained in the operation of the house can offset much of this cost. Indeed, annual savings can be as much as $560, which would deliver a payback period of less than 6 years [5]. Most importantly, energy efficient design will deliver houses that are more comfortable to live in, cheaper to run and help stem the rising levels of energy consumption and the consequent production of greenhouse gases.

References

[1] Commonwealth Bureau of Meteorology (2003) Climate averages for Brisbane regional office., Commonwealth Bureau of Meteorology Website, http://www.bom.gov.au/climate/averages/tables/cw_040214.shtml, Canberra.
[2] Tucker, S. N., Newton, P. W., Delsante, A. E., Ambrose, M. D., Johnston, D. R., Allen, S, Rasheed, B. and Remmers, T. R. (2002) *AGO-CSIRO Greenhouse Efficient Design*, (BCE DOC 02/118), CSIRO Building, Construction and Engineering, Melbourne.

[3] Eco Dezign Houses (2003) Eco Dezign Houses Website, http://www.edh.com.au/north_lakes_project.html, Landsborough, Queensland.

[4] *Australian Financial Review* (2002) House building costs will soar, says HIA, Newspaper article (09/08/2002).

[5] Reardon, C. (2001) *Your House, Design for Lifestyle and the Future – Technical Manual*, Australian Greenhouse Office, Canberra.

Keywords index

The following keywords are indexed to the first page of each chapter.